Rural Politics in Contemporary China

This collection provides an overview of China's rural politics, bringing scholarship on agrarian politics from various social science disciplines together in one place. The twelve contributions, spanning history, anthropology, sociology, environmental studies, political science, and geography, address enduring questions in peasant studies, including the relationship between states and peasants, taxation, social movements, rural-urban linkages, land rights and struggles, gender relations, and environmental politics. Taking rural politics as the power-inflected processes and struggles that shape access and control over resources in the countryside, as well as the values, ideologies and discourses that shape those processes, the volume brings research on China into conversation with the traditions and concerns of peasant studies scholarship. It provides both an introduction to those unfamiliar with Chinese politics, as well as in-depth, new research for experts in the field.

This book was originally published as a special issue of the *Journal of Peasant Studies*.

Emily T. Yeh is Associate Professor of Geography at the University of Colorado, Boulder. She conducts research on nature-society relations in Tibetan parts of the PRC, including the political ecology of pastoral environment and development policies, the relationship between ideologies of nature and nation, natural resource commodity chains, indigenous knowledge about climate change, and emerging environmental subjectivities.

Kevin J. O'Brien is Alann P. Bedford Professor of Asian Studies, Director of the Institute of East Asian Studies, and Professor of Political Science at the University of California, Berkeley. His books include *Grassroots Elections in China* (Routledge, 2011) (with Suisheng Zhao).

Jingzhong Ye is Professor of Development Studies and Deputy Dean at the College of Humanities and Development Studies (COHD), China Agricultural University. His research interests include development intervention and rural transformation, rural 'left behind' populations, rural education, land politics, and sociology of agriculture.

Critical Agrarian Studies

Series Editor:
Saturnino M. Borras Jr.

Critical Agrarian Studies is the new accompanying book series to the *Journal of Peasant Studies*. It publishes selected special issues of the journal and, occasionally, books that offer major contributions in the field of critical agrarian studies. The book series builds on the long and rich history of the journal and its former accompanying book series, the Library of Peasant Studies (1973-2008) which had published several important monographs and special-issues-as-books.

Critical Perspectives in Rural Development Studies

Edited by Saturnino M. Borras Jr.

The Politics of Biofuels, Land and Agrarian Change

Edited by Saturnino M. Borras Jr., Philip McMichael and Ian Scoones

New Frontiers of Land Control

Edited by Nancy Lee Peluso and Christian Lund

Outcomes of Post-2000 Fast Track Land Reform in Zimbabwe

Edited by Lionel Cliffe, Jocelyn Alexander, Ben Cousins and Rudo Gaidzanwa

Green Grabbing: A New Appropriation of Nature

Edited by James Fairhead, Melissa Leach and Ian Scoones

The New Enclosures: Critical Perspectives on Corporate Land Deals

Edited by Ben White, Saturnino M. Borras Jnr., Ruth Hall, Ian Scoones and Wendy Wolford

Rural Politics in Contemporary China

Edited by Emily T. Yeh, Kevin O'Brien and Jingzhong Ye

Rural Politics in Contemporary China

Edited by
**Emily T. Yeh, Kevin J. O'Brien and
Jingzhong Ye**

LONDON AND NEW YORK

First published 2015
by Routledge
2 Park Square, Milton Park, Abingdon, Oxfordshire OX14 4RN

and by Routledge
711 Third Avenue, New York, NY 10017, USA

First issued in paperback 2015

Routledge is an imprint of the Taylor & Francis Group, an informa business

British Library Cataloguing in Publication Data
A catalogue record for this book is available from the British Library

ISBN 13: 978-1-138-79230-2 (pbk)
ISBN 13: 978-1-13878-700-1 (hbk)

Typeset in Times New Roman
by Taylor & Francis Books

Publisher's Note
The publisher accepts responsibility for any inconsistencies that may have arisen during the conversion of this book from journal articles to book chapters, namely the possible inclusion of journal terminology.

Disclaimer
Every effort has been made to contact copyright holders for their permission to reprint material in this book. The publishers would be grateful to hear from any copyright holder who is not here acknowledged and will undertake to rectify any errors or omissions in future editions of this book.

Contents

Citation Information

The chapters in this book were originally published in the *Journal of Peasant Studies*, volume 40, issue 6 (November 2013). When citing this material, please use the original page numbering for each article, as follows:

Chapter 1
Rural politics in contemporary China
Emily T. Yeh, Kevin J. O'Brien and Jingzhong Ye
Journal of Peasant Studies, volume 40, issue 6 (November 2013) pp. 915-928

Chapter 2
A century of rural self-governance reforms: reimagining rural Chinese society in the post-taxation era
Alexander F. Day
Journal of Peasant Studies, volume 40, issue 6 (November 2013) pp. 929-954

Chapter 3
Debating the rural cooperative movement in China, the past and the present
Yan Hairong and Chen Yiyuan
Journal of Peasant Studies, volume 40, issue 6 (November 2013) pp. 955-982

Chapter 4
Chinese discourses on rurality, gender and development: a feminist critique
Tamara Jacka
Journal of Peasant Studies, volume 40, issue 6 (November 2013) pp. 983-1008

Chapter 5
Finance and rural governance: centralization and local challenges
John James Kennedy
Journal of Peasant Studies, volume 40, issue 6 (November 2013) pp. 1009-1026

Chapter 6
Measurement, promotions and patterns of behavior in Chinese local government
Graeme Smith
Journal of Peasant Studies, volume 40, issue 6 (November 2013) pp. 1027-1050

Chapter 7
Rightful resistance revisited
Kevin J. O'Brien
Journal of Peasant Studies, volume 40, issue 6 (November 2013) pp. 1051-1062

Chapter 8
Violence as development: land expropriation and China's urbanization
Sally Sargeson
Journal of Peasant Studies, volume 40, issue 6 (November 2013) pp. 1063-1086

Chapter 9
In defense of endogenous, spontaneously ordered development: institutional functionalism and Chinese property rights
Peter Ho
Journal of Peasant Studies, volume 40, issue 6 (November 2013) pp. 1087-1118

Chapter 11
The politics of industrial pollution in rural China
Bryan Tilt
Journal of Peasant Studies, volume 40, issue 6 (November 2013) pp. 1147-1164

Chapter 12
The politics of conservation in contemporary rural China
Emily T. Yeh
Journal of Peasant Studies, volume 40, issue 6 (November 2013) pp. 1165-1188

Chapter 13
The politics of water in rural China: a review of English-language scholarship
Darrin Magee
Journal of Peasant Studies, volume 40, issue 6 (November 2013) pp. 1189-1208

Please direct any queries you may have about the citations to
clsuk.permissions@cengage.com

Rural politics in contemporary China

Emily T. Yeh, Kevin J. O'Brien and Jingzhong Ye

We examine overarching themes in the contributions, including critiques of neo-liberalism, rural-urban linkages, the relevance of mixed methods and cross-disciplinary approaches, the need to engage social theory, and variation across space and time. At the same time, we provide an overview of rural Chinese politics and explain that the goal of the collection was to bring findings that have appeared in area studies or disciplinary outlets into conversation with peasant studies research. After discussing intellectual debates about the peasantry, everyday practices of governance, contentious politics, the mutual constitution of the rural and urban, and environmental politics, we conclude that work on the Chinese countryside needs 'lumping' (to discover unexpected similarities) and 'splitting' (to uncover patterns and forks in the road). Chinese rural politics is neither 'turtles all the way down' and baffling complexity, nor one master story that applies in all times and places. Instead we must continue to navigate the path between exoticizing China and treating its rural transformation as a tale many times told.

Introduction

Much news about today's China focuses on the urban. A milestone was reached in 2011, when the proportion of the People's Republic of China (PRC)'s 1.34 billion citizens living in cities reached 50 percent, the result of a remarkably rapid 'great urban transformation' (Hsing 2010) that began in the 1980s. By 2025, China is projected to have 221 cities with over one million inhabitants. Still, with hundreds of millions moving to urban areas, hundreds of millions more will continue to live in the countryside and work in agriculture. The fact that more people in China make their home in cities than villages marks a historic shift. At the same time, it is the product of long-standing dynamics through which the urban and rural are mutually constituted by processes, politics and ideologies that link, transgress and span both (Murdoch and Lowe 2003, Davis 2004, McCarthy 2005). Even as China becomes more urban, the politics of its countryside will continue to be central to the PRC and around the world.

We would like to thank Professor Jingzhong Ye and the College of Humanities and Development at the Chinese Agricultural University for convening the conference at which six of the papers in this collection were first presented. In addition to many helpful comments from the audience, we'd like to single out James C. Scott for joining us during a sparkling, late September weekend in Beijing and serving as a discussant. Jun Borras also provided unstinting support throughout, from writing the original call for papers, to helping select the contributors, to ushering the contributions through the refereeing process.

This collection addresses China's rural politics, broadly construed as the power-inflected processes and struggles that shape access to and control over resources in the countryside, as well as the values, ideologies and discourses that shape those processes and struggles. Though scholarship on agrarian politics in China has taken off over the past three decades, the literature has tended to appear in area studies journals, or disciplinary outlets in which questions central to a single field are placed front and center. Our intention here is different. In commissioning a set of review essays on themes in critical agrarian-environmental studies, we sought to bring what China experts have uncovered into conversation with the traditions and concerns of peasant studies scholarship. Toward this end, we assembled an international group of established researchers who span the social sciences, including political science, sociology, anthropology, geography, history and environmental studies, to address enduring questions in peasant studies, including the relationship between states and peasants, taxation, social movements, rural-urban linkages, land rights and struggles, gender relations, and environmental politics.

Rural China in brief

As the world's largest developing country, China's success in reducing child mortality, promoting primary education, eliminating infectious diseases and lessening hunger has contributed significantly to global progress in meeting Millennium Development Goals. In a remarkable accomplishment, the number of rural people living in poverty was brought from 85 million in 1990 down to 36 million in 2009. China itself has set a goal of not just eliminating extreme poverty, but of achieving an 'all-around *xiao kang*' or 'moderately well-off' society by 2020. This objective is to be met by a combination of targeted government investment and the country's rise as an economic power. Still, while China's turn toward capitalism has brought prosperity to many, leading some analysts to wonder whether 'peasants' continue to exist as a social category,[1] it has also exacerbated income disparities, transforming the PRC into one of the most unequal societies on earth. Peaking at 0.49 in 2008, China's current Gini coefficient of 0.47 ('Gini out of the bottle,' 2013) approaches those of Nigeria and Brazil, and is higher than that of the United States. The wealth gap is regional as well as spatial, with average per-capita income in rural areas less than one third that in cities.

China's rise has been fueled by more than 250 million migrant workers, members of the 'floating population' (*liudong renkou*), whose labor in export processing zones, cities and better-off villages has turned China into 'the world's factory'. The 'household registration' (*hukou*) system, which has tied citizens to their place of birth since the 1950s, was relaxed in 1984 to allow peasants to move to urban areas. As the township and village enterprises that spurred economic growth and absorbed rural labor after 'opening up and reform' (*gaige kaifang*) went bankrupt or were privatized in the late 1980s, the flow of migrant laborers increased. To this day, however, the *hukou* system denies 'peasant workers' (*nongmin gong*) state services, such as access to education, health care and housing, which are reserved for urban citizens.[2] In addition, migrants continue to be looked down upon by

---

[1]This was the theme of a Presidential panel on 'The Persistence of the Peasant' at the 2012 Association of Asian Studies Conference.

[2]Note, however, that this varies by city, with some municipal governments (for example, Shanghai and Chengdu) providing more services than others (for example, Beijing). Thanks to Alexsia Chan, and her forthcoming PhD dissertation at the University of California, Berkeley, for this point.

urban residents, blamed for crimes, paid salaries late or not at all and discriminated against (Solinger 1999, L. Zhang 2002, Yan 2003, Ngai 2005).

As migration exploded in the 1990s, and the countryside was emptied of working age men and women, so too did a national ideology that valorized the urban and denigrated the rural, positing cities as the primary site of political, cultural and economic worth (Bulag 2002, Cartier, 2003, 2005, Ma 2005, Yeh 2013a). Cities became metonyms for development, and urbanization became a top goal of China's modernization strategy. Along with this, city dwellers were deemed to be of higher quality, or *suzhi*, than rural residents (O'Brien and Li 1993–94, Bakken 2000, Anagnost 2004, Murphy 2004, Kipnis 2006). This privileging of the urban and disparaging of the rural led to what has been called the 'spectralization' (Yan 2003) of agriculture and the countryside, as villages became ghostly reminders of the past, a wasteland inhabited only by the 'left-behind', particularly children and the elderly (Jacka 2013, this collection, Ye 2013, this collection).

The 1990s also witnessed an overhaul of the fiscal system that shifted many expenditures to local governments. In poorer, agricultural villages this often led to spiraling taxes and fees, including illegal ones. These 'peasant burdens' (*nongmin fudan*) were a major cause of resentment and contention, ranging from everyday resistance (Scott 1985) to 'rightful resistance' (O'Brien and Li 2006), to thousands of sometimes violent 'mass incidents' (*qunti shijian*). As the new century unfolded, the frequency and intensity of protest grew. Whereas 8700 mass incidents were reported in 1993, the figure for 2011 reached 180,000 (Zheng 2012, 30), or nearly 500 every day.

A mounting sense of crisis, referred to as the 'three rural problems' (*san nong wenti*, or rural people, society and production), drew the attention of intellectuals concerned with the peasantry and state leaders worried about social stability and regime legitimacy. Debates emerged between liberals, some of whom argued that land rights should be privatized, and the new left, which critiqued neoliberalism and called for protections from the market and retaining equal distribution of farmland. One current among left-leaning intellectuals centered on calls for a New Rural Reconstruction Movement, modeled after a Chinese populist program of the early twentieth century. These scholar-activists urged fellow intellectuals to lead a rural cultural revival that would remake the countryside, in part through the formation of cooperatives.

The state also took steps to combat the 'three rural problems'. In 2002, a tax reform abolished most local fees, foreshadowing the complete elimination of agricultural taxes in 2006. That same year, Beijing launched the New Socialist Countryside program, an initiative designed to spark rural development, reduce income inequality and check unrest by redistributing resources and income to rural areas. Its components included expanding the cooperative medical system, eliminating school fees, enhancing water conservancy and completing the electric power network. Despite a shared aim of addressing rural problems, the thrust of the state's program is quite different from that advanced by leftist intellectuals, insofar as it calls for further urbanization, consumption and market-driven growth. The New Socialist Countryside program, as Elizabeth Perry (2011) has noted, also harkens back to an earlier Maoist campaign in its goals and because it has been implemented through propaganda and work teams. Political campaigns have also become a prime means by which environmental targets, whether for reducing pollution or afforestation, have been addressed (Economy 2002, van Rooij 2006). The reliance on mobilization and 'education and ideology work' (*jiaoyu sixiang gongzuo*), in areas as different as environmental protection, village election implementation (Schubert and Ahlers 2012), protest policing (Deng and O'Brien 2013), and population control and crisis management (Perry 2011) speaks to the Party's enhanced presence at the grassroots level since the late 1980s.

The end of the agricultural tax has also produced a number of other far-reaching consequences. One has been a hollowing out of the township, the lowest level of formal government. This has led some Chinese observers to call for the township to be eliminated, while others have proposed that it become little more than a service provider (Kennedy 2013, this collection, Day 2013, this collection). Perhaps an even more important result of revamped fiscal relations has been the turn to land appropriation by local authorities to generate revenue to compensate for lost taxes and fees. In China's complex property rights system, local officials can expropriate farmland, transfer it to state ownership, and then sell it to real estate developers. But rural collectives cannot sell their land or move it to non-agricultural uses. Peasants are compensated based on the average value of the land's agricultural output, typically only a small fraction of the market price. As a result of land takings, roughly 88 million peasants became landless between 1990 and 2008, with another 50 million expected to join them by 2030 (Sargeson 2013, this collection). Not surprisingly, whereas high taxation inspired much discontent in the 1990s and early 2000s, land expropriation has come to the fore now, accounting for some 65 percent of rural 'mass incidents'.

Environmental protests are also becoming more common. With five of the 10 most polluted cities in the world, and pollution reducing gross domestic product (GDP) by 8–15 percent not including health care costs (Hilton 2013, 8–9), grievances run deep. Though much attention has been paid to urban pollution, as in coverage of Beijing's 2013 'airpocalypse', unhealthy air and water are also common in rural areas, including the notorious 'cancer villages' where death rates far exceed the national average (Lora-Wainwright 2010). Despite growing concern with environmental degradation and stringent regulations, enforcement remains lax, largely because local governments are financially dependent on polluting factories. Coal mining accidents, particularly at small, unregulated and often illegal mines, illustrate these dynamics well. Fearful of protest, the state hesitates to condone anything that interferes with economic growth. This necessitates continuing use of coal to avoid brownouts and keep industry humming, and reduces incentives to enforce worker safety and environmental measures (Weston 2007).

The environmental politics of China's capitalist transformation are not limited to air and water pollution. The building of large hydropower dams has led to the involuntary resettlement of millions of peasants. Plans for new dams, particularly along the Nu River, have also been targets of local protest and mobilization by environmental non-governmental organizations (ENGOs) (Büsgen 2006, Litzinger 2007, Yang and Calhoun 2007) as well as the source of much bureaucratic infighting (Mertha 2008). Other development efforts, including those designed to improve the environment, have also led to large-scale resettlement of farmers and pastoralists (Yeh 2009). The rapid expansion in the number of nature reserves, for example, has reduced access to crucial livelihood resources, leading to various forms of peasant resentment and resistance (Yeh 2013b, this collection).

The essays

All these issues and more are discussed by the 12 contributors to this collection. Though written from a variety of disciplinary perspectives, the papers are all grounded in Chinese culture and society, and an approach to politics that rejects essentialist understandings of a reified 'Chinese culture'. Many of the reviews include both English-language and Chinese sources; this is true of the contributions that discuss Chinese debates on rural society, and those on taxation and land expropriation. Furthermore, though access makes long-term fieldwork in the countryside difficult, most of the contributions are grounded in just such experiences. This engagement with daily life and the quotidian is reflected in the 'view from below' found in many of the essays.

Although a single volume cannot take on all of agrarian politics in China, this collection covers substantial territory. Drawing on a wide range of topics, five themes emerge: intellectual debates about the peasantry, everyday practices of local governance, contentious politics, rural-urban linkages, and environmental politics.

Intellectual debates and the New Rural Reconstruction Movement

Our first set of essays, by Alexander Day, Yan Hairong and Tamara Jacka, explores Chinese debates about past, present and future state-peasant relationships. Both Day and Yan argue that there are resonances between contemporary discussions, particularly within the New Rural Reconstruction Movement, and those of the early twentieth century. Jacka, on the other hand, offers a feminist critique of the New Rural Reconstruction Movement. All three find that, even while criticizing neoliberal proposals to further integrate agriculture and rural labor power into national and global markets, the movement's intellectual leaders tend to view peasants as a homogenous whole, ignoring class, gender and other forms of differentiation crucial to understanding rural society. This complicates their efforts to protect the peasantry from economic exploitation and to ameliorate a growing income gap, the aging and feminization of agriculture, the fragmentation of family life and livelihood challenges.

Alexander Day's essay provides a historical backdrop for issues discussed by other contributors, including the debate over the future of the township (Kennedy 2013, this collection) and rural property rights (Ho 2013, this collection). He argues that in both the decades following the end of the Qing dynasty and in the 1990s, rural China was characterized by 'state involution': intensifying economic crisis, predation by entrepreneurial brokers and weakened central capacity. Drawing on ways earlier intellectuals reimagined rural society and politics, particularly the notion of 'self-governance', contemporary analysts associated with the New Rural Reconstruction Movement have championed alternative models of organization, such as cooperatives (Yan 2013, this collection), peasant associations and democratic self-governing villages. Their vision for how society should be organized following the abolition of the agricultural tax differs markedly from those of liberal intellectuals who favor privatizing land. However, even without privatization and full proletarianization of labor, capital has penetrated the countryside and new class relations have developed (Q.F. Zhang and Donaldson 2008, 2010, Huang 2011).

Yan Hairong's essay examines recent efforts to promote rural cooperatives. Like Day, Yan hears echoes of debates from the 1930s Rural Reconstruction Movement. The crux of the 1930s controversy concerned the nature of agrarian China. Liang Shuming, a prominent advocate of cooperatives as a 'third way' alternative to Western capitalism and communism, stressed the uniqueness of Chinese culture and society, which he argued lacked class divisions. This led him to support reform rather than revolution. Mao Zedong, of course, thought otherwise. Mao believed that Chinese society was riven by class opposition, contradiction and struggle, and that revolution was the only way out. Today's supporters of a New Rural Reconstruction Movement also promote cooperatives as a possible 'third way'. Like Liang Shuming, who lamented villagers' inertia, contemporary intellectuals lament 'fake' cooperatives, which exist in name only, or are dominated by a few rich farmers or 'dragon-head' (*longtou*) enterprises. Yan shows that many supporters of today's movement downplay class differentiation because, as Day (2013, this collection) also argues, they conceptualize rural society as an undifferentiated whole. Despite this blind spot, supporters of rural reconstruction appropriately, in Yan's view, challenge modernization theory as well as the dominant and unsustainable neoliberal vision of capitalist agriculture.

Whereas Day and Yan highlight the absence of class in New Rural Reconstruction Movement discourse, Jacka's feminist critique focuses on its elision of gender inequalities. Discussing the movement as a form of alternative development, she finds that it gives insufficient attention to the gendered nature of institutions, and so reproduces injustice. Thus, as an alternative globalization movement, New Rural Reconstruction is less able to ally with women's movements, in the way the Zapatistas in Mexico or the MST (Movimento dos Trabalhadores Rurais Sem Terra, or Landless Rural Workers' Movement) in Brazil have. Jacka examines the work of influential scholars Wen Teijun and He Xuefeng, and their writings about the left-behind population. Both, she argues, implicitly equate 'the peasant' with a man, neglecting the presence and agency of women, the significance of a gendered division of labor, and gender-specific challenges common in contemporary rural Chinese society. Moreover, their appeal to Chinese culture assumes an essentialized view of the patriarchal family as unquestioned norm. Drawing on Nancy Fraser's work, Jacka concludes that rural reconstruction could have transformative potential, but is currently better characterized as taking an affirmative approach to economic and cultural injustice that fails to tackle global capitalism head-on. This means it risks compounding the inequities experienced by poor rural women.

Everyday practices of governance

The next contributions by Kennedy and Smith address the nitty-gritty politics of county and township governments. Kennedy traces some of the same transformations in governance as Day does, though from a political scientist's vantage point. He argues that the trajectory of tax reform has arced toward centralization. In 1994, county and township officials' main responsibilities shifted from service provision to collection of taxes and fees, and then, after further reform in 2002 and 2006, back to service provision. However, these changes looked quite different to villagers and to analysts who viewed them from a broader institutional viewpoint. Villagers in poorer regions tended to blame township officials for rising fees and taxes, whereas from an institutional perspective it was a structural bias against villagers, rather than township officials per se – who often had to go without salaries to meet county quotas – that encouraged excessive extraction and sparked protest across rural China. The cadre management system, analyzed in depth by Smith, was also a crucial factor that shaped county-township relations and created the rural tax burden.

Noting that too little of the literature on rural governance addresses 'how the local state actually get things done', Smith examines informal and sub rosa practices, including the role of the 'shadow state'. His contribution suggests several logics of Chinese rural governance, such as 'the party trumps the government, and everyone wants to live in town', while also cautioning against assuming uniformity or overly general rules of thumb. Other principles he identifies include localism, the weak role of townships, and the importance of informal networks and patronage. His contribution, like many others (for example, O'Brien and Li 1999, Whiting 2000, Edin 2003),[3] underscores performance targets

---

[3]For critiques of what might be called the 'fetishism of targets', see Ahlers and Schubert (2013) and Mei and Pearson (forthcoming). Ahlers and Schubert, in particular, suggest that recent research on the local state sometimes turns cadres into narrow *Homo economicus* interested only in promotion (and in disguised defiance of their superiors). Cadre evaluation matters, they acknowledge, as do unfunded

embedded in the cadre evaluation system. These, as seen in Tilt's (2013, this collection) essay, are also central to understanding why China's environmental regulations have not translated into reduced pollution, and also to changes that may be underway with the introduction of high-priority, quantitative targets that may finally motivate cadres to pay more attention to the environment (Wang 2013).

Contentious politics

One way that research on rural China has opened a dialogue with peasant studies has been through O'Brien and Li's (2006) notion of 'rightful resistance'. Like James Scott's concept of 'everyday resistance', rightful resistance identified a form of contention between quiescence and rebellion that enables peasants to 'act up' more than otherwise seems prudent. Rightful resistance, in particular, involves using the language of power to challenge local cadres who fail to implement the policies, laws and commitments of the Center. Its defining characteristics include operating near the boundary of authorized channels, employing the rhetoric of the powerful, exploiting divisions within the state and mobilizing community support. Kevin O'Brien's contribution to this collection constitutes a 'self-criticism' (*ziwo piping*) that engages various critiques of the concept and its application to rural China. Though O'Brien and Li's book, *Rightful resistance in rural China*, had an apparent 'ground-level orientation', O'Brien agrees that there is room for more ethnographic, less state-centric studies of protest that focus on peasant subjectivities, rural communities, cultural norms and local histories. Moreover, though the book offered a view of the state from below and unpacked the hierarchy level by level, he suggests it is time to go beyond disaggregating the state vertically to a horizontal disaggregation that examines conflicts within each level. O'Brien also notes that, as a conceptual effort, the book spent little time exploring variation by region or issue, and that doing so is the next task for survey researchers and others to dive into. Finally, he concludes that a number of vexing questions about rightful resistance suffer from overly stark either/or conceptualizations that confuse more than they inform. In particular, he argues that it probably does not make sense to insist that rightful resistance is either sincere or strategic, reactive or proactive, or reflects rights versus rules consciousness. A more open reading of contention in rural China allows for ironies and paradoxes, revels in the intentionally oxymoronic term 'rightful resistance', and encourages us to hesitate before leaping to conclusions about the relationship between local protest and systemic change.

As noted above, the most prevalent bone of contention in the countryside today is land appropriation, which now accounts for around half of local government revenue. Sargeson's contribution examines land expropriation violence.[4] Like O'Brien and Li's rightful resistance, Sargeson's application of Arturo Escobar's concept of 'violence as development' has relevance to peasant studies beyond the PRC. Coming from a radically different analytical approach, Sargeson's essay addresses some of the same critiques of rightful resistance that O'Brien examines. In particular, it leans hard against overly rationalist assumptions about individuals weighing costs and benefits that O'Brien (2013, this

---

mandates, bureaucratic collusion and factionalism, but they also find a striking amount of cooperation between cadres and their superiors to achieve 'good enough' implementation of initiatives such as the New Socialist Countryside program.

[4]Some suburban farmers, however, treat land-taking as an opportunity to shake down the state. Andrew Kipnis, personal communication, 17 March 2012. See also Paik and Lee (2012) on rural people who want to be expropriated.

collection) writes have been justifiably criticized in *Rightful resistance in rural China*, and instead attaches more weight to memories and a community's historical relationship to a place, as well as to how subjectivities are transformed. Sargeson locates problems in each of three discussions of violence that accompanies land expropriation: game-theoretic, rational choice approaches in which violence is instrumental; a spatial ecology approach in which there is an implicit spectrum between a rational, propertied urban citizenry and inchoate resistance by peasants in the hinterlands; and a 'political maturation' approach in which violence is seen as a catalyst that transforms villagers into politically astute citizens. All three of Sargeson's critiques echo O'Brien's concern that some research on rural protest is too rooted in neoclassical assumptions about individual behavior and reflects 'developmental thinking', or at least can fall prey to teleological readings. Instead, Sargeson argues that understanding violence as constitutive of development better accounts for non-instrumental, spatially dispersed and socially complex violence, including episodes where villagers become disillusioned with the possibilities of collective rightful resistance.

Rural-urban linkages

All of the contributors recognize the interconnectedness of the rural and urban, but two essays highlight this linkage explicitly. Peter Ho's contribution on the institutional structure of rural-urban property rights argues that 'the functioning of rural land tenure cannot be understood separately from the urban land market', while Ye *et al.*'s essay focuses on the people left behind as a result of China's massive rural-urban migration.

Ho mounts a critique of neoliberal arguments that insist that China's insecure and opaque property rights mean that the institutional structure must change or collapse. Reviewing rural and urban property from the formation of the PRC to the present, with particular attention to land markets, Ho suggests that the persistence of the institutional set-up, in which rural land is untitled, informal and frequently reallocated, confirms that it is functional and credible and in no danger of falling apart. Instead, it will most likely change from within. Rather than viewing flexibility and ambiguity as weaknesses, Ho sees them as drivers of China's capitalist development. He stresses that 'frictions and distributional conflicts' are inevitable. Though he mentions land expropriation, his focus, unlike Sargeson's (2013, this collection), is not on 'structural violence' or its traumatic effects. That peasants support frequent reallocation of land (Kong and Unger 2013) is evidence of the flaws in neoliberal thinking on property.

Like Ho, Ye *et al.* challenge tenets of mainstream economics, in this case the belief that China's rural-urban migration is nothing more than an efficient reallocation of labor that serves national economic development. Their contribution presents a comprehensive review of the literature on internal migration, placing the Chinese case alongside studies around the globe. The authors suggest that research on migration typically proceeds from three starting points: neoclassical economics, a social and cultural perspective, or neo-Marxism. They favor the neo-Marxist approach. Worldwide, most studies find significant variation in the effects of migration on education, physical and psychological health, and marriage and gender roles of people left behind, at both the individual and community level. For China, most research on left-behind children, women and the elderly ignores political, economic and structural forces, frames their problems as moral failings, and seeks solutions in the form of charity, care and 'love'. Instead, Ye *et al.* propose a framework that builds on Foucault's and Agamben's work on biopolitics to explain why both migrants and those left behind exist in a precarious 'state of exception'. Like Sargeson, Ye *et al.* are critical of the structural violence of development. Like Ho, they stress urban-rural linkages, though they are much less optimistic than Ho about the current state of affairs in rural China. Rather than seeing institutions governing internal migration as endogenous, autonomous and credible, they treat them as forms of biopolitics, which discipline, exclude and sometimes kill migrants. Still, all is not hopeless. Ye *et al.* suggest that a biopolitical perspective on migration could one day help an organized working class emerge 'out of the present cocoon' of semi-proletarian migrant laborers.

Environmental politics

The last group of contributions focuses on issues in rural environmental politics: industrial pollution, conservation, and water. Tilt, Yeh and Magee all adopt a political ecology approach and advocate expanding the study of rural politics to include human-environment relations, as well as bringing politics to the forefront of literatures often dominated by managerial approaches.

Tilt proposes a framework that makes sense of the current state of industrial pollution in rural China – perhaps the largest cause of protests after land appropriation – by conceptualizing it as a domain consisting of rural citizens' knowledge of environmental harm, the actions that they take and their results, and regulations that shape pollution. Though industrial pollution is often assumed to be an urban concern, air, water and soil contamination are severe in the countryside, largely due to the spread of rural factories and industrial parks. Information and data about pollution, however, remain scarce, and causal links to health effects are difficult to prove. Because science cannot provide indisputable answers, it is important to understand perceptions, which arise from specific socio-political contexts. In terms of action, rightful resistance is a particularly apt concept for environmental protest, given the state's own discourse about environmental protection. At the same time, most grievances end up being addressed through compensation, which routinizes pollution rather than reduces it. The fiscal constraints discussed by Kennedy and Smith further limit enforcement.

Yeh's contribution shows that much of the literature on forest protection and rehabilitation, nature reserves and grasslands is technocratic and managerial and fails to consider the politics of access. A number of researchers have examined the effects of reforestation – under the massive Sloping Land Conversion Program – on off-farm labor participation, rural income and forest ecology. Their studies have generated mixed results, though many suggest that the program is unlikely to benefit marginalized communities. The political ecology literature on Chinese nature reserves shows that conservation enclosures have often sparked peasant resistance and that biodiversity goals are compromised by the revenue imperative and performance criteria by which cadres are judged (see Smith 2013, this collection). With reference to grasslands, Yeh reviews studies that show how cadre evaluation and pastoralists' efforts to maintain their livelihood have foiled various programs to rehabilitate rangelands. At the same time, she argues that pastoralism deserves more attention from the peasant studies community. She points out that environmental politics in ethnic minority areas is a form of rural politics, and that ethnicity is one of many types of differentiation that should not be neglected. Omitting struggles over resources in minority areas from the literature on peasant politics mirrors and reinforces their exclusion from repertoires of resistance available to other rural people.

Finally, Darrin Magee's examination of the limited western scholarship on the politics of rural water overlaps with Tilt's discussion of water pollution. Both focus on the obstacles peasants face accessing China's legal system and how activists and non-governmental organizations (NGOs) provide legal aid to victims of water contamination. Beyond this,

Magee also explores discourses, power relations and institutions that govern irrigation, household water use, power generation and in-stream flows. These discussions pose the deeply political question of whether water should be treated as a basic right or saleable commodity, and reflect a growing interest in neoliberal solutions such as Payment for Ecosystem Services. In his treatment of irrigation, Magee cites research that shows how uncertainty about land tenure may have contributed to a decline in effective water use, a downside to the property rights regime that Ho finds 'credible'. Magee's analysis of dam-building touches on several themes that reappear throughout the collection, including protests, inequality and the relationship between cities and the countryside. Dams produce (mostly urban) winners and (mostly rural, often ethnic minority) losers, as well as contention both before and after they are constructed.

Common messages

Several common themes stand out in these contributions. First, many critique neoliberalism, though they take different tacks in doing so. Ho attacks the idea that institutions are the products of intention, which he considers a fundamental tenet of neoliberalism. Sargeson focuses on the human cost of capitalist development, and how its structural violence disrupts the way of life, social standing, self-respect and sense of self and place of many rural people. Ye *et al.*, Yan, Day and Jacka all criticize theories of rural society that do not take structural forces and class differentiation into account. Finally, Yeh and Magee discuss problems that arise when Payments for Ecosystem Services programs are used to manage China's forests and water.

Second, these studies suggest the need to think about rural politics relationally, highlighting rural-urban linkages and expansive conceptions of both 'rural' and 'urban'. This is made explicit in Ye *et al.*'s study of migration and Ho's argument that the division of rural and urban land markets is complicated and more fluid than it appears. It is also evident in Sargeson's and Ho's discussions of land-taking as a driver of urbanization and a strategy for rural governments to generate income. Moreover, as Magee emphasizes, hydropower development is often tied to discourses about rural poverty, but peasants generally lose out from projects that expropriate land to provide urban residents with electricity. Protest also draws the urban and the rural together. In the contributions by Day, Yan, O'Brien and Jacka, we see urban activists, intellectuals and even officials working together with rural people to improve life in the countryside.

Third, a number of the papers demonstrate the relevance of mixed methods and cross-disciplinary approaches (Borras 2009), while clarifying the relative strengths of qualitative and quantitative research. Kennedy, for example, finds that quantitative studies of the 1994 tax reforms revealed regional differences across China, in particular showing that the effects were not as uniformly negative as is often believed. Ho supports the use of regression and factor analysis to help make sense of the relationship between economic growth and land institutions. These authors remind us that qualitative research by itself, divorced from a sense of frequencies and diversity, can be misleading, particularly if a single village is taken to stand in for the whole of China. As William Hurst (2009) has argued concerning Chinese workers, we are long past the time when studies of Chinese peasants in one locale at one time can stand in for their circumstances, hopes and struggles everywhere.

Qualitative analysis, on the other hand, shines in offering insight into informal and illegal revenue collection, as well as villagers' resistance to it. In-depth field research also offers a fuller understanding of how rural people experience and cope with pollution day to day. Tilt, in his essay, argues for combining the regional and national-level data that dominates work on pollution with qualitative work, which shows us how and why peasants respond to environmental degradation. Sargeson and Smith, too, make strong arguments for long-term 'soaking and poking' (Fenno 1978) in the countryside. Smith notes that such fieldwork is often dismissed by quantitative social scientists as unrepresentative and lacking in rigor. He maintains, however, that it is only through extended stays and the development of trust that phenomena such as the selling of government posts and the 'shadow state' can be understood. Sargeson also makes a forceful appeal for longitudinal and qualitative research, which alone can reveal 'how people bring unique histories to bear when participating in political-economic processes that span long periods of time and space'.

Sargeson's and Ye *et al.*'s essays highlight another theme threaded through the contributions: the need to engage social theory, including Marxian and poststructural approaches to development (see also Borras 2009). Other authors, particularly Day and Yan, also underscore the relevance of classic theories of political economy, particularly ones that emphasize class politics,[5] by pointing out the neglect of class differentiation in analyses of rural society by left-leaning Chinese intellectuals.

Another issue that appears throughout this collection is how to make sense of geographic differences in China. A number of contributors point to sources of disagreement in the literature that are in fact the product of regional and local variation. Smith, for example, suggests that the degree to which the local rural state has been characterized as predatory, developmental, corporatist or mafia-like hinges on the presence and ability of firms to apply pressure, which in turn depend on location. Similarly, Kennedy demonstrates that the effects of local tax and fiscal policies were different for poorer, inland regions, which had to impose more informal fees and levies, and wealthier coastal townships, which experienced little change. Thus, geography, understood as the 'path-dependent', relational production of places, matters a great deal for understanding rural governance in China.

O'Brien also notes that many readers of *Rightful resistance in rural China* wanted to know more about variation across space and time. Though the intent of the book was to show that seemingly disparate phenomena could be thought of within a single framework, O'Brien points out that the next task is understanding why rightful resistance appears to be more common in some provinces than others, and in some parts of the countryside (for instance, is it more common in suburban or more remote villages?). We also need to know more about whether the form resistance takes varies by issue, and how different elite allies (for example, officials, journalists, lawyers, entrepreneurs, roving scholars) affect the course and outcomes of protest. By contrast, Ye *et al.*, while examining a bewildering amount of variation in how migration affects sending areas, emphasize the need to go beyond the conclusion that 'it depends' to uncover underlying patterns and processes.

Peasant studies clearly needs both 'lumping' (to discover unexpected similarities) and 'splitting' (to uncover patterns and forks in the road). It's neither 'turtles all the way down' and baffling complexity, nor one master story that applies in all the places and times. This holds true for peasant studies and also for China studies and comparative research that includes China as a case. China is sometimes held up by area specialists as unique and exceptional. At the same time, its recent capitalist turn has been treated by some grand theorists as a story whose outline is already known, with only minor, empirical details to be filled in. This collection suggests that it is neither. Instead, we must continue to navigate the path between exoticizing China and treating its rural transformation as a tale many times told.

---

[5]On class in contemporary China see Hanser (2008) and Yan (2008).

References

Ahlers, A.L. and G. Schubert. 2013. Effective policy implementation in China's local state. Unpublished paper.

Anagnost, A. 2004. The corporeal politics of quality (*suzhi*). *Public Culture*, 16(2), 189–208.

Bakken, B. 2000. *The exemplary society: human improvement, social control, and the dangers of modernity in China*. Oxford: Oxford University Press.

Borras, S.M. Jr. 2009. Agrarian change and peasant studies: changes, continuities and challenges – an introduction. *Journal of Peasant Studies*, 36(1), 5–31.

Bulag, U. 2002. From yeke-juu league to Ordos municipality: settler colonialism and alter/native urbanization in Inner Mongolia. *Provincial China*, 7(2), 196–234.

Büsgen, M. 2006. *NGOs and the search for Chinese civil society: environmental non-governmental organizations in the Nujiang campaign*. Institute of Social Studies Working Paper Series No. 422. The Hague: Institute of Social Studies.

Cartier, C. 2003. Symbolic city/regions and gendered identity formation in south China. *Provincial China*, 8(1), 60–77.

Cartier, C. 2005. City-space: scale relations and China's spatial administrative hierarchy. *In*: L. C. Ma and F. L. Wu, eds. *Restructuring the Chinese city: changing society, economy and space*. London: Routledge, pp. 21–38.

Davis, M. 2004. Planet of slums: urban involution and the informal proletariat. *New Left Review*, 26, 5–34.

Day, A. 2013. A century of rural self-governance reforms: reimagining rural Chinese society in a post-taxation era. *Journal of Peasant Studies*, 40(6).

Deng, Y.H. and K.J. O'Brien. 2013. Relational repression in China: using social ties to demobilize protesters. *The China Quarterly*, 215, 533–52.

Economy, E. 2002. China's Go West campaign: ecological construction or ecological exploitation? *China Environment Series*, 5, 1–11.

Edin, M. 2003. State capacity and local agent control in China: CCP cadre management from a township perspective. *The China Quarterly*, 173, 35–52.

Fenno, R.F. Jr. 1978. *Home style: house members in their districts*. Boston: Little, Brown.

Gini out of the bottle. 2013. *The Economist*. 26 January http://www.economist.com/news/china/21570749-gini-out-bottle [Accessed 24 June 2013].

Hanser, A. 2008. *Service encounters: class, gender and the market for social distinction in urban China*. Stanford, CA: Stanford University Press.

Hilton, I. 2013. Introduction: the return of Chinese civil society. *In*: S. Geall, ed. *China and the environment: the green revolution*. London: Zed Books, pp. 1–14.

Ho, P. 2013. In defense of autonomous, endogenous development: the institutional structure of China's rural-urban property rights. *Journal of Peasant Studies*, 40(6).

Hsing, Y.T. 2010. *The great urban transformation: politics of land and property in China*. Oxford: Oxford University Press.

Huang, P.C.C. 2011. China's new-age small farms and their vertical integration: agribusiness or co-ops? *Modern China*, 37(2), 107–34.

Hurst, W. 2009. *The Chinese worker after socialism*. Cambridge: Cambridge University Press.

Jacka, T. 2013. From spectralization to participation and reconstruction: a feminist critique of contemporary Chinese discourses on rurality, gender and development. *Journal of Peasant Studies*, 40(6).

Kennedy, J.J. 2013. Finance and rural governance: centralization and local governance. *Journal of Peasant Studies*, 40(6).

Kong, S.T. and J. Unger. 2013. Egalitarian redistributions of agricultural land in China through community consensus: findings from two surveys. *The China Journal*, 69, 1–19.

Kipnis, A. 2006. *Suzhi*: a keyword approach. *The China Quarterly*, 186, 295–313.

Litzinger, R. 2007. In search of the grassroots: hydroelectric politics in northwest Yunnan. *In*: E. J. Perry and M. Goldman, eds. *Grassroots political reform in contemporary China*. Cambridge: Harvard University Press, pp. 282–99.

Lora-Wainwright, A. 2010. An anthropology of 'cancer villages': villagers' perspectives and the politics of responsibility. *Journal of Contemporary China*, 19(63), 79–99.

Ma, L.J. 2005. Urban administrative restructuring: changing scale relations and local economic development in China. *Political Geography*, 24, 477–97.

Magee, D. 2013. The politics of water in rural China: A review. *Journal of Peasant Studies*, 40(6).

McCarthy, J. 2005. Rural geography: multifunctional rural geographies – reactionary or radical? *Progress in Human Geography*, 29(6), 773–82.

Mei, C.Q. and M.M. Pearson. Forthcoming. Killing a chicken to scare the monkeys? Deterrence failure and local defiance in China. *The China Journal*.

Mertha, A. 2008. *China's water warriors: citizen action and policy change*. Ithaca: Cornell University Press.

Murdoch, J. and P. Lowe. 2003. The preservationist paradox: modernism, environmentalism and the politics of spatial division. *Transactions of the Institute of British Geographers*, NS28, 318–32.

Murphy, R. 2004. Turning Chinese peasants into modern citizens: population quality, demographic transition, and primary schools. *The China Quarterly*, 177, 1–20.

Ngai, P. 2005. *Made in China: women factory workers in a global marketplace*. Durham, NC: Duke University Press.

O'Brien, K.J. 2013. Rightful resistance revisited. *Journal of Peasant Studies*, 40(6).

O'Brien, K.J. and L.J. Li. 1993–94. Chinese political reform and the question of 'deputy quality'. *China Information*, 8(3), 20–31.

O'Brien, K.J. and L.J. Li. 1999. Selective policy implementation in rural China. *Comparative Politics*, 31(2), 167–86.

O'Brien, K.J. and L.J. Li. 2006. *Rightful resistance in rural China*. New York: Cambridge University Press.

Paik, W. and K. Lee. 2012. I want to be expropriated!: the politics of *xiaochanquanfang* land development in suburban China. *Journal of Contemporary China*, 21(74), 261–80.

Perry, E.J. 2011. From mass campaigns to managed campaigns: 'constructing a new socialist countryside'. *In*: S. Heilmann and E. J. Perry, eds. *Mao's invisible hand: political foundations of adaptive governance in China*. Cambridge: Harvard University Press, pp. 30–61.

Sargeson, S. 2013. Violence as development: land expropriation and China's urbanization. *Journal of Peasant Studies*, 40(6).

Schubert, G. and A.L. Ahlers. 2012. *Participation and empowerment at the grassroots: Chinese village elections in perspective*. Lanham, MD: Lexington Books.

Scott, J.C. 1985. *Weapons of the weak: Everyday forms of peasant resistance*. New Haven: Yale University Press.

Smith, G. 2013. Measurement, promotions and patterns of behavior in Chinese local government. *Journal of Peasant Studies*, 40(6).

Solinger, D. 1999. *Contesting citizenship in urban China: peasant migrants, the state and the logic of the market*. Berkeley: University of California Press.

Tilt, B. 2013. The politics of industrial pollution in rural China. *Journal of Peasant Studies*, 40(6).

van Rooij, B. 2006. Implementation of Chinese environmental law: regular enforcement and political campaigns. *Development and Change*, 31(1), 57–74.

Wang, A. 2013. The search for sustainable legitimacy: environmental law and bureaucracy in China. *Harvard Environmental Law Review*, 37(2), 365–440.

Weston, T. 2007. Fueling China's transformations: the human cost. *In*: T. Weston and L. Jensen, eds. *China's transformations: the stories behind the headlines*. Boulder, CO: Rowman & Littlefield, pp. 68–89.

Whiting, S. 2000. *Power and wealth in rural China: the political economy of institutional change*. Cambridge: Cambridge University Press.

Yan, H.R. 2003. Spectralization of the rural: reinterpreting the labor mobility of rural young women in post-Mao China. *American Ethnologist*, 30(4), 1–19.

Yan, H.R. 2008. *New masters, new servants: migration, development, and women workers in China*. Durham, NC: Duke University Press.

Yan, H.R. 2013. Rural cooperative movement in China: the past and the present. *Journal of Peasant Studies*, 40(6).

Yang, G.B. and C. Calhoun. 2007. Media, civil society, and the rise of a green public sphere in China. *China Information*, 21(2), 211–36.

Ye, J.Z., *et al.* 2013. Internal migration and the left-behind population in China. *Journal of Peasant Studies*, 40(6).

Yeh, E.T. 2009. Greening western China: a critical view. *Geoforum*, 40, 884–94.

Yeh, E.T. 2013a. *Taming Tibet: landscape transformation and the gift of Chinese development*. Ithaca, NY: Cornell University Press.

Yeh, E.T. 2013b. The politics of conservation in contemporary rural China. *Journal of Peasant Studies*, 40(6).

Zhang, L. 2002. *Strangers in the city: reconfigurations of space, power and social networks within China's floating population*. Stanford, CA: Stanford University Press.

Zhang, Q.F, and J.A. Donaldson. 2008. The rise of agrarian capitalism with Chinese characteristics: agricultural modernization, agribusiness, and collective land rights. *The China Journal*, 60, 25–47.

Zhang, Q.F, and J.A. Donaldson. 2010. From peasant to farmer: peasant differentiation, labor regimes and land-rights institutions in China's agrarian transition. *Politics and Society*, 38(4), 458–89.

Zheng, Y.N. 2012. China in 2011: anger, political consciousness, anxiety, and uncertainty. *Asian Survey*, 52(1), 28–41.

Emily T. Yeh is an Associate Professor of Geography at the University of Colorado Boulder. She conducts research on nature-society relations, primarily in Tibetan parts of the People's Republic of China, including projects on conflicts over access to natural resources, the relationship between ideologies of nature and nation, the political ecology of pastoral environment and development policies, vulnerability of Tibetan herders to climate change, and emerging environmental subjectivities. Her book, *Taming Tibet: landscape transformation and the gift of Chinese development* (Cornell University Press, 2013), explores the intersection of the political economy and cultural politics of development as a project of state territorialization.

Kevin J. O'Brien is the Alann P. Bedford Professor of Asian Studies, Director of the Institute of East Asian Studies, and Professor of Political Science at the University of California, Berkeley. His books include *Grassroots elections in China* (2011) (with Suisheng Zhao), *Popular protest in China* (2008), *Rightful resistance in rural China* (2006) (with Lianjiang Li), and *Engaging the law in China: state, society and possibilities for justice* (2005) (with Neil J. Diamant and Stanley B. Lubman).

Jingzhong Ye is a professor of development studies and deputy dean at the College of Humanities and Development Studies (COHD), China Agricultural University. His research interests include development intervention and rural transformation, rural 'left behind' populations, rural education, land politics, and sociology of agriculture.

A century of rural self-governance reforms: reimagining rural Chinese society in the post-taxation era

Alexander F. Day

In 2006, the Chinese central government abolished the agricultural tax. This came after several years of intense focus on the growing rural crisis, sparking a new debate on the shape of rural society. Putting these contemporary debates in the context of 100 years of rural governance reforms, this paper argues that contemporary rural advocates find themselves in a situation similar to that of the first half of the twentieth century, and their reimagination of rural governance draws on the ideas of that time as well. It focuses on the contemporary visions for rural reform of Xu Yong, Dang Guoying, Yu Jianrong, Wu Licai, Li Changping, Cao Jinqing and He Xuefeng.

The crisis of peasant welfare in the late 1990s and new state policies leading to the 2006 abolition of the agricultural tax in China altered the state-rural society relationship, sparking a new debate on rural governance and peasant organization among intellectuals that advocated for the peasantry. As many have noted, the abolition of the agricultural tax brings to an end thousands of years of the Chinese state taxing agriculture for revenue. But perhaps a more telling periodization can be constructed out of the last one hundred years of state-making, state involution, state extraction from the rural sphere and the politics of peasant advocacy.[1] In the last decade of the Qing Dynasty (1644–1911), rural governance underwent a significant transformation, beginning a century-long experiment in modernizing rural governance that produced decidedly mixed results for China's peasants, though the successful industrialization of China. These experiments in state-rural society relations combined an attempt to modernize rural economic, political and social life with a new extractive power – a combination of modernization and central-state extraction that the Chinese state only brought to an end in 2006. The contemporary political debate over the social, economic and political organization of the peasantry that coincided with the

The author would like to thank Gail Hershatter, the editors of this collection and the anonymous reviewers for comments and suggestions on this paper. Guolin Yi was helpful in obtaining sources.
[1] By 'politics of peasant advocacy' I am referring to the attempt by intellectuals to influence the official politics of rural reform of the state and party as well as attempts by intellectuals to directly organize peasants. Benedict Kerkvliet (2009) argues that there are broadly three forms of peasant politics: official, advocacy and everyday politics.

transformation of the state-rural society relationship is clearly marked by this century of rural state-making and visions of reform.

This paper looks at the central state's competition with local elites and local governments for control over rural surplus in order to place contemporary debates by intellectual advocates for the peasantry into perspective, for many advocates see the contemporary moment as an opportunity to create a new central state-peasant relationship but also as a potential crisis in which the pace of rural social disintegration might quicken. In other words, this essay asks about the meaning of the present moment, of 'opportunity' or 'crisis', and about the context within which we want to understand it.

The debates at the turn of the new millennium are eerily reminiscent of those of the first half of the twentieth century. Both periods were characterized by a disintegration of rural social relations and governance, increasing rural economic crisis, intensifying predatory extraction and exploitation, and the weakening of the central state's capacity to control the rural sphere. Prasenjit Duara calls this combination of expanding state structure with its weakening power, which resulted in intensifying exploitation of rural surplus by entrepreneurial brokers, 'state involution' (1988). And this paper argues that a similar situation began to emerge in the 1990s. What divides these two moments was a period of strong state action in the countryside during the Mao period, and, not coincidentally, the successful industrialization of the Chinese economy, a situation in which state involution ended and entrepreneurial brokerage was eliminated. It is this trajectory of state involution – its emergence, disappearance and return – that provides a context for contemporary debates on rural governance and peasant organization.

At heart, these debates concern whether the end of central-state extraction of rural surplus, symbolized by the abolition of the agricultural tax, could mean the peasantry's democratic control over rural society and the surplus it produces, and what kind of rural organization or governance could facilitate such an outcome. Contemporary rural advocates ask: should the state continue to create institutions that facilitate the integration of rural production and labor power into the national/global economy, or should it support rural political and economic institutions that protect the peasantry from economic exploitation and persistent subsistence crises? Focusing on the politics of peasant advocacy, this paper notes how advocates responded to the state's (often failed) attempt to reach deeper into the rural sphere by proposing new models of organizing rural social life – whether as cooperatives, peasant associations, communes or democratic self-governing villages – shaping state policy and grassroots rural organizing alike.

A century of modernization and extraction

The Qing state was forced by the Boxer Protocols of 1901 to undergo serious reform in the first decade of the twentieth century. These modernizing administrative reforms, known as the New Policies, changed the relationship between the central state and local governance, increasing the intrusion of government into local society to the advantage of county-level elites who gained new powers to extract wealth from the peasant masses. In response, China saw an upsurge of rural revolts and incidents of social disruption in the last years of the Qing, leading up to the 1911 Republican Revolution (Prazniak 1999, 17–8, Bianco 2001).

A main point of these Qing New Policies was an attempt to formalize and centralize village governance, which up to that time was largely decentralized in various village associations (C. Liu 2007, 58). Before this time, the imperial state ended at the county level, which was ruled by a magistrate appointed from the center. That magistrate had to rely on unofficial local elites as well as less well-off agents to help it collect taxes and police society. While this was a rather frugal system of administration, it also opened up the state to problems of tax avoidance and corruption. With the New Policies, however, the Qing state began to promote a mild form of local self-government (*difang zizhi*), a central component of the New Policies, in order to deepen the reach of the state and modernize and mobilize rural society (Kuhn 1975, 276). But these reforms ended up only partially extending the reach of the state, and they had diverse effects on rural governance and society. As Roxann Prazniak argues, these self-government reforms ended up augmenting the power of the rural elite (Prazniak 1999, 32), who continued their class-based extraction through rents, usury and control over local taxation. Chang Liu argues that, while new village heads of North China were to be elected under these reforms, they were also supposed to act as a government representative in the village (C. Liu 2007, 58). Li Huaiyin sees these reforms as the 'formalization of local leadership' (H. Li 2005, 17). County level self-government bureaus were formed through elections of elite representatives from the subdivisions of the county, and the councils were mandated to create new schools and police offices, take a census and gather funds for modernizing projects (Prazniak 1999, 31, 34). These projects placed a new fiscal burden on the rural population in the form of surcharges and levies. Furthermore, new taxes known as the *tankuan* were levied on villages as a whole, instead of on individual property, increasing the importance of village political organization (Duara 1988, 3–4, C. Liu 2007).

The term *zizhi* (self-governance or self-government), through which these reforms came to be understood, continues to be a key category of contemporary rural politics and remains a flexible political category. For example, it could indicate an increase in the state's top-down reach through the mobilization and management of local initiative or, alternatively, a bottom-up non-state form of political and social organization. By the 1930s, the term reached beyond the political sphere, with the ability to indicate forms of cooperative economic organization as well. Moreover, with the end of the central state's rural extraction for modernization and industrialization in 2006, the questions of organizing the peasant and reimagining rural social and political life took on new importance, and contemporary advocates use the term *zizhi* to blur the boundary between state and society, opening the space for new and innovative forms of village organization. In other words, contemporary debates on rural governance clearly resonate with those from the first half of the twentieth century, and the vicissitudes of rural governance reforms and of the category *zizhi* take on new significance in this moment in which the central state has abolished its extraction from rural society.

The category *zizhi* entered late-Qing discourse from Japan through the efforts of Huang Zunxian, a Qing diplomat and official who served in Japan from 1877 to 1882.[2] As an official in Hunan in the late 1890s, Huang argued against the law of avoidance, which barred officials from holding office within their home locality, and believed that the local elite should take a greater role resolving local problems (Kuhn 1975, 270, Gao 2003, 112). The indigenous discourse of *fengjian*, indicating a 'feudal' governing system in which leaders are chosen from the local population, helped set the stage for adopting the *zizhi* concept of self-governance (Kuhn 1975, Min 1989, Thompson 1995). In fact in 1893, well before the New Policies were introduced in the last decade of the Qing, Chen Chi suggested that township officials (*xiangguan*) should be elected (Min 1989, 118–20, Ma 1997, 188). Furthermore, the term was used to convey Confucian ideals, and, in a

---

[2]The concept was brought to Japan from Prussia and was written into Japanese imperial law in 1888 (Kuhn 1975, 272, Thompson 1995, 12–6).

speech before the Southern Study Society in Changsha in 1898, Huang 'told his listeners to both "cultivate themselves" (*zizhi qishen*) and to "attend to (*zizhi*) local affairs (*xiang*)"' (Thompson 1995, 8). As Philip Kuhn argues, the new stress on local leadership was a response to the disintegrating effects that nineteenth-century rebellions had on Qing rule and was a route not to 'the dispersal of power but rather the building of nationhood', based on 'local activism'. This elite activism would both bolster the local government and also work to initiate commerce and industry (Kuhn 1975, 270, Rankin 1997).

In announcing the local self-governance reforms, the Qing state clearly stated that self-government 'refers to that which accompanies bureaucratic rule without contradicting it; it is absolutely not a term referring to that which moves ahead on its own regardless of bureaucratic rule' (quoted in Ma 1989, 202). Yet the category *zizhi* was ambiguous enough to mean different things to different people. Its coordinates included local autonomy – enlarging local power vis-à-vis the center – and social and popular participation. It could mean the mobilization of societal interest groups – especially local elite and leaders – to participate in local governance as a 'supplement to bureaucratic administration', a top-down ideal or a bottom-up 'vehicle for societal initiative' (Rankin 1997, 269). As Mary Rankin argues concerning the early Republic, 'elite participatory ambitions often put a higher premium on taking part in government than on safeguarding autonomy and defining rights against the state' (1997, 269). North China governors Zhao Erxun in Shanxi and Yuan Shikai in Zhili and Shandong developed New Policy *zizhi* reforms (Thompson 1988, 1995). The concept was picked up by reformist intellectuals such as Kang Youwei, who in 1902 argued that only self-government would bring about local development and national survival (Kuhn 1975, 272–74).

The New Policies, however, also sparked a negative and violent reaction among the peasantry that spread around China shortly before the end of the Qing Dynasty. Interpreting these local revolts against the modernizing reforms and the new local-state extractions that funded them, Chinese intellectuals became critical advocates for reforming rural social relations and state-society relations. Late-Qing anti-Manchu and anarchist revolutionary Liu Shipei interpreted local revolts against the New Policies as the emergence of a revolutionary peasantry capable of transforming China.[3]

Liu believed the New Policies, with their combination of national assembly (*guohui*) and local self-governance (*difang zizhi*), would lead to a strengthening of elite rule over the people as well as increasing economic inequality (S. Liu 1996a, 205–7). Instead, he proposed a new, higher level of organization for rural life and work, believing that peasants should be organized into *xiang*, free associations of about a thousand people (S. Liu 1996b). All property and capital would be abolished and become public. With mechanization, manual labor would diminish over time. The key for Liu was that there would be no division of labor: everyone would take part in every job during the course of a lifetime, and thus there would be no division between mental and manual labor or between ruler and ruled. The *xiang* form would recreate the unified agrarian-industrial system of production that Liu associated with traditional rural society but at a higher level of collective organization, one befitting human nature, human morals and the direction of social evolution (S. Liu 1996c, 285). From the time of Liu's 1907 proposal down to today, intellectuals have acted as advocates of new forms of peasant organization, often becoming very critical of the role of the central state and the predatory local state in the countryside.

---

[3]For more on Liu see Dirlik 1991 and Zarrow 1990.

Other Chinese intellectual advocates for the peasantry took a less revolutionary approach, and saw in both the Qing New Policies and Republican period village self-government reforms opportunities for social reform in the countryside, reforms that would push the definition of self-government in a more grassroots direction than the state intended. Some reformist gentry-intellectuals, for example, believed that the gentry would play a key role in organizing the harmony of the rural sphere, and that national strength could only come through rural and class unity. In 1903, Mi Digang of Zhaicheng in present-day Dingxian, Hebei Province, returned from study abroad in Japan, where he had been influenced by the Japanese new village movement (J. Liu 2008). Mi's father, Mi Jiansan, had been promoting village education reform in Zhaicheng since the 1890s and had opened a group of schools in the village beginning in 1902 (C. Liu 2007, 65). Upon his return to Zhaicheng, Mi Digang initiated what he called *cunzhi* (village governance) reforms, building on the late-Qing New Policies by centralizing school finance. Working with the county magistrate, Sun Faxu, Mi initiated reform projects that included schools, village banking, purchasing and marketing cooperatives, the reform of customs and tax collection, transforming village governance by making it both more centralized and more democratic (Alitto 1979, 146, C. Liu 2007, 65–6).[4] Together with Mi in Hebei, Wang Hongyi, who returned from study in Japan to open rural schools in Caozhou, Shandong, formed the core of the *cunzhipai* (village governance group) (Zheng 1999, 182, Dai 2007, 28–9). *Cunzhi* was understood as heightening the organization of village life through increased participation in local governance and through the use of economic cooperatives. But it was only later in the Republican period that the *cunzhi* reforms that Mi and Wang promoted became a national model.

The Republican period

After the 1911 fall of the Qing, Yuan Shikai, the new president of the Republic of China whose 1907 Tianjin self-government reforms were influential on the Qing New Policies, ended self-government by decree in order to limit the power of the rising gentry. Cooperation between local elites and the central government over the control of rural institutions and the extraction of rural surplus increasingly turned to competition. The subcounty ward (*qu*), the lowest level of the state structure at the time, was not a self-governing body but an extension of the county government, used by Yuan to attempt to bring local leaders into the state structure. Kuhn argues that the *qu* or *zizhi qu* (self-governing ward) was an institution created to control and harness the power of various gentry organizations that emerged in response to the Taiping and other late-imperial rebellions to take on activities once under state charge (1970, 215–20). While Yuan later began to promote limited experimentation in local self-government, his new self-government code of 1914 was only instituted to any extent in Shanxi, although it had some influence in other areas, and again its purpose was to bring local men under the magistrate (Kuhn 1975, 276–80, Rankin 1997, 270). In the early years of the Republic, the category *zizhi* could mean local autonomy or elite and even popular participation, thus '[p]rovincial self-government became a convenient slogan for the militarists, but was also used by civilian reformers opposed to Yuan's control' (Rankin 1997, 270). Yet the developing

---

[4]Sun Faxu is listed as Sun Chunzhai in Alitto (1979). Sun later left to work with Shanxi warlord Yan Xishan – for whom Liu Shipei had also acted as an advisor – helping to implement village self-governance reforms (Alitto 1979, 146, C. Liu 2007).

relationship between imperial officials and progressive elite activists of the late Qing, Rankin argues, was 'eroded' by the politics of the early Republic, and the repression of the Yuan period made a participatory form of state-society relationship 'impossible' (Rankin 1997, 267). Failed centralization descended into repression, and local autonomy turned into warlord rule: the peasantry was more exploited than ever by local and regional elites.

At the time of the 1926 Northern Expedition, in which the Chinese Communist Party (CCP) worked with the Nationalist Party to end warlord rule and reunify China, the CCP organized peasants into peasant associations aimed at breaking the rural power structure and mobilizing peasants for revolution.[5] In his pre-Marxist days, Mao used the concept of *zizhi* to mean radical popular self-rule (Duara 1995, 191). As the Northern Expedition pushed north from Guangdong, the peasant movement grew rapidly, especially in Hunan. Mao arrived in the area and soon began to investigate the movement, writing his famous 'Hunan Report' in early 1927 (Mao 1965). Mao used the report, which was his strongest endorsement of peasant revolutionary activity and of organizing the peasantry into peasant associations, to argue forcefully that the party should fully support the rural revolution and guide the development of peasant organizations. In the text, Mao argues for a notion of revolution in which peasants had to enter the revolution on their own terms, and only by actively participating, even in violence, would the peasantry free itself from the power relations that enforced the rural class structure. Self-rule became self-activity. A main target of this self-activity would be the old gentry organizations formed during and following the Taiping Rebellion, organizations that the weak Republican state had attempted to incorporate into the state structure as subcounty administrative units.[6] This rural organization work was largely new for the young CCP, and its radicalism helped push Chiang Kai-shek to violently break with the Communists in 1927.

After the 1927 split with the CCP, the Nationalists increasingly utilized military means to root out radicalism and communist organization, and the party had little in the way of an effective rural reform program. Once the Nationalists had gained control over North China in 1928, they attempted to formalize local government at the subcounty level, though it remained strongly controlled by the county (Kuhn 1975, 286, Duara 1988, 62, H. Li 2005, 18). As the work of Li Huaiyin (2005, 16), Chang Liu (2007), and Kenneth Pomeranz (1993, 272) show, in different areas, under different conditions, this state-making process had different results. In some areas there was a tendency towards village breakdown, in other areas villages were able to increase internal solidarity in resistance to state-making pressures. Yet in some areas one finds a combination of successful state-making and a strengthening society.

Prasenjit Duara argues that incomplete state making in North China villages from the time of the late-Qing New Policies through the Republican period led to the emergence of 'predatory state-brokers' who exploited the peasantry for personal gain – the competition for rural surplus between state and local elites continued, with the central state losing ground. While the state attempted 'to deepen and strengthen its command over rural society', it was never able to fully formalize its administrative power (Duara 1988, 1).

---

[5]There were earlier attempts at forming peasant associations or unions. On the organizing work of Peng Pai, see Hofheinz (1977) and Marks (1984); on Fang Zhimin, see Sheel (1989); and, on the Nationalist Shen Dingyi see Schoppa (1995).

[6]Mao (1965) talks of the *du* and the *tuan*, which became formalized as levels of governance as the *qu* and *xiang* during the Republican period, for example (see Kuhn 1970, 220).

Duara argues that there were two types of brokerage that mediated the relationship between rural society and the state from the late Qing. 'Protective brokerage' existed where communities collectively attempted to gather taxes or carry out other demands of the state in order to avoid predatory middle men, or 'entrepreneurial brokers'. The latter inserted themselves in between the state and rural society in order to make a profit. Throughout this time, according to Duara, as the state increased its demands for local resources without matching those demands with an increase in its capacity for state making, protective brokerage was gradually replaced by entrepreneurial brokerage: in other words, 'the ability of the state to control local society was less than its ability to extract from it' (Duara 1988, 73).

While the Nationalist regime did attempt to institute a system of *xiang* (township) – between 100 and 1000 households – and *zhen* (municipality) as the highest level of local self-government, in reality it still put the burden on villages, which grew in importance as a result, to pay the new levies it demanded. Informal forces played a greater and greater role as state demands increased. Duara calls this 'state involution', in which there is both a growing state structure and its disintegration, and the result is the increasing power of the entrepreneurial broker, accountable neither to the state nor to the local population.

In a comparison of the social and political make-up of the countryside in South and North China, Chang Liu (C. Liu 2007) argues that many North China villages became more unified in response to state-making pressures, creating an opportunity for CCP organizing that did not exist in the south. In the south, a weakening landlord class, which had largely moved into the cities by the twentieth century, relied more heavily on the government to collect its rent and protect its property. Mediating class relations there allowed the state to gain the upper hand through its 'infrastructural power' and ability 'to penetrate local society, and to coordinate the activities of different social groups and classes for its own purposes, without employing a despotic, coercive force' (C. Liu 2007, 47). Alongside the state involution that Duara finds in the north, however, Liu argues that many villages grew stronger in response to state intrusion and its new levies, giving the villages a new unity – a process to which, Liu argues, Duara does not pay enough attention. The CCP was able to base its revolution against the Nationalists in these resistant village communities.

In Huailu County, Hebei Province, a core rural area of North China, Li Huaiyin (2005) also finds tight-knit communities that defended their villages against growing outside pressures. There, in the administration of the village, informal village cooperative institutions dominated. Yet this was not a simple dichotomy of society versus the state, as the county 'magistrate often acted as a mediator between the provincial government and local elites, rather than as a representative of the state' (H. Li 2005, 18). Interdependence and cooperation were the norm between state and village society in Huailu County. These informal village institutions persisted into the twentieth century, despite the inroads of the state, although the state attempted to formalize them. In 1928, however, the Nationalists were able to weaken the power of county-level urban elites and successfully extend the formal bureaucracy to the subcounty level. Nonetheless, subcounty leaders still primarily acted as protectors of the village community. Thus, in Huailu, while Li finds 'a concurrence of tighter state control over local society and the formalization of local administration that enhanced the leadership of rural elites' (H. Li 2005, 18), cohesive tight-knit communities remained strong protectors of local interests.

Non-communist peasant advocates, like the state itself, intervened in the complex conditions of rural society, attempting to 'reconstruct' the rural social, cultural and economic life of the involutionary North China village. In 1924, *cunzhipai* advocate Wang Hongyi convinced Liang Shuming, a conservative Beijing University scholar who promoted traditional culture as the foundation to rebuilding China, to move to Shandong, leading him to support the *cunzhipai*'s reform projects (Alitto 1979, 145–9). This led to the Rural Reconstruction movement (*xiangcun jianshe yundong*), led by Liang and Yan Yangchu (also known as James Yen). Following the *cunzhi* approach of village democratization and cooperatives, Liang argued that Chinese culture needed democratization (*minzhuhua*) and socialization (*shehuihua*), the essence of rural reconstruction. Collectivized villages would form the basis for the complete reorganization of Chinese society, bringing about development without the conflicts of class society and urban industrialization (Alitto 1979, 174–5, Dai 2007, 51–3). Instead of emphasizing the importance of rural class divisions, as Mao did, Liang stressed that rural and cultural reconstruction could lead to rural society acting as a unified whole. It was Liang's view of the peasantry as a unitary political subject that returned at the end of the twentieth century, and Mao's stress on class differentiation dropped out.

Yan, a Christian educated in Hong Kong and the US who saw education as the most important method for ending rural poverty, formed the Mass Education Association in 1923 with other Chinese rural-focused reformers.[7] Originally active among the urban poor, the society moved its activities to the countryside in 1926, basing itself in Dingxian, where the Mi family's Zhaicheng model of village reform was located (Douw 1991, 35). Yan believed that peasant ignorance was a root cause of rural poverty, and his Dingxian reforms focused on the 'four weaknesses' of villages: 'poverty, disease, ignorance, and misgovernment' (Hayford 1990, *x*). While education remained its basis, as Yan grew to be more involved in transforming rural society he also promoted the economic cooperatives of the Zhaicheng model. His Dingxian experiment came to be seen as one of the most important models of the Rural Reconstruction movement.

Some members of the *cunzhipai* grew very critical of Liang's views, including Lü Zhenyu, later an important CCP historian. In 1928, Lü went north from Hunan to Hebei and Shanxi to investigate local self-government efforts, and, impressed by the Zhaicheng experiment, he decided to learn more about the *difang zizhi* and *cunzhi* movements. From the beginning of his time in the group, Lü believed that *cunzhi* should mean real self-determination, self-governance and self-activity of the people, for only they would 'destroy the contradiction of ruler and ruled' (Dai 2007, 35). Opposed to *zizhi* (self-government) was *guanzhi* (official rule), which was the expropriation of the people's power; but even the recent *difang zizhi* movement was really the rule of the people (*zhimin*) by local gentry. Yet Lü still foresaw a role for the state to 'assist' and 'guide' the people in fostering a self-governing capacity (Dai 2007, 32–40).

Lü, however, argued that Liang mis-analyzed the roots of rural poverty, which by the early 1930s he saw as a class issue: peasants did not have access to land or the tools of agricultural production and were often forced to become unemployed migrants in the cities and countryside. Peasant cooperatives were helpful for small landowners, but could not solve the land question. Likewise, while rural education might be provided to peasants by humanitarian gentry, education was not a solution but itself would only truly be improved if the rural economic problem was solved first, by giving 'land to the tiller' (Dai 2007, 56–8). Once land was in the hands of the tillers, then the socialization of agriculture could begin, increasing the scale of production and management by unifying the land, capital and labor of each village (Dai 2007, 45–8).

---

[7]For a comprehensive biography of Yan, see Hayford (1990).

With the 1937 Japanese invasion of North and East China, most rural reconstruction experiments ended, although a new, more forceful and brutal form of governance appeared. In 1941, the Japanese introduced the *daxiang* (large township), 1000 households in size, in order to break what power the natural village had left. Yet this only furthered the involution of local state control, as more revenue was extracted to pay for this new level of government, and entrepreneurial brokers took on new positions of power as the headmen of the township (Duara 1988, 223–34).

In communist-controlled areas a different process took place, one that Chang Liu's work details well (2007). There, the more unified villages created under the pressures of modernization and state extraction – especially the *tankuan* levies – since the late Qing became a fundamental building block of communist power, and, as Liu argues, the CCP took 'the village as the basic unit' of taxation and mobilization (C. Liu 2007, 98). In fact, the class politics of the party was used to produce a more egalitarian and solidified village community, and it was on the village community as a whole that taxes were usually levied. As Liu concludes, '[t]he institution of the unified progressive tax in the base area [Jin-Cha-Ji] brought an end to the transformation of taxation from a mainly state-individual affair to the one of entirely state-village intercourse', a transformation that began in the late Qing (C. Liu 2007, 99). The CCP thus built on a preexisting process to strengthen the village community, which became the focus of CCP mobilization. As the CCP attempted to construct the village government as its lowest level of administration with the important task of taxation, however, a tense relationship between the party and village often returned (C. Liu 2007, 110–2).

The People's Republic

After 1949, the party moved from the revolutionary seizure of power to reconstruction. In one of the most momentous transformations of rural China, land reform was completed by 1953, creating largely egalitarian villages and eliminating one set of rivals for the control of rural surplus – class-based surplus extraction was largely ended. In some areas, however, as the post-war economy improved, inequality began to return. Yet the state was caught between seemingly contradictory goals of rapidly growing the economy and maintaining economic equality in the countryside. With Mao's forceful support, the policy moved more swiftly towards collectivization, which Jonathan Unger rightly calls 'one of the largest efforts in world history to reorganize people's lives and livelihoods' (2002, 7). Collectivization would accomplish several goals according to Mao: on the one hand, it would protect against the return of economic inequality within the villages; on the other hand, it would also facilitate the development of agricultural production, so vital to the newly industrializing nation. At the same time, the CCP eliminated the 'entrepreneurial state brokers' that had siphoned off so much of the rural surplus (Duara 1988, 254). While this paper places the state-rural society relationship of the Mao period into a narrative of the transformation of rural governance and state control over rural surplus, the significance and impact of the CCP's elimination of rural class-based exploitation should not be underestimated. The new Chinese state was certainly looking to eliminate what it saw as a wasteful and illegitimate use of rural surplus through the introduction of new institutions, but it also attempted to radically transform rural power relations and rural life.

Introduced in 1953, 'unified purchasing and marketing' (*tonggou tongxiao*) gave the state the sole right to buy and sell grain (and later other agricultural goods), and at fixed prices and quotas. Not only did this remove the middlemen of the old system and guarantee low-cost grain supplies to the cities, state control over the price and quota meant the system acted like a hidden tax by underpaying agricultural producers (Sicular 1988, 253–4, Wen 1999a, Kuhn 2002, 106, Unger 2002, 12). From the mid- to late 1950s, organizing the peasantry into enlarged collectives and even communes the size of marketing towns – the old *xiang* – made it easier to extract both agricultural taxes and below-market-price agricultural resources from the countryside (during the Great Leap Forward, this ease of extraction led to catastrophic consequences). Reducing the transaction costs of this peasant-state exchange was, of course, not the only reason for collectivization, but recently this rationale for collectivization has become central to left-leaning critiques of current state policy in China (Wen 1999a, 1999b, 2001b, He 2004a, 2008, Cao 2010). Wen Tiejun's narrative of the Mao period, for example, argues that collectivization was primarily a way to facilitate the extraction of rural surplus at low costs in order to rapidly industrialize the Chinese economy, and, thus, the nation now owes the peasantry for their earlier sacrifices (Wen 1999a and 1999b).[8] Nonetheless, this narrative tends to depoliticize the radical transformation that the CCP and rural activists helped to produce in the countryside; although clearly Wen and other contemporary left-leaning scholars who employ this narrative do so in order to stress the positive, and often forgotten, role the peasantry played in the industrialization of China and its current economic success.

With land reform, state-controlled purchasing and marketing, and collectivization the state brought over 50 years of 'state involution' to an end in rural China.[9] None of this would have been possible without the party's grassroots organizing efforts in the countryside – based in the village – which took place during the revolution and land reform. Nor would it have been possible if the party had not built up goodwill among the peasantry. No earlier state was able to reach so deep into rural Chinese society to produce an effective structure such as this.

Following the disastrous Great Leap Forward (1958–1961), the party changed the direction of rural organization, laying the basis for rural governance over the next couple of decades (Y. Xu 2003). The unit of economic planning was reduced from the commune (the size of a market town – *xiang* – together with its villages) to the production team (*shengchan dui*), which was the size of a hamlet or part of a village with 10 to 50 households. They were led by a head chosen by the members and over time were able to provide a basic level of important public goods such as basic health care, elder care for those without progeny to support them, and rudimentary schooling (Unger 2002, 10–11). The hamlet or village neighborhood became more tightly knit, continuing a trend of the pre-1949 period. The production team head, chosen by team members, attempted to protect the team's interests, a form of 'protective state brokerage', to use Duara's terms (1988).

Above the teams was the production brigade (*shengchan dadui*), the size of a large village and the lowest level of administration in which the party branch operated with the production brigade party secretary. Though brigade leaders were originally from the village, they were chosen by the party, and this, as Unger shows, put them in a difficult position: 'politics within Chinese villages often revolved around an intrinsic conflict between the production teams – economic entities that were somewhat democratic – versus the brigade level, a political entity, essentially autocratic and under pressure to uphold and enforce Party policy' (Unger 2002, 17).

---

[8]For an English language version of Wen 1999b, see Wen 2001a. For a detailed discussion of these political narratives, see Day 2013.
[9]Philip Huang (1993), however, argues that 'economic involution' continued, as rural labor productivity did not markedly increase during the Mao period.

Higher up were the commune and county governments, through which the party controlled the brigades and teams. For Vivienne Shue, the political structure of rural governance during the Mao period was defined by its 'honeycomb pattern', in which power was segmented by the various levels of the bureaucracy and economic and social life grew increasingly self-contained. [10] This characterized the commune level as it did the production brigade or team. Philip Kuhn views the commune as a bureaucratic extension of the state and 'a substantial extension of state control', when compared to the Nationalist period (2002, 110). For Kuhn, this system 'blended small-scale socialism with intensified bureaucratic control', in effect, 'a deeper state penetration into village society and a more rigorous system of extraction from it' (Kuhn 2002, 110).

This system, Kuhn argues, was 'confirmed' during the reform period (1978 to the present) by the 1983 'separation of township administration from the rural economy. Bureaucratic penetration had survived, even as socialism disintegrated' (Kuhn 2002, 110). While confirming the general trend, Shue sees a bigger break in the form of rural governance, arguing that the Deng regime sought to shatter the Mao-period honeycomb structure of rural governance, as it had the capacity to resist the Dengist reform program. A new 'weblike' pattern emerged, in which networks, largely built around rural-urban commercial exchange, transcended the old cells of the honeycomb structure (Shue 1988, 131, 147–52). In other words, the little protection the honeycomb structure of governance had provided rural society was, with the reforms, losing its viability. In fact, Shue sees the rural market reforms as being designed to attack the honeycomb structure and the localism of the Mao period (1988, 148). In the long run, however, these reforms did not lead to stronger state control so much as the reappearance of state involution.

Over the 1980s, therefore, the reforms took apart the Mao-period structure that was built from the 1950s. Land was contracted to households, which became the basic economic unit in the countryside. In 1983, communes were disbanded as political and economic functions were separated, with township (*xiang*) governments taking over their administrative function (Unger 2002). In 1985, the unified purchasing and marketing system was transformed into a contract procurement (*hetong dinggou*) system, and over time, market forces increasingly controlled the agricultural exchange (Oi 1986). There were scattered experiments in village self-government (*cunmin zizhi*) in the early 1980s, with villagers' committees (*cunmin weiyuanhui*) replacing production brigades. While some were critical of village self-government, calling instead for the village to be governed by top-down 'administrative offices' (*cungongsuo*), such proposals lost the debate within policy circles (O'Brien and Li 2000, 470–1). By the late 1980s the creation of democratic villagers' committees had become law, although implementation was slow and uneven. The disruption caused by the crackdown on the 1989 Tiananmen protests put democratizing reforms on hold through much of the 1990s, and only after the passing of a revised law on rural elections in 1998 did institutionalization of rural democracy really begin to take hold. Even so, the local party branch was still the main power in the village (O'Brien and Li 2000).

---

[10]Unger argues that 'the village now faced the state on its own, naked as it were, as it was largely isolated from neighboring villages' (Unger 2002, 22). Vivienne Shue, too, argues that peasant communities became more bordered and internally unified at this time (Shue 1988, 132–47), continuing the trend of CCP reliance on unified villages that Chang Liu noted (C. Liu 2007).

The crisis of the 1990s, *sannong wenti*, and the return of state involution

Although the rural political and economic structure was markedly reformed in the 1980s, the state at various levels continued its extractions from peasants and agriculture. Furthermore, competition over rural surplus and the control of rural labor power – between the central state, the local state and capital – intensified through the reform period, with the amount the local state ate up and the burden on the peasantry mushrooming during the 1990s. Shifting state taxation policy, more local autonomy concerning revenues and expenditures, increasing regional inequality, and the failure and privatization of local revenue-generating enterprises led to worsening rural economic conditions. While many scholars, especially in the west, blame the increasing burden on the peasantry on a lack of political capacity or political will of the central state (for a discussion of these issues see Kennedy 2013), many of the Chinese rural advocates discussed in the next section point to the structure of the rural political economy as the fundamental problem. Nonetheless, by the 1990s, central and local state extractions, and the burden on the peasantry they produced, brought about decreasing returns for local society over time, a return of the 'state involution' that Duara (1988) found in the early twentieth century.

After years of decline in the central government's proportion of tax revenues, the fiscal reforms in 1994 separated the revenue streams of central and local governments. In an influential report published at the time, Wang Shaoguang and Hu Angang (1994) noted that the dropping percentage of central state revenues was severely weakening the state's capacity to affect the economy. The reforms increased both total government revenues and the central government proportion, but they also created a number of problems and unintended effects. The reforms increased revenue inequality between poor and rich provinces. It also gave more fiscal autonomy to provincial and sub-provincial governments, giving local governments more leeway and incentive to squeeze more revenues out of the local population. Local extra-budgetary funds in the form of various fees grew tremendously in the 1990s as a result – a problem the central state has attempted to deal with since that time (Herschler 1995, S. Wang 1997, Göbel 2010). The tax reforms of the 1990s through to the 2006 abolition of the agricultural tax, therefore, should be seen as a central-state response to the reemergence of state involution in the countryside.

After two decades of rural reforms, commentators on peasants and agriculture began to talk about the crisis of rural society, and many began to consider that the rural situation was deteriorating in the 1990s (see, for example, Wen 1999a, C. Li 2002, 2003, Kuang 2005).[11] In the late 1990s, rural industries began to go bankrupt or were privatized – often in manner that allowed local cadres to plunder village and township resources – leading to a decline in rural nonfarm employment; peasant incomes stagnated (Whiting 2001, 289–90, Bernstein and Lü 2003, 48–9, Keidel 2007, 86–7). There were increasing peasant protests, usually concerning taxation and the financial 'burden' (*fudan*) on the peasantry brought about by the involutionary rural government (Bernstein and Lü 2003, O'Brien and Li 2006, Walker 2006, 2008). This growing rural crisis became known as *sannong wenti* (the three rural problems: peasants, rural society and agriculture), a phrase popularized by Wen Tiejun (Day 2013).

From the late 1990s on, there was an increase in advocacy for peasants within academic circles, not only because rural crisis became more visible from that time, but also because of the intensifying politics of inequality that emerged in the 1990s. As peasant incomes fell

---

[11]For a detailed discussion of this shift in understanding, see chapter four of Day (2013).

further behind the urban population (a difference of at least one to three around the turn of the millennium) many rural scholars shifted to the left, and began to advocate for better conditions for the peasantry. This new meaning of 'left' in Chinese politics that emerged during the 1990s, particularly around debates on rural issues, indicates various political stances taken against the privatization of land, for the protection of the peasantry from market forces, and for the building of cooperative relations in the countryside (Day 2013). The most prominent formulation of this trend is 'New Rural Reconstruction' (*xin xiangcun jianshe*), an attempt to bring together rural experiements in democratic cooperatives and social organization modeled in part on the Rural Reconstruction movement of the 1930s (Day 2008a, 2008b, 2013, Day and Hale 2007; see also Jacka and Yan 2013). Others maintained a more liberal view that focused on defending the rights of the peasant to control their property, democratic governance and the retreat of the state.

This increasingly visible rural crisis put more political pressure on the central state to adjust its rural policy for the benefit of peasants. The tax-for-fee (*shuifei gaige*) reform of 2002 continued earlier efforts to rein in the proliferating fees at the local level in rural China. The reform aimed to transform local fees into a regulated and limited tax. Christian Göbel (2010) contends that the reforms hit the poor regions hardest, and local resistance led to their failure. A new, and more dramatic, policy was announced in 2004–2005 that led to the abolition of agricultural taxation altogether in 2006. Some see this as more of an evolution of the policy than as a sign of its failure (Kennedy 2007, 2013). The state itself pronounced the tax-for-fee reform a success and suggested that shifting to the abolition of the agricultural tax a few years later was simply a continuation of that policy. Although the abolition had an immediate goal of reining in the involutionary rural state, without further rural governance reforms many local governments could continue their extractions through other means, such as the sale of agricultural land for development. For many localities, central-state transfers have been inadequate to make up for the loss of local revenue, creating an incentive for local governments to continue to find other ways to tap rural surplus. Advocates for the peasantry took this moment to push for further reform of rural governance, aiming to finally bring the involutionary process to an end.

For the central state, too, the goal of abolishing the agricultural tax by 2006 signalled the beginning of a new intensity of rural policy formulation, with grander visions for transforming the countryside. This more comprehensive direction for rural policy was encompassed within the policy framework of 'constructing a New Socialist Countryside' (*jianshe shehuizhuyi xin nongcun*, hereafter NSC), which both brought together many older development projects, such as the creation of a rural cooperative medical system, and created new ones (Pan and Li 2006, W. Wang 2006, X. Xu 2006, Ahlers and Schubert 2009, Perry 2011). Following the announcement of the abolition of agricultural taxes, the idea of the NSC entailed increasing central transfers to rural governments to make up for the loss of taxes, comprehensive urban-rural planning (*chengxiang tongchou*), accelerated construction of rural infrastructure, especially roads and electricity lines, improved education facilities and the resettlement of peasants into better housing, among other projects. While the NSC has meant different things in different areas, one important aspect is the encouragement of in-place urbanization and commercialization, making use of specialized supply-and-marketing cooperatives, with the recognition that migration to large cities was not an immediately viable path for the majority of China's peasants. The NSC framework signals a recognition of the changing shape of rural governance in the post-taxation era. With fiscal transfers from the center to the local level taking precedence over local revenue, counties gained more power over local development projects, as under the NSC framework villages and townships had to apply to their county for project funding (Ahlers and Schubert 2009).

27

At the same time the central state was attempting to confront rural state involution from the 1990s through to the abolition of the agricultural tax in 2006, it was also attempting to further integrate agriculture into the national and global economy, facilitating non-state forms of rural surplus extraction. In some areas this has meant large-scale capitalist farming and class differentiation among the peasantry (Zhang and Donaldson 2008, 2010). But this has been somewhat limited, in part because of the resistance of many peasants to the privatization of land. As Philip Huang has shown, the family farm has persisted in much of rural China and the state has been attempting to use specialized cooperatives and contract farming to vertically integrate peasant farming into the national economy (Huang 2011, Huang *et al.* 2012; see also Day 2008b and 2013 on contemporary debates over rural cooperatives). Burak Gürel, in contrast, argues that China's incomplete privatization of land has in fact facilitated the development of agribusiness and the appropriation of rural surplus labor by capital without full proletarianization (Gürel 2012, 2013). Nonetheless, following the rise of the involutionary local state in the 1990s and reduction of central-state extraction, rural China also began to undergo a transformation in class relations, with the penetration of capital and agribusiness in various forms.

The debate on rural governance and organizing the peasantry

It was within this complex context of an involutionary rural state, disintegrating rural social relations, rural economic crisis, changing central state policy, increasing market integration and emerging class differentiation that the academic debate among advocates for the peasantry turned again to the issue of rural organization and governance at the beginning of the new millennium. Those attempting to reimagine and advocate for a new structure of rural governance faced several problems. How could the local government be reined in to make it both more efficient and less of a burden on the peasantry? Was the main problem the state trampling on the rights of the peasantry, or were capital and the market economy the primary threat to the subsistence and livelihood of the peasantry? Had the state's new rural investment (including agricultural price support, new health and education spending, new infrastructure spending and various projects under the NSC rubric) and the abolition of the agricultural tax solved rural problems or simply opened up an opportunity for further structural reform? Was thinking of rural governance in terms of a state-society dichotomy useful or did it hinder understanding? Should rural organization focus on democratization and the representation of interests in the political sphere or collective organization in the economic sphere as well? The main problem that these advocate-intellectuals seek to address is that of the involutionary rural state, and when they do turn their focus to the rural economy they often treat the issue as one in which an undifferentiated peasantry faced the market. Class differentiation is rarely addressed.

Intellectuals hailed the abolition of the agricultural tax as the beginning of a new era of state-peasant relations. In fact, one of the remarkable aspects of the recent debates on rural governance is the degree to which all participants view the present moment as a major historical turning point, although the meaning of this moment is different for different participants. Notable in almost all of these periodizations is the dismissal of the significance of the Mao period. Advocates either ignored the Mao period, characterized it as a historical mistake, or reduced its effects to establishing successful state extraction of rural surplus. Left-leaning scholars, who had grown more critical of the disintegration of rural society since the 1990s, perhaps had the most coherent, if reductionist, view of the Mao period. Viewing the Mao period as a moment of heavy extraction, they periodized the present as a shift from a century-long era of state extraction from the countryside for industrialization to a new era in which industrial society could pay back the peasantry for their previous sacrifices. For Wen Tiejun, this moment was characterized as the end of the primitive accumulation of capital (*ziben yuanshi jilei*) from the countryside after four attempts at industrialization: the Qing self-strengthening movement that began in the 1860s, modernization under the nationalists from the 1920s on, the Mao period, and the reform period (T. Wen 1999a, 1999b, 2001a, 2001b, 2006). He Xuefeng, constructing a similar historical narrative, characterized the moment as one in which agriculture no longer had to support industry, but instead should be organized to provide the rural population with a stable existence (He 2004a, 218–9, 2008, 219, 227). Cao Jinqing also saw this as a major historical transition in which industry now supported rural society, allowing for a new relationship between the state and the peasantry and a true peasant self-governance to develop (2010, 238–9, 257). Commenting on the involutionary rural state of the 1990s, Li Changping called the present moment the end of the era in which 'agriculture supported government' (*yinong yangzheng*) (C. Li 2004b, 86).

Yet, while intellectuals of all political stripes largely supported the abolition of the agricultural tax, there was much less agreement on how peasants should be organized under the new conditions of the post-tax contryside and, in particular, whether collective economic organizations were legitimite and beneficial. The crux of the issue was an atomized peasantry operating within a market economy and often facing an undemocratic local government that extracted rural surplus for its own ends – a situation with clear similarity to that of the first half of the twentieth century. Was the countryside going to simply continue to disintegrate socially and slide into and out of economic crisis? Could rural society be rebuilt, either through new formal political organizations or through collective cultural and economic reconstruction?

In this debate, some advocates for transforming rural governance, often political liberals, viewed the problem in a state/society framework focusing on separating formal rural government organizations from the economy. This separation and the political reform of formal rural governance would remove the local state from the economy and allow for democratic control over it by the rural populace. Others, however, were critical of the focus on formal state organizations (He 2004a, 105) and the use of the state/society dichotomy (Wu 2004a). These advocates, often leaning to the left, used the political category *zizhi* (self-governance) to blur the distinction between state and society. *Zizhi* – as we have seen a key political category of early twentieth century rural reform movements – was neither state (*guojia*) nor government (*zhengfu*), but a form of rural social organization in between state and society. In fact, it could encompass political, social and economic organization. In other words, as the central state ended its focus on rural extraction, a political space was opened up in which to imagine a variety of new forms of organization befitting rural society that escaped a modernization paradigm modeled on the capitalist west. New possibilities emerged under the category *zizhi* for a genuine self-governance, one that could shake off the top-down state domination that had smothered it in earlier times.

Xu Yong (2003, 137–8), representing the liberal state/society perspective, argues that a different model of rural governance must match up with each historical stage of rural social, economic, and political development. After the Mao period, which Xu characterizes as a period in which rural communes unified state and society, the first wave of rural institutional innovation in the early 1980s produced a split between state and society when the commune system was abolished and replaced by the township government and household-based agricultural production – a separation that Kuhn viewed as the survival of 'bureaucratic penetration' (Kuhn 2002, 110). Yet after that time, tensions increased between the representative of the state (the township government) and rural society, leading to bureaucratism, top-down governance practices and an increasing financial burden on the peasantry. The township government and village self-governance system (*xiangzheng cunzhi*) created at the time, in other words, was no longer suitable for the development of rural society and the market economy, thus necessitating a 'second wave of institutional innovation in township and village administration' (*xiangcun guanli de dierci zhidu chuangxin*). With the tax-for-fee reforms, a new period could arise, in which these state-society tensions or state-peasant tensions could be resolved through further governance reforms at the township level, the key link between the state and rural society. Such reform would entail a shift from 'township government and village self-governance' (*xiangzheng cunzhi*) to 'county government, township branch, and village self-governance' (*xianzheng xiangpai cunzhi*), a reform that would enhanced county-level powers (Y. Xu 2003, 120, 137–45).

Under Xu's influential formulation, the state would retreat and the county would become the basic level of state power in the countryside. While it would have a higher degree of independence, its power would be restricted by state regulation and law, and not by direct administrative intervention. The township would become a county administrative branch office (*xian de paichu jigou*), financed by the county, reducing its primary functions to managing family planning, public security and various social service tasks. It would also work to guide (*zhidao*) the activities of the self-governing villages. In turn, the township people's representatives would check the activities of the township. Such a township reform would shrink the size of its personnel and lighten the burden on the peasantry. Underneath the township, Xu argues, village self-governance should become a real self-governing organization that would carry out tasks decided upon by the villagers' committee. The goal for Xu was to create an efficient and cheap governance system in the countryside that facilitated 'positive interaction between state and society' (*guojia yu shehui liangxing hudong*) and raised the self-organization of the peasantry (Y. Xu 2003, 121, 144–5). As rural urbanization (*chengzhenhua*) and depeasantization-deagriculturalization (*feinonghua*) accelerated, according to Xu, small cities (*xiaocheng*) and towns (*zhen*) should become self-governing organizations separate from townships (*xiang*) (Y. Xu 2003, 166–9).

Dang Guoying (2004), following a similar modernization paradigm, argues that most discussions on reforming the township government mistakenly focused on merging townships to reduce administrative costs. Dang argues, however, that with these reforms some townships became too big, leading to inefficiency and lack of services for the peasantry. According to Dang, however, the point should be to provide more and more efficient services. Dang advocates shrinking the size of government organizations above the township level, allowing for more rapid urbanization, and separating urbanizing towns (*zhen*) to become more independent of the townships. As labor migrates, the township level would increasingly take over the service function of the village level (*cun*), which would lose meaning over time. The township would then also become democratic. Furthermore, in the mainstay of the liberal argument, Dang advocates that peasants should be given more legal land rights and be able to represent their interests through popular organizations like peasant associations (*nonghui*).

The liberal-populist Yu Jianrong (2010a, 2010b, 2010c, Yu and Cai 2008) argues that rural governance reform must begin at the county level. Thus, while some, such as He Xuefeng (2004b), worry that the abolition of agricultural taxes could weaken rural governance, especially at the township level, Yu argues that that was not a bad thing (2010c). In general, Yu argues that the source of power for the county must shift from above to below, leading to an independent and democratic county government. County-level people's congresses should have the authority to oversee, investigate and even dismiss county administrators. Furthermore, administrative power should be concentrated in the county executive, making it more efficient and easily accountable (2010c, 2011). Over time, the county head could be directly elected as well (2010c). The township government should be abolished and transformed into a self-governing organization (*xiangzhen zizhi*) (Yu 2004).

According to Yu, however, such a democracy could not turn into reality if peasants remained second-class citizens (*erdeng gongmin*). Here we see peasants treated as a unitary class, produced in its relationship to the state, masking class differentiation and non-state forms of extraction and exploitation. Peasants need to be 'liberated' (*jiefang nongmin*) and turned into 'modern citizens' (*xiandai gongmin*). Yu's remedy to long-term state discrimination is to tie democracy to property rights. If peasants are restricted in their ability to create wealth (*chuangzao caifu*) by the collective land ownership system, they will not be able to realize local democracy. Clarifying land rights and interests could mean turning the collective into a shareholding system, in which peasants could sell their shares if they move out of the village (2010c). The most important aspect of Yu's proposals is the establishment of peasant associations, modeled in part on the experience of Taiwanese peasant associations, which would represent peasant interests to the government and prevent various vanity projects of local cadres that waste resources and open the government up to corruption (2006). Peasant associations would be 'of the peasants, by the peasants, and for the peasants' (*nongyou, nongzhi he nongxiang*) and not state organizations. They could aid in the development of marketing cooperatives, helping peasants to 'go into the market' (*zouxiang shichang*) (2010c).

Breaking with the liberal state/society framework, Wu Licai (2004a, 2004b) focuses on the historical relationship between the county, the township, and the village, noting that the township was caught between higher levels of government and village self-government. Wu argues that the state/society framework is not useful, as there are multiple levels in relation with each other. In reality, the township mediates the relationship between the peasantry and the state; it is a 'third realm' between state and society, citing Philip Huang's formulation (Huang 1993). We could note that the mediating role of the township that Wu observes places it in the position of the state brokers central to Duara's (1988) analysis of the early twentieth century, and thus the township could play either a protective or an entrepreneurial role. Wu's goal is to figure out how to institutionalize this mediating role in a way that turns the township into a productive and protective institution instead of an involutionary one.

Turning the township government into a county-directed administrative office (*gongsuo*), as Xu Yong and others advocated, would not solve this problem, but would merely extend the bureaucracy downward – it is a bureaucratic, not democratic, instrument. Wu instead argues for the importance of recognizing and strengthening the mediating role of the township level, turning the township government into '*xiangzheng zizhi*' (township government self-governance). This would help to create true cooperation between the officials and the people (*guanmin hezuo*). The power of the higher level government would allow the *xiangzheng zizhi* to have independence from the county, making it a real unit of rural governance; whereas direct elections at the *xiangzheng zizhi* level would make it a popular organization. In other words, Wu advocates for both the strengthening of state backing at the township level and the strengthening of democracy.

Li Changping (2004a, 2004b), a former rural cadre and critic of liberal proposals for land privatization, argues that overly strong official power, originating in a commune system set up to extract resources for industrialization, was the cause of rural governance problems. As long as the local government lives off the surplus of the peasantry, reform would be impossible. With the abolition of the agricultural tax, true reform becomes possible. Yet popular power still has to grow to counter official power. The dominance of official power over popular power in the countryside meant that rural reforms continued to fail through the 1990s. Because of this, simply changing the *xiangzhen* government into a *xianggongsuo* (township administrative office) will also fail if popular power is not strengthened. Likewise, even *cunmin zizhi* cannot counter official power unless rural popular interest groups are better organized. Li advocates turning the township and town (*xiangzhen*) governments into self-governing organizations (*zizhi zhengfu*) and argues that existing rural social organizations, such as youth and women's organizations, should become more popular based. In addition, the state needs to create space for new rural organizations to come into existence that can represent different interest groups, such as civil associations (*xiehui*), peasant associations (*nonghui*) and cooperatives (*hezuoshe*), especially community financial cooperatives (*shequ jinrong hezuo zuzhi*). Under these conditions, one of the prime functions of the township-level government would be to support those rural social organizations (C. Li 2005).

For Li, therefore, unlike many liberals, rural interests are not individual interests; higher-level social and economic organization is necessary to match higher-level political organization, and ideally they should merge. The political organization of interests cannot simply bring together the individual interests of an atomized peasantry, in other words, but would also need to raise the level of village economic and social organization in order to be truly strong enough to match official power. In addition, peasants need greater political and economic organization as a collective in order to defend themselves within the market economy. Whether village collectives could really resist the destabilizing effects of the entry of capital and agribusiness, or whether collectives would lead to their further integration into the market economy, is a problem not fully addressed by Li, in part because of his tendency to view the peasantry as a unitary, undifferentiated class. Yet according to Li, a peasantry organized in this fashion could protect its interests, and rural democracy could be expanded upward to the *xiangzhen* level. With the end of the agricultural tax, village self-governance could now become truly independent and democratic, as village leadership did not matter as much to the township government, which no longer had to extract revenue from the village.

While the abolition of the agricultural tax reduced the burden on the peasantry and changed the power dynamics within rural governance, Li does not believe that peasants should pay nothing for the use of land; in fact, he feel it is crucial that peasants pay fees for agricultural land use if the village is to remain a collective in any real sense. Li's most unique proposal, therefore, is for 'taxes to be turned into rents' (*shui zhuan zu*) paid to the self-governed village collective (C. Li 2004b, 2005, 2006). At least part of the rural surplus that is no longer being collected in taxes should be controlled by the democratic collective, in other words, and not just revert back to individual households. Giving the village collective its own financial resources, generated by the collective itself and used for projects the collective decided upon, would strengthen the collective and form its material basis.

Furthermore, as peasants give up their land and leave the collective to migrate into the city, the collective could use its funds to pay for their entry into the urban social insurance system, facilitating the smooth urbanization of the surplus rural population. At present, many urbanizing or urbanized rural migrants hold on to their land, no longer participating in collective activities and acting as landlords renting their land out to landless collective members. A land rent paid to the collective would incentivize the transfer of land to landless members and benefit poorer villagers while maintaining a functioning collective. The transfer of collective land from agriculture to non-agricultural uses or to non-collective members would be up to the democratic decision of the collective as a whole. Countering the main liberal argument for land privatization, Li argues that the power of the collective would be a much stronger protection against the illegal or unfair expropriation of land than a simple privatization of land would be. The democratic collective would make the decision and gain the benefits. For much of central and western China, Li argues, land is the only collective resource, so such a system would be particularly important there.

Cao Jinqing (2010) suggests that provinces should be shrunk in size, increasing their number from 28 to about 50, and that counties and cities should be directly under the province. This would shrink the size of the government structure and the number of its personnel. He also argues that for agricultural regions the township should be self-governed (*xiangzhen zizhi*) along with the village level, with free elections. This would turn the township and the village into administrative offices (*xianggongsuo* and *cungongsuo*) controlled by the populace and protective of their interests – the local entrepreneurial state, the result of state involution, would be eliminated. Over time, the county could move towards democratic governance as well, beginning with the right of the populace to dismiss or recall county officials. With the agricultural tax abolished, the prime function of the county would be to guarantee the small peasant economy based on the land contract system and to continue birth planning. The township and village would be service organizations, and, as they would be self-governed, their functions would vary from region to region depending on local conditions.

Self-governed villages and townships, however, depend on an organized peasantry in order to function. As Cao and others argue, peasants are a weak or vulnerable social group (*ruoshiqunti*), and when unorganized and separate (*fensan*) they have little real negotiating power within the market or with the local government. Traditional rural social organizations, although they exist, are too weak and insufficiently generalized to counter the power of the market and local state. Only through organization could peasant power and rural self-governance have any meaning. Like Yu Jianrong, Cao suggests that county- or township-level peasants' associations (*nongmin xiehui*) are necessary. However, for Cao, these associations could have a stronger economic function, forming rural cooperatives of various types, and not just represent peasant interests to the state. As part of a 'new cooperative movement' (*xin hezuo yundong*), these peasant-association cooperatives would need financial and organizational help from the state, but should be allowed to develop in the direction members see fit.

While He Xuefeng (2004a, 105) is critical of studies that focus solely on the reform of formal political organizations in the countryside, he also stresses that the retreat of the state and the formation of grassroots organizations or self-government as its replacement is not sufficient to remedy rural social disintegration. What is needed is a strengthening of self-governance (*zizhi*) and rural government (*zhengfu*), as well as a deeper organization of the peasantry. He (2004b, 2007) points to the dangers of rural tax reform, arguing that it could weaken rural governance and lead to further disintegration of an already fractured and atomized rural society. According to He, taxation was one of the primary reasons that township cadres worried about villagers and their relationship to villagers (He 2004a, 2004b). With tax reform, that important link is gone, and cadres can largely ignore villagers, leaving development up to the peasants and the market. He and others fear that the provision of public goods, most importantly irrigation works, will further disappear. On the other hand, He points out that despite these dangers, a less contentious relationship is at least possible (He 2004b, 2007). Cadres would worry less about the outcome of village elections, and thus they have a better chance of being independent, for example (He 2007, 229).

Most important for He, the end of taxation means the end of the 'rural interests struc-ture' (*xiangcun liyi gongtongti*), in which the prime interest for county and township cadres, even township cadres who were responsible for providing public goods, was the collection of taxes (He 2007, 229–35, 2008, 224–6). This interest structure that emerged in the 1990s, produced by the relationship between the governance organizations of the county, township and village, had a similar effect to that which appeared in the first half of the twentieth century, and which likewise increased the burden on the peasantry in extracting resources for its own good (2008, 225). In fact, He implies that the 1990s were a lost decade for rural society. While industry no longer needed the support of agricultural extraction, and the era of 'primitive accumulation' had come to an end, rural extraction continued largely to benefit rural cadres and the county and township interest structure. In other words, the central state was too slow in bringing about the reform of rural governance to bring it into line with the new period produced by this shift in the Chinese political economy, and the longer it waits the more entrenched this interest structure will become and the more rural social relations will disintegrate. Rural governance was thrown into crisis, an involutionary situation similar to that of the early twentieth century. It is no surprise, therefore, that He's response draws on the work of rural reformers from the first half of the twentieth century, in particu-lar those involved in rural reconstruction projects such as Liang Shuming.

With the abolition of the agricultural tax, the institutional roots of this structure were dismantled, but the results are ambiguous for He. The abolition of the tax only created the opportunity for real meaningful reform that could solve this century-old problem. While township cadres could now focus on the provision of public goods, they have also become more passive, and a new institutional structure that will push them to work for the benefit of peasants has not replaced the old. For the New Socialist Countryside to truly create better rural conditions for the peasantry, the rural government needs to become more active, not less. Here, He is critical of those who believe that the withdrawal of the state (*guantui*) will automatically create the conditions for village self-governance and economic development. He argues that a stronger state and a more self-governed and active peasantry are both necessary for rural construction to continue. The peasantry needs to be better organized, but at present, with the outflow of rural people and resources, it is very difficult for the peasantry to self-organize. A peasantry atomized by the market will not be able to step forward (*minjin*) as the state retreats (*guantui*); instead, various 'graying social forces' (*hui shehuihua*), illegal and disruptive, will fill the void (He 2007, 235, 2008, 125, 137).[12] Rural governance needs to be institutionally redesigned under these new conditions, transformed from a top-down to a bottom-up structure, in order to actively focus on the provision of public resources and raise the cooperative capacity of peasants (*nongmin de hezuo nengli*) (He 2007, 236, 2008, 138).

In other words, neither opening the space for more self-organization of the peasantry, whether in the form of villager self-governance (*cunmin zizhi*) or peasant associations (*nongmin xiehui*), nor increasing central restraint on county or township governments, would alone solve rural governance problems (He 2008, 127). What is necessary is an inter-mixing of both, blending state and social forces together. The county should be democra-tized, in order to restrain and limit the power of top-down forces, but not weakened. He suggests district competitive elections for a county people's congress and allowing those representatives to choose, as well as supervise, county and township leaders and adminis-trators. Like Xu Yong, He argues that the township government should be turned into a

---

[12]On the 'graying' of rural society, see Tan Tongxue (2007).

county administrative office (*xianggongsuo* or *xianji paichujigou*). As a simple administrative office, the township would lose its party organization, government, congress and political consultative conference. At the village level, village self-governance would be strengthened as village-level competitive elections developed alongside county level elections (He 2004a, 102–4, 138–9).

He is not as optimistic as Xu Yong, however, about the power of contemporary village self-governance to counteract both the market and the political interests of the counties and the townships (He 2004a, 129). What is necessary, says He, is an increase in the cooperative capacity of the village. Cooperative labor on cultural, social and economic projects by mutual consent could build this capacity over time and overcome the atomization of the peasantry within the market economy (2004a, 138). These bottom-up 'construction' (*jianshe*) projects would not succeed, however, without state support and protection. This stress on cultural and social cooperation at the village level separates He's approach from Xu Yong's view on rural governance reforms. Like the other scholars discussed in this section, however, He largely sees the peasantry as a unitary class, despite their atomization, and this puts him much closer to the Republican-period Liang Shuming than to Mao. He believes that organization can overcome atomization. Also like Liang, He suggests that the basis for organization is rebuilding rural culture, including drawing on the resources of traditional culture, for neither democratic politics nor economic cooperation alone can provide a strong enough foundation for the regeneration of cooperative capacity. Emerging forms of class-based extraction disappear in this formulation. In the long run, according to He, a combination of political forces, both formal democracy at the county level and a strengthened self-governance and cooperative labor and culture at the village level, would lead to a resurgence of rural governance.

Conclusion

If most explicit in the work of Li Changping and He Xuefeng, almost all of the participants of recent debates on rural governance view the 1990s as a lost decade of reform, a moment in which state involution – so characteristic of the first half of the twentieth century – returned. If the 1990s saw the return of the state involution, as this paper argues, it also saw a resurgence of the political advocacy of the earlier period as well. Rural reconstruction of the 1930s has again become a model, as has *cunzhi* activism, and the category *zizhi* is again a site of debate and innovation. While at the beginning of the twentieth century the explicit goal of the Qing state was 'using self-government to aid bureaucratic rule' (Ma 1997, 204), 100 years later, intellectual advocates for the protection of the peasantry have utilized the ambiguity of the category *zizhi* to question the state's role in rural society. With the end of central-state extraction, the forms of rural governance that characterized this century-long trajectory are being transformed. In the present moment, *zizhi* is no longer understood as a way to increase the reach of the state. For liberal advocates, the goal is the withdrawal of the state and its replacement with democratic self-governance, market institutions and civil organizations to represent the interests of peasant farmers. For those on the left, the category *zizhi* is used to blur the lines between state and society in order to increase rural construction projects and the capacity of peasants to work together. Thus, while the need to organize the peasantry is universally acknowledged, the state's role in this process is more circumscribed than ever.

Contemporary liberals tend to see peasant organization as a question of political and civil organization; in other words, peasants who were atomized in the market – acting as individual producers with property rights – should come together politically in peasant associations and through the exercise of democratic rights to express their interests as a social group. For liberals, *zizhi* implies the introduction of democratic institutions into the countryside. For those on the left, social atomization has to be overcome not just through political and civic organizations, but in the economic and productive spheres as well. The basis for this organization could be economic collectives or, as some stress, its real basis should be in the sphere of culture – leading us back to the early twentieth-century work of rural reconstructionists such as Liang Shuming. *Zizhi* and *cunzhi* can indicate far more than formal political institutions, therefore, but are seen by some to allow for a new democratic village community to emerge that could gain more control over its land and surplus, protecting these resources from those who wished to exploit them, whether the local government or the forces of capital. These terms name the unified cultural, social, economic and political project to reconstruct rural village communities.

Yet the end of central-state extraction has not meant the end of extraction of rural surplus, but continued local-state extraction (especially through the requisitioning of land) and increasing class-based extraction. Nonetheless, almost all scholars and peasant advocates involved in these debates, no matter how they view the state-society relationship, tend to see the peasantry as an undifferentiated class. Atomization, in other words, is viewed primarily as a problem of organization and governance, not an issue of class differentiation and class politics – a politics that is largely proscribed, it should be said, from reform-period intellectual discussion. For liberals, that organization happens at the formal political level, through democratic institutions, where the interests of individuals can be mediated. For those leaning to the left, that organization happens through building a culture of cooperation. If advocates continue to pay much less attention to the issue of class differentiation than they do to state involution, they will potentially be blindsided by the emerging class fractures developing in the countryside. Needed here is a rethinking of the meaning of the peasant and rural society in the present moment commensurate with the momentous rethinking that occurred in the first half of the twentieth century.

At issue in these contemporary debates is the meaning of the rural sphere itself. The participants base their stances on differing long-term historical narratives of modernization, yet all view the present moment as a historical break necessitating new forms of governance. If the rural sphere during the Mao period was shaped by the process of industrialization via the state extraction of rural surplus, what is the role of the rural sphere today now that that extraction has ended? Has its position been pushed to the margins of society, a place to store China's surplus population to be dipped into for cheap labor when necessary, whose government wrings out the little surplus that is produced? Can it be smoothly integrated into the national economy and global capitalism or should it be given institutional protection from the economy? How one answers these questions largely depends on whether one believes that the market economy will create enough non-agricultural and urban employment to significantly shrink the rural population or not. A growing skepticism in China surrounding the potential for continued high economic growth rates and rapid urbanization has led to the emergence of left-leaning critics of the reform process and the intensification of political debate around the role of the rural sphere. Nonetheless, advocates for the peasantry – including those on the left – have seized on the reforms surrounding the abolition of the agricultural tax as a moment of opportunity to deal decisively with the involutionary rural government and find new ways to organize the peasantry, goals of rural reformers since the time of Liu Shipei at the very beginning of the twentieth century.

References

Ahlers, A.L. and G. Schubert. 2009. 'Building a new socialist countryside' – only a political slogan? *Journal of Current Chinese Affairs*, 38(4), 35–62.

Alitto, G. 1979. *The last Confucian: Liang Shu-ming and the Chinese dilemma of modernity*. Berkeley: University of California Press.

Bianco, L. 2001. *Peasants without the party: grass-roots movements in twentieth-century China*. Armonk, NY: M.E. Sharpe.

Bernstein, T., and Xiaobo Lü. 2003. *Taxation without representation in contemporary rural China*. Cambridge: Cambridge University Press.

Cao, J. 2010. *Ruhe yanjiu Zhongguo* [How to study China]. Shanghai: Shanghai renmin chubanshe.

Dai, K. 2007. *Lü Zhenyu zaoqi sixiang yu shijian yanjiu* [Research in the early thought and practice of Lü Zhenyu]. Changsha: Hunan shifan daxue chubanshe.

Dang, G. 2004. *Xiangzhen jigou gaige yao jiejue shenme wenti* [What problems does township organization reform want to solve]. *In:* Li Changping and Dong Leiming, eds. *Shuifei gaige Beijing xia de xiangzhen tizhi yanjiu*. Wuhan: Hubei renmin chubanshe, pp. 66–70.

Day, A. 2008a. The central China school of rural studies: guest editor's introduction. *Chinese Sociology and Anthropology*, 41(1), 3–9.

Day, A. 2008b. The end of the peasant? New rural reconstruction in China. *boundary 2*, 35(2), 49–73.

Day, A. 2013. *The peasant in postsocialist China: history, politics, and capitalism*. Cambridge: Cambridge University Press.

Day, A, and M.A. Hale. 2007. New rural reconstruction. *Chinese Sociology and Anthropology*, 39(4), 3–9.

Dirlik, A. 1991. *Anarchism in the Chinese revolution*. Berkeley: University of California Press.

Douw, L. 1991. The representation of China's rural backwardness: 1932–1937: a tentative analysis of intellectual choice in China, based on the lives, and the writings on rural society, of selected liberal, Marxist, and nationalist intellectuals. PhD Dissertation, Department of History, University of Leiden.

Duara, P. 1988. *Culture, power, and the state: rural North China, 1900–1942*. Stanford: Stanford University Press.

Duara, P. 1995. *Rescuing history from the nation: questioning narratives of modern China*. Chicago: Chicago University Press.

Gao, W. 2003. *Wan Qing Zhongguo de zhengzhi zhuanxing: yi Qing mo xianzheng gaige wei zhongxin* [Late Qing China's political transition]. Beijing: Zhongguo shehui kexue chubanshe.

Göbel, C. 2010. *The politics of rural reform in China: state policy and village predicament in the early 2000s*. London and New York: Routledge.

Gürel, B. 2012. Land ownership in capitalist agriculture: a critique of Zhang and Donaldson. Unpublished paper.

Gürel, B. 2013. Changing relations of production in Chinese agriculture from decollectivization to capitalism. Unpublished paper.

Hayford, C.W. 1990. *To the people: James Yen and village China*. New York: Columbia University Press.

He, X. 2004a. *Xiangcun yanjiu de guoqing yishi* [Rural research and national conditions consciousness]. Wuhan: Hubei renmin chubanshe.

He, X. 2004b. Dangqian nongcun zhengce tiaozheng guo meng [Abrupt adjustments in contemporary rural policy]. *Sannong Zhongguo*, 3, 21–23.

He, X. 2007. *Xiangcun de qiantu: xin nongcun jianshe yu Zhongguo daolu* [The future prospects of the countryside: new rural construction and the Chinese path]. Jinan: Shandong renmin chubanshe.

He, X. 2008. *Shenme nongcun, shenme wenti* [What is the countryside, what is its problem]. Beijing: Falü chubanshe.

Herschler, S.B. 1995. The 1994 tax reforms: the center strikes back. *China Economic Review*, 6(2), 239–45.

Hofheinz, Roy. 1977. *The broken wave: the Chinese communist peasant movement, 1922–1928*. Cambridge: Harvard University Press.

Huang, P.C.C. 1993. 'Public sphere'/'civil society' in China: the third realm between state and society. *Modern China* 19(2), 216–40.

Huang, P.C.C. 2011. China's new-age small farms and their vertical integration: agribusiness or co-ops? *Modern China* 37(2), 107–34.

Huang, P.C.C., Gao Yuan and Yusheng Peng. 2012. Capitalization without proletarianization in China's agricultural development. *Modern China* 38(2), 139–73.

Jacka, T. 2013. From spectralization to participation and reconstruction: A feminist critique of contemporary Chinese discourses on rurality, gender and development. *The Journal of Peasant Studies*, 40(6), DOI 10.1080/03066150.2013.855723

Keidel, A. 2007. *China's economic fluctuations: implications for its rural economy*. Washington, DC: Carnegie Endowment for International Peace.

Kennedy, J.J. 2007. From the fee-for-tax reform to the abolition of agricultural taxes: the impact on township governments in northwest China. *The China Quarterly*, 189(March), 43–59.

Kennedy, J. 2013. Finance and Rural Governance: Centralization and Local Challenges. *The Journal of Peasant Studies* 40(6). DOI 10.1080/03066150.2013.866096

Kerkvliet, B.J.T. 2009. Everyday politics in peasant societies (and ours). *Journal of Peasant Studies*, 36(1), 227–43.

Kuang, X. 2005. Sannong heyi weiji [Why is *sannong* a crisis?]. *Sannong Zhongguo*, 6, 12–24.

Kuhn, P.A. 1970. *Rebellion and its enemies in late imperial China: militarization and social structure, 1796–1864*. Cambridge: Harvard University Press.

Kuhn, P.A. 1975. Local self-government under the republic: problems of control, autonomy, and mobilization. *In:* F. Wakeman, Jr. and C. Grant, eds. *Conflict and control in late imperial China*. Berkeley: University of California Press, pp. 257–98.

Kuhn, P.A. 2002. *Origins of the modern Chinese state*. Stanford: Stanford University Press.

Li, C. 2002. *Wo xiang zongli shuo shihua* [I told the premier the truth]. Beijing: Guangming ribao chubanshe.

Li, C. 2003. The crisis in the countryside. *In:* C. Wang, ed. *One China, many paths*. London: Verso, pp. 198–218.

Li, C. 2004a. Xiangzhen tizhi gaige: guanbenwei tizhi xiang minbenwei tizhi zhuanxing [Township-town system reform: the transition from a system of officials at the basis towards the people at the basis]. *In:* Li Changping and Dong Leiming, eds. *Shuifei gaige Beijing xia de xiangzhen tizhi yanjiu*. Wuhan: Hubei renmin chubanshe, pp. 24–32.

Li, C. 2004b. Quxiao nongyeshui jiang yinfa yixilie shenke biange [Abolishing the agricultural tax will lead to a series of profound changes]. *Dushu*, 6(June), 86–91.

Li, C. 2005. Xiangzhen tizhi bianqian de sikao: zai tan 'houshuifei shidai' de xiangzhen tizhi gaige [Think about township-town system change: another discussion on 'post-taxation era' township-town system reform]. *Sannong Zhongguo*, 5, 42–8.

Li, C. 2006. Sannong wenti zai jinyan: nongyeshui yinggai zhuanwei dizu [Offering another opinion on the triple rural problem: the agricultural tax should become land rent]. *Sannong Zhongguo*, 7, 103–7.

Li, H. 2005. *Village governance in North China: Huailu County, 1875–1936*. Stanford: Stanford University Press.

Liu, C. 2007. *Peasants and revolution in rural China: rural political change in the North China plain and the Yangzi Delta, 1850–1949*. New York: Routledge.

Liu, J. 2008. Zhaicheng: Zhongguo cunminzizhi de faxiangdi. *Zhongguo hezuo jingji* (3) [online]. Available from: http://www.zh-hz.cn [Accessed 10 February 2010].

Liu, S. 1996a. Lun xinzheng wei bing min zhi gen. *In:* Li Shaogen, ed. *Guocui yu xihua: Liu Shipei wenxuan*. Shanghai: Shanghai yuandong chubanshe, pp. 205–12.

Liu, S. 1996b. Renlei junli shuo. *In:* Li Shaogen, ed. *Guocui yu xihua: Liu Shipei wenxuan*. Shanghai: Shanghai yuandong chubanshe, pp. 170–8.

Liu, S. 1996c. Lun nongye yu gongye lianhezhi kexing yu Zhongguo. *In:* Li Shaogen, ed. *Guocui yu xihua: Liu Shipei wenxuan*. Shanghai: Shanghai yuandong chubanshe, pp. 282–5.

Ma, X. 1997. Local self-government: citizenship consciousness and the political participation of the new gentry-merchants in the late Qing. *In:* J.A. Fogel and P.G. Zarrow, eds. *Imagining the people: Chinese intellectuals and the concept of citizenship, 1890–1920*. Armonk, NY: M. E. Sharpe, pp. 183–211.

Mao, Z. 1965. Report on an investigation of the peasant movement in Hunan. *In:* Committee for the Publication of the Selected Works of Mao Tse-tung, ed. *Selected works of Mao Tse-tung*. Beijing: Foreign Language Press, pp. 23–59.

Marks, R.B. 1984. *Rural revolution in South China: peasants and the making of history in Haifeng County, 1570–1930*. Madison: University of Wisconsin Press.

Min, T. 1989. *National polity and local power: the transformation of late imperial China*. Cambridge: Harvard University Press.

O'Brien, K. and L. Li. 2000. Accommodating 'democracy' in a one-party state: introducing village elections in China. *The China Quarterly*, 162, 465–89.

O'Brien, K. and L. Li. 2006. *Rightful resistance in rural China*. Cambridge: Cambridge University Press.

Oi, J. 1986. Peasant grain marketing and state procurement: China's grain contracting system. *The China Quarterly*, 106, 272–90.

Pan, W. and Li Changping, eds. 2006. *Shehuizhuyi xin nongcun jianshe de lilun yu shijian* [The theory and practice of new socialist countryside construction]. Beijing: Zhongguo jingji chubanshe.

Perry, E. 2011. From mass campaigns to managed campaigns: 'constructing a new socialist countryside'. *In:* S. Heilmann and E. Perry, eds. *Mao's invisible hand: the political foundations of adaptive governance in China*. Cambridge: Harvard University Asia Center, pp. 30–61.

Pomeranz, K. 1993. *The making of a hinterland: state, society, and economy in inland North China, 1853–1937*. Berkeley: University of California Press.

Prazniak, R. 1999. *Of camel kings and other things: rural rebels against modernity in late imperial China*. Lanham: Rowman, Littlefield Publishers.

Rankin, M.B. 1997. State and society in early republican politics, 1912–18. *The China Quarterly*, 150 (June), 260–81.

Schoppa, K. 1995. *Blood road: the mystery of Shen Dingyi in revolutionary China*. Berkeley: University of California Press.

Sheel, K. 1989. *Peasant society and marxist intellectuals in China: Fang Zhimin and the origin of a revolutionary movement in the Xinjiang region*. Princeton: Princeton University Press.

Shue, V. 1988. *The Reach of the State: Sketches of the Chinese Body Politic*. Stanford: Stanford University Press.

Sicular, T. 1988. Grain pricing: a key link in Chinese economic policy. *Modern China*, 14(4), 451–86.

Tan, T. 2007. Paths and social foundations of rural graying: the case of two townships in southern Hunan. *Chinese Sociology and Anthropology*, 39(4), 39–49.

Thompson, R. 1988. Statecraft and self-government: competing visions of community and state in late-imperial China. *Modern China*, 14(2), 188–221.

Thompson, R. 1995. *China's local councils in the age of constitutional reform, 1898–1911*. Cambridge: Harvard University Press.

Unger, J. 2002. *The transformation of rural China*. Armonk, NY: M.E. Sharpe.

Walker, K.L.M. 2006. 'Gangster capitalism' and peasant protest in China: the last twenty years. *Journal of Peasant Studies*, 33(1), 1–33.

Walker, K.L.M. 2008. From covert to overt: everyday peasant politics in China and the implications for transnational agrarian movements. *Journal of Agrarian Change*, 8(2–3), 462–88.

Wang, S. 1997. China's 1994 fiscal reform: an initial assessment. *Asian Survey*, 37(9), 801–17.

Wang, S. and Hu Angang. 1994. *Zhongguo guojia nengli baogao* [A report on China's state capacity]. Hong Kong: Oxford University Press.

Wang, W., ed. 2006. *Jianshe shehuizhuyi xin nongcun de lilu yu shijian* [The theory and practice of building a new socialist countryside]. Beijing: Zhonggong zhongyang dangxiao chubanshe.

Wen, T. 1999a. *Zhongguo nongcun jiben jingji zhidu yanjiu: 'sannong' wenti de shiji fansi*. Beijing: Zhongguo jingji chubanshe.

Wen, T. 1999b. 'Sannong wenti': shiji mo de fansi. *Dushu* 12, 3–11.

Wen, T. 2001a. Centenary reflections on the 'three dimensional problem' of rural China. *Inter-Asia Cultural Studies*, 2(2), 287–95.

Wen, T. 2001b. Bainian Zhongguo, yibo sizhe [China's 100 years: a river with four bends]. *Dushu*, 3, 3–11.

Wen, T. 2006. Xin nongcun jianshe de beijin, jingyan, yu jiaoxun (daixu) [The background, experience, and lessons of new countryside construction (introductory article)]. *In:* Wen Tiejun, ed. *Xin nongcun jianshe lilun tansuo* [Theoretical exploration of rural reconstruction]. Beijing: Beijing chubanshe, pp. 1–25.

Whiting, S. 2001. *Power and wealth in rural China: the political economy of institutional change*. Cambridge: Cambridge University Press.

Wu, L. 2004a. Guanmin hezuo tizhi: 'xiangzheng zizhi' – xiangzhen zhengfu gaige silu shentao [A system of official-popular cooperation: 'township government self-governance' – a discussion of the path of township-town government reform]. *In:* Li Changping and Dong Leiming, eds. *Shuifei gaige Beijing xia de xiangzhen tizhi yanjiu*. Wuhan: Hubei renmin chubanshe, pp. 33–66.

Wu, L. 2004b. Nongdi zhidu gaige: shichanghua [Reform of the agricultural land system: marketization]. *Sannong Zhongguo*, 3, 64–8.

Xu, X., ed. 2006. *Jianshe shehuizhuyi xin nongcun: xuexi duben* [Building a new socialist countryside: study book]. Beijing: Xinhua chubanshe.

Xu, Y. 2003. *Xiangcun zhili yu Zhongguo zhengzhi* [Township-village governance and Chinese politics]. Beijing: Zhongguo shehui kexue chubanshe.

Yu, J. 2004. Xiangzhen bushe zhengfu, huifu nongcun zizhi [The township is not a government, restoring rural self-governance]. *Jiangsu nongcun jingji*, 1, 18.

Yu, J. 2006. Shehuizhuyi xin nongcun jianshe xuyao jianli xinxing nongmin zuzhi [On the necessity of establishing new types of peasant organizations for building a New Socialist Countryside]. *Henan shehui kexue*, 14(3), 17–21.

Yu, J. 2010a. Dangqian Zhongguo jiceng zhengzhi gaige de kunjing he chulu [The predicament and way out for current Chinese basic-level political reform]. *Dangdai shijie shehuizhuyi wenti*, 2, 3–12.

Yu, J. 2010b. Zhenggai buneng qi xiwang yu xiangzhengfu [Political reform should not place hope in the township government]. *Zhengzhi ganbu cankao*, 5, 40.

Yu, J. 2010c. Hou shuifei shidai: jiceng quanli 'xuanfu' zhi you [The post-taxation era: worry over the 'suspension' of base level power]. *Renmin luntan*, 277, 18–9.

Yu, J. 2011. Xianzheng yunzuo de quanli beilun ji qi gaige tansuo [The paradox of the operation of county power and an exploration of its reform]. *Tansuo yu zhengming*, 7, 27–30.

Yu, J. and Cai, Yongfei. 2008. Xianzheng gaige shi Zhongguo gaige xin de tupokou [County government reform is the new break-out point of Chinese reform]. *Dongnan xueshu*, 1, 45–50.

Zarrow, P. 1990. *Anarchism and Chinese political culture*. New York: Columbia University Press.

Zhang, Q.F. and J.A. Donaldson. 2008. The rise of agrarian capitalism with Chinese characteristics: agricultural modernization, agribusiness and collective land rights. *The China Journal*, 60, 25–47.

Zhang, Q.F. and J.A. Donaldson. 2010. From peasant to farmers: peasant differentiation, labor regimes, and land-rights institutions in China's agrarian transition. *Politics and Society*, 38(4), 458–89.

Zheng, D. 1999. *Minguo xiangcun jianshe yundong* [The republican rural reconstruction movement]. Beijing: Shehui kexue wenzhai chubanshe.

Alexander F. Day is an Assistant Professor of History at Occidental College, with a research focus on modern China. His *The peasant in postsocialist China: history, politics, and capitalism* was published by Cambridge University Press in 2013. The book examines contemporary debates on China's emerging rural crisis and its relationship to intellectual politics in the reform era.

Debating the rural cooperative movement in China, the past and the present

Yan Hairong and Chen Yiyuan

Rural cooperatives appear to be flourishing in China. Yet this blossom has been controversial. Some contest whether specialized farmer cooperatives should be promoted. They are opposed to the implications and consequences that derive from the growth of such cooperatives. Many criticize that most of the cooperatives thus far developed are 'fake' cooperatives. Some propose comprehensive peasant associations in Japan, South Korea and Taiwan as a model for emulation. These contestations are about rural cooperatives, but also go quite beyond them. For those passionately involved in the support and critique of rural cooperatives, what is at stake is both rural sustainability and the possibility of China pursuing a third-way development. In the 1930s, rural cooperatives also blossomed in China, and it was accompanied by heated intellectual debates about the future of China. This paper will examine intellectual perspectives and debates both in the past and at present about rural cooperative development in China. Not only are there some remarkable intellectual parallels between the two, but also both movements have their own structural difficulties. In the face of the rapid agrarian change in China, the 1930s debate might still shed a light on today's conundrum.

1. Introduction

Rural cooperatives appear to be flourishing in China. With the Law on Specialized Farmer Cooperatives formally implemented in July 2007, rural cooperatives having industrial-commercial business registration stood at 100,000 in 2008. They grew to 689,000 by the end of 2012 and are expected to reach 900,000 by 2015 (China Review News 2013). For the purpose of such cooperatives, the law defines specialized farmers as 'the producers and operators of the same kind of farm products or the providers or users of services for the same kind of agricultural production and operation' and allows agro enterprises to join as cooperative members. According to the law, specialized farmer cooperatives should 'mainly serve their members, offering such services as purchasing the means of agricultural production, marketing, processing, transporting and storing farm products, and providing technologies and information related to agricultural production and operation (Falüjie 2007).' In 2008, the central

The authors wish to acknowledge the support of the project 'Forging New Trans-border Links, Social/ Community Economies (SCEs) in Hong Kong and the Pearl River Delta (PRD)' (K-QZA2). We would like to thank Alex Day, Matt Hale, Burak Gurel, as well as two anonymous reviewers for their valuable comments and suggestions.

government encouraged a closer linkage between agro enterprises and rural producers (X. Zhang 2009, 14). A number of government or quasi-government organs are involved in promoting cooperatives, including the Ministry of Agriculture, All China Federation of Supply and Marketing Cooperatives, China Association for Science and Technology, People's Bank, etc. At local levels, more government agencies are also involved in promoting cooperatives (Tong and Wen 2009, 16). Concerned intellectuals and student-youth groups who take rural support positions are pioneer promoters of cooperatives.

Yet the development of cooperatives so far has been rather controversial. Some contest the promotion of specialized farmer cooperatives and see the promotion as following a Euro-American model that promotes the interests of capitalist farmers. Many criticize that most of the cooperatives thus far developed are 'fake' cooperatives. Some propose comprehensive peasant associations in Japan, South Korea and Taiwan as a model for emulation. These contestations are about rural cooperatives, but also go quite beyond them. For those passionately involved in the support and critique of rural cooperatives, what is at stake is both rural sustainability and the possibility of China pursuing a third-way development.

We should remember that this is not the first time that rural China saw a rising tide of the rural cooperative movement. Rural cooperatives first blossomed in the 1930s in the context of the Rural Reconstruction Movement (RRM) that enjoyed much official promotion and intellectual participation. At the same time, the flourishing of the movement then was accompanied by heated intellectual debates about the future of China, epitomized in the conversation between Liang Shuming (1893–1988) – one of the leading figures of the 1930s RRM – and Mao Zedong about the generality and particularity of Chinese Society. The current growth of rural cooperatives also enjoyed much official promotion and intellectual engagement. Contemporary intellectuals in China who take rural support positions also see rural cooperatives as a key component of the new RRM. Despite many differences between 1930s and today, both the old and new RRM have been involved in a search for an alternative national development, and both have seen a lack of organization among rural producers to be a fundamental problem for rural and even national reconstruction. Both old and new RRM have determined that rural cooperative organization is the way to get small rural producers organized. For today's leading rural advocates,[1] such as Wen Tiejun, the earlier movement is an inspiration and a legacy. However, the cooperative movements in the 1930s and today both have experienced conundrums whose conditions are intractably structural.

In our view, the debate surrounding the 1930s RRM and its cooperative movement is highly relevant today, for two reasons. First, there some remarkable intellectual parallels between the two RRMs and it is from the earlier RRM that some promoters of the RRM today by and large draw their inspiration (Li Zhonghua et al. 2008). The factors contributing to the contemporary significance of the earlier movement will be examined later in this contribution. Second, both movements have their own structural difficulties. The old debate on the nature of agrarian China, particularly between Liang Shuming and Mao Zedong, is little heeded today, but is – as will be argued – worth revisiting for our own critical reflection on the problems of contemporary rural cooperative movement. This paper will thus examine intellectual perspectives and debates both in the past and at present about rural cooperative development in China. In the face of the rapid agrarian change in China, the 1930s debate might still shed a light on today's conundrum.

---

[1]The rural advocates discussed in this paper are rural support intellectuals who show concerns for the wellbeing of the rural population and who often declare their pro-rural positions and their appreciation of rural values in their intellectual works and policy proposals.

2. The debate in the 1930s: commonality or particularity of Chinese society

Chinese peasants have long had traditional practices of mutual help on an ad hoc basis, typically involving no more than a few families.[2] It was in the early twentieth century's context of intellectual fermentation that some Chinese intellectuals became attracted to the idea of the cooperative (Du *et al.* 2002).[3] One of the earliest promoters of cooperatives, Tang Chang yuan, was fairly representative in embracing cooperativism as an alternative. In his view, 'Cooperation is opposed to capitalism. Its impact will more than undermine economic imperialism. But its approach is different from that of Marxism. Cooperativism does not emphasize revolution, but emphasizes construction. It does not rely on the state, but is based on organizations. Its approach is gradual and its action is far-reaching' (Y. Chen 1983, 97, quoted in Bo 1994, 122). This view was considered by Yu Shude, an early Communist and secretary of Sun Yat-Sen, himself also an important proponent of cooperatives, to be a misconception of cooperative organization. Yu unambiguously pointed out in 1927 that cooperative organization is not socialism, but belongs in the realm of social policy (Yu 1929, Foreword). This insight was borne out in the reality of the first half of the twentieth century when different political forces – the Nationalist Party (KMT), the Chinese Communist Party (CCP), non-partisan intellectuals of different political positions, and even Japanese colonial force in Northeast China – all promoted cooperatives as a matter of social policy, but in different contexts and for different ends.

CCP's promotion of cooperatives was part of its larger class-based political mobilization. It first began with workers, but was soon extended to the peasantry, when the peasant movement rose in the 1920s. In 1922, while organizing a famously successful strike among mine workers in Anyuan, Jiangxi province, Mao and his colleagues also organized workers to form a consumer cooperative. The cooperative's rapid growth was met with a warlord military crackdown in 1925. Similar worker-centered cooperatives were also organized in parts of Hunan and Guangdong (Du *et al.* 2002, 35–6). In 1925, the CCP's 'Letter to Peasants' (*zhongguo gongchandang gao nongmin shu*) encouraged the emerging peasant associations to be active in forming cooperatives. In 1926 and 1927, cooperatives was one of the topics taught at the Guangzhou Institute of Peasant Movement that was jointly run by both the CCP and KMT (Du *et al.* 2002, 37). From 1925 to 1927, peasant associations in Guangdong, Hunan, Hubei and Jiangxi, where peasant movements were strongest, all issued resolutions on cooperatives (Shi 1957, 73–8).

Organizing cooperatives was part of the overall class-based peasant movement in the 1920s that included suppression of landlord political and military power, rent reduction, anti-imperialist mobilization, literacy movements, etc.[4] These resolutions on cooperatives stated that they were for 'poor peasants' or 'small peasants', to avoid extortion by 'landlords', 'rich people' (*fuweng*) or rich households (*fuhu*) (Shi 1957, 74). These resolutions promoted cooperation in marketing, supply and credit. After the KMT broke away from the

---

[2]Shi Jingtang *et al.* (1957, 3–69) has a wonderful collection of investigative reports produced in 1940s and 1950s on traditional forms of mutual help found in China's different rural regions.

[3]Some Chinese intellectuals, who studied in or visited Japan, Germany, France and the USA, introduced the idea of cooperative economy to the Chinese reading public via publications and university teaching. The earliest promoters include Qin Shougong (1877–1938), Xue Xianzhou (1878–1927), Xu Changshui (1895–1925), Dai Jitao (1891–1949), Tang Cangyuan (1881–1931), Zhu Jinzhi (1888–1923) and Yu Shude (1894–1981). See Du Runsheng *et al.* (2002, 11–14) for a description of their publications and activities.

[4]Mao's well-known 1927 investigative report in Hunan detailed the kinds of struggles engaged by the peasant movement then (1991, Vol. 1, 12–44).

two-party coalition in 1927, the CCP began to build its own rural bases where land reform was carried out. In the context of land reform and the need for men to join the Red Army and the local militia to defend these bases, initiative existed among peasants both to use traditional and to invent new forms of cooperation in sharing tools and animal power and mobilizing women in production.[5] The CCP promoted cooperatives, particularly in production. In 1933, the CCP produced a policy outline on the organization of labor cooperatives, which was based on the principle of 'relying on poor peasants and uniting middle peasants', as well as keeping landlords, rich households and capitalists out of cooperatives (Shi 1957, 35–6, Mei 2004a, 105). In the 1940s when the CCP was in the midst of organizing national resistance against Japanese invasion, it enthusiastically promoted cooperatives in its Jin-Cha-Ji bases (Mao 1943). By this time, CCP-led class-based mobilization had achieved land reform in half of these bases and rent reduction in the other half (Mao 1943). In this context, cooperatives served the purpose of not only stimulating cooperation and productivity, but also organizing a united front of all those who resisted Japanese invasion, including anti-Japanese rich peasants, landlords and capitalists (Liu Qingli 2010, 14).

The KMT began to organize rural credit cooperatives in 1928 (Bo 1994, 127).[6] Cooperatives grew in the 1930s in KMT-controlled areas. In 1932, the KMT government issued its first policy to support cooperatives in areas where it was combating the Communist presence. In the ensuing years, the KMT government continued to support cooperatives with policies and funds. The number of cooperatives grew, with a great many being credit cooperatives. The presence of credit cooperatives could not help poor peasants with their production. Nor did it relieve them from predation by loan sharks. Instead, it made them vulnerable for extortion by landlords and merchants who took control of credit cooperatives and acted as their guarantors for credit (Mei 2004b, 87).

In the context of the wide-spread agrarian crisis in China in 1932 and the imminent national crisis brought on by Japanese occupation of China's Northeast, some non-partisan intellectuals dedicated themselves to the RRM and promoted cooperatives as the RRM's key component.[7] These intellectuals saw the cooperative movement as a way of people's self-organization and a key to national salvation. They also saw it as an alternative to the cropping up of Soviet bases in rural counties (1930–1934) founded by the CCP (Bo 1994, 129). The cooperative movement grew quickly, owing in part to the endorsement by the Nationalist government. 1933 saw the establishment of the first national network of rural reform organizations. By its third convention in 1935, the network had recruited 99 organizations from 10 provinces, including civil associations, universities and governmental departments, as well as some major newspapers (Jiang and Jia 2008, 76, 80).

---

[5]Mao's 1933 rural investigation reports on Changgang in Jiangxi and Caixi in Fujian detailed post-land reform rural cooperative initiatives (Mao 1982). In 1932, the Communist government issued a guideline for cooperative organization (Wei and Zeng 2010, 34). See Selden (1971) for cooperative economy promoted by the CCP in the Yan'an period.

[6]Chen Guofu, a KMT leader, established a 'Zhongguo hezuo yundong xiehui' [Chinese association for the cooperative movement] in 1924 and the KMT passed a resolution on 'the peasant movement' in 1926, which mentioned promotion of peasant cooperatives. In 1928, Jiang Jieshi and Chen Guofu again called for a resolution on the cooperative movement. In 1931, the KMT government issued 'A preliminary regulation on rural cooperatives' and in 1934 it formally issued a 'Law on Cooperatives' (Bo 1994, 130). Bo Guoqun argues that without the KMT government's promotion, rural cooperatives would not have been possible (130).

[7]It was widely believed that rural China was facing collapse or bankruptcy. It was partly impacted by the world economic crisis from 1929 to 1931 and partly brought on by the 1931 flooding of the Yangtze River and Japanese invasion (Bo 1994, 129).

At its peak, the end of 1936, there were over 1000 rural experiment sites (Yan Yangchu 1989, 305 quoted in Jiang and Jia 2008, 79), 37,318 cooperatives in 16 provinces that covered over 1.6 million members (Bo 1994, 127).[8] These experiments were financed either by the KMT government or by foreign funders (Liang 1989, vol. 2, 580). Renowned sociologist Fei Hsiao-tung (Fei Xiaotong) proposed in 1939 that cooperative rural workshops and rural industry could play a very important role in reconstructing rural China (Fei 2002, 238–9).

Despite the fact that both the CCP and the KMT promoted cooperatives for different ends, intellectual proponents of the 1930s cooperative movement continued the earliest proponents' vision for it to be a 'third way' alternative to both Western capitalism and communism. Liang Shuming's experiment in Zouping of Shandong province, running from 1931 to 1936, was the largest, having 307 cooperatives that involved 8828 members in its peak year 1936 (Qiu 2002, 101–2). It became one of the KMT government-endorsed experimental counties in 1933 and Liang himself was briefly the head of the county in 1935 (Bo 1994, 129). An established neo-Confucian thinker and social reformer,[9] Liang Shuming gave his experiment great significance by clearly linking it with national reconstruction. Due to Liang's influence both in the past and at present, we focus on his view as the most important in the 1930s RRM.

A variety of socialisms became popular in China around the May 4th Movement in 1919 and attracted a large number of Chinese intellectuals. Influenced by Peter Kropotkin's writing on mutual aid, cooperation and guild socialism (F. Yang 1999), Liang believed that RRM could build a new social structure with which China could embark on a path neither capitalist nor communist (Liang 1989, vol. 2, 151). Like many of his contemporaries, Liang saw China experiencing deterioration in the political (harassment by soldiers, banditry and severe taxations, etc.), the economic (economic invasion by foreign countries, etc.) and the cultural dimensions (social disorder). The demise of China in the past century, in his view, is 'a history of rural destruction' (Liang 1989, vol. 2, 150). Not unlike Mao, he held the countryside was the key to solving China's problem (Liang 1989, vol. 2, 161, vol. 5, 374), but these two figures differed in their diagnosis, as will be shown below. Disappointed with political-military powers that had been socially destructive since the fall of the last dynasty, Liang turned to social-cultural structure for diagnosis and cure (Alito 1986, 280, Liang 1989, vol. 2, 162–4). He stated, 'China's problem today is that the social organization and structure, which had lasted for thousands of years, has collapsed, but the new one has not yet been established. RRM is a movement that reconstructs China's social organization and structure' (Liang 1989, vol. 5, 375).

Liang based his RRM on two major assumptions. First, rural reconstruction in itself is a self-adequate foundation for the reconstruction of the entire Chinese society. On the one hand, Liang emphasized that China's problems were ignited from outside and were not autogenic (Liang 1989, vol. 2, 233–4, 577). On the other hand, he believed that the solution to China's problems lies in the RRM, which 'naturally contains solutions to all problems', as it includes progress in productive technologies, development of cooperative organizations, improved levels of education and a growth in peasant power (Liang 1989, vol. 5, 374). Liang envisioned 'rural school programs' [xiangnong xuexiao] to provide two

---

[8]Of all the organizations facilitating and overseeing cooperatives, 87 percent were governmental and 12.7 percent were social (He Jianhua 2007, 20)

[9]Alitto (1986) and Lynch (1989) provide the most comprehensive scholarly works in English on Liang Shuming.

things that he saw sorely lacking in rural China, knowledge of science and techniques and a new form of social organization (Liang 1989, vol. 2, 191). The schools treated all villagers as students (*xue zhong*) and taught cooperation, agriculture skills and knowledge, literacy, etc. It also organized more than 300 cooperatives in production, sales, credit, procurement, etc. More than educational functions, Liang's rural schools interfaced with the political, economic and military defense functions of the locality. Liang thought that this experiment served to develop a model for the political system of the entire state. The presumed self-adequacy of the RRM is based on the assumption that it could be autonomous from the dominant imperialist and local warlord political-economic powers. Liang's confidence in this assumption was to be devastated in 1938 (F. Yang 2001, 216, 221).

Liang's assumption of rural self-adequacy also includes his vision that China's agriculture production could serve as a foundation for its industrialization (Liang 1989, vol. 2, 496). Liang reasoned that compared with China's feeble industry which was oppressed by Western industrial interests, agriculture was China's comparative advantage (508). Liang envisioned that China's path of development would have to be an agricultural one, based on cooperation. Need-based non-profit industrialization could be developed with enterprises owned not by individuals, but by cooperatives, social organizations and the state (509). It would be a new civilization, different from both the old agrarian civilization and Western urban civilization. Liang planned a route to national construction:

> inland rural areas can make use of surplus funds from elsewhere to restore and increase [agricultural] production, [rural folks'] increased purchasing power can then spur the development of national industry. Then industry and all other sectors can in turn prosper. (Liang 1989, vol. 5, 367–8)

As discussed below, this vision of a positive circularity between agriculture and industry and its promise to bring about national reconstruction was criticized for ignoring the predominance of imperialism in China.

The RRM was unfolding alongside a series of heated intellectual debates, between the late 1920s and the first half of 1930s, about the nature of Chinese society. These debates covered the nature of Chinese society (1927–1928), Chinese social history (1932–1933) and rural society (1934–1935), with the last debate being the most important.[10] This series of debates was driven by grave concern for the future of China. As observed by a prominent editor at the time, 'The question about the nature of the Chinese economy now demands an answer from scholars of all classes. Intellectuals of any class, for the sake of their class's future, will have to address this question' (Guo 2003, 50–1). These debates – in the wake of the 1927 violent split between the KMT and the CCP – also dovetailed with the debates within international and Chinese communist movements about aims and strategies of Chinese revolution. The outcome of the debates was a much wider intellectual reception of the characterization that China was not a capitalist, but a semi-colonial, semi-feudal society.

The China Agrarian Group (Zhongguo nongcun pai), a network of Marxist and left-wing intellectuals, which had earlier engaged in the debate about the nature of China's rural society, launched critiques of RRM, including experiments led by Liang Shuming and James Yan, another leading figure known for his reformist experiment in rural

---

[10]For summaries of these debates in the English language, see Chapters 6–8 in Chiang (2001) and Chapter 4 in Han (2005). For a comprehensive summary in Chinese, see G. He (1937).

education. In their 1936 publication, *A critique of China's rural reconstruction* [Zhongguo xiangcun jianshe pipan] (Qian and Li 1982), the editors challenged the assumed autonomy or self-adequacy of RRM, 'Can the RRM or the construction of Chinese national economy be accomplished independently of the political task of national liberation?' (Qian and Li 1982, 64). More specifically, they asked, 'can development of agriculture bring about industrialization and salvage the cities?' While Liang intended a non-capitalistic development, the actual situation was indeed that these 'funds' [*zijin*] – as he referred to them – from commercial banks or government-owned financial institutions – behaved in stereotypically capitalistic manner when going to the countryside; they concentrated in localities that had easy transportation, benefitted the rich and middle peasants and excluded poor peasants (Bo 1994, 132–3).[11] Citing cases of how cotton grown by rural cooperatives in Hebei and Shandong fed imperialist enterprises in China, the editors questioned the relationship between rural reconstruction and imperialist expansion in China. A few years later, an influential study by Chen Hansheng [Chen Han-seng], the leading figure of the China Agrarian Group, showed how the British-American Tobacco Company, through comprador merchants, local gentry and rural cooperatives, trapped tens of thousands of Chinese peasants in Anhui, Henan, and Shandong provinces in unfavorable terms of production and sales of tobacco leaf to the company (H. Chen 1980 [1939]).

Liang's second major assumption about rural construction is his basic outlook on Chinese social particularity, which became the core of his 1938 debate with Mao. In Liang's view, Chinese society was unlike Western society in two respects: Chinese society had a division of labor, but no division of class; Chinese society was held together by ethics – duties and obligations towards each other, not based on individualism. Liang argued China lacked class dynamics for revolution, as none of the classes – peasants, workers and capitalists – could serve as a class base for revolution. On the one hand, Liang recognized disparity in land ownership in China, due to the system of private ownership, and thought that equality in land could be achieved either by public ownership or by equalization of land ownership. Yet he dismissed the possibility of either, as he saw China lacking the necessary political conditions (Xu and Zhao 2011). On the other hand, Liang particularly refuted the relevance of class for rural China and saw rural China lacking a radical separation between land and rural producers. Liang's evidence was that land holding could both concentrate and disperse, as land could be freely bought and sold and family assets are equally divided among all sons (Liang 1989, vol. 3, 146, Xu and Zhao 2011). In a similar vein, Liang denied the existence of a ruling class in China and argued that as there was social mobility, rulers and the ruled could exchange places in China. He thus famously stated, 'Chinese society has rulers, but no ruling class' (Liang 1989, vol. 3, 155). As opposing classes could not be formed in China, Chinese society

---

[11]A couple monographs published in the 1990s about Liang or his experiment in Shandong have the view that the cooperative members were mostly landlords and rich peasants (Ma 1992, 205–6, Zhu 1996, 147, 150). Guy Allito [Ai Kai] argued that the cooperative mainly included the middle peasants, even though there was a tendency of rich peasant domination (Ai 1996, 262, 260). Yang Feirong reexamined the cooperative records and argued that most cooperative members were likely to be middle and poor peasants (F. Yang 2001, 201–4). Although cooperative records then had information on the size of land held by member households, they had no information on the size of each household, thus making it difficult to estimate the class status of the member households based on per capita land ownership. However, even if cooperative members were mostly middle and poor peasants, available cooperative records cannot tell us the decision-making process within cooperatives and who benefited most from them.

had no class struggle, a condition Liang thought conducive to cooperation. Liang once advocated avoiding words such as '*nongmin*' [peasantry] and '*bei yapo de minzu*' [the oppressed nation] because of their class connotations, and instead using '*xiangcun jumin*' [rural residents] to create inclusiveness (Liang 1992 [1933], 215).

Although often regarded as conservative (Webb 2008) or Confucianist (Alitto 1986), Liang Shuming's RRM was nevertheless about finding a new social organization beyond the age-old institutions such as family, lineage or village. As Liang explained, the mission of the RRM is 'to build a *normal* (*zhengchang xinatai de*) human civilization so that economic 'wealth' and political 'power' are in the hands of the society and shared by everyone' (Liang 1989, vol. 2, 76). This 'social movement' – as Liang called it – should be led by intellectuals and should be based on villagers themselves (Liang 1989, vol. 2, 377). Regarding villagers as an undifferentiated group, what Liang wanted to construct might be called a rural civil society that would be jointly headed by the old/rural and new/urban elites. Rural schools in the RRM became the 'public sphere' where social reformers from the outside mobilized villagers (Thogersen 1998, 147). Schools were overseen by school boards whose members were always drawn from among rural elites, while teaching was led by social reformers. Liang and his followers considered their organization to represent an autonomous society independent from the state, but in reality it functioned as 'a mediator between the Chinese state and a rural hinterland which had so far been almost inaccessible to state officials' (159). Liang was more a prototype modernizer in practice than what he appeared in his writing (158).

In their critique of the RRM, the China Agrarian Group also took on Liang's RRM philosophy and practice for failing to address the ownership of means of production. They questioned rhetorically whether the movement could alleviate the pains of Chinese peasants if it focused only on agricultural skills, transportation, marketing and finance, but left the problem with the ownership of means of production, particularly land, untouched (Qian and Li 1982, 64). In 1935, Liang himself admitted that the RRM could neither relieve peasants of their burden of taxes, nor help them with land redistribution (Liang 1989, vol. 2, 581). Moreover, Li Zixiang, a member of the China Agrarian Group, indicated in his critique that cooperatives based on small producers not only do not contradict large enterprises, but are actually dominated by large enterprises and banks (Z. Li 1982, 69). Li predicted that Liang's 'socialization of production' and 'socialization of distribution' would thus be blank checks that will never be cashed. As to what Liang touted as a positive cycle between production and consumption, it would end up facilitating the expansion of imperialist industry and commodity markets in China. In practice, the RRM also almost solely depended on intellectuals, even though Liang thought it should be otherwise (70). While Liang in his RRM methodology emphasized objective conditions, Li pointed out that the objective conditions in Liang's eyes are immutable and thus his theory and methodology is inevitably a form of conservatism which succumbed to these 'objective conditions' (70). Liang's philosophy of RRM was therefore idealist, his methodology was conservative and his 'new civilization' was only to be a semi-colonial civilization (69). Li Zixiang sharply concluded that the movements (including the one led by James Yan) 'intended to use "cultural work" to mend the fundamental contradictions of the entirety of China's social-political-economic problems' and that these efforts, like pushing a cart uphill, would collapse one day (74).

The RRM ended with Japan's invasion of China in 1937. In 1938, searching for hope to defeat the Japanese invasion and intrigued by CCP's united front policy, Liang made a trip to Yan'an where he had some lengthy exchanges with Mao. These candid and friendly talks allowed them to find some common ground in national liberation and in their disapproval of

Western-style constitutional democracy, but they also made most apparent their different views of Chinese peasantry and society. When Mao asked Liang what he found difficult in his RRM, Liang frankly stated, 'The biggest difficulty is that peasants prefer passivity to action [*hao jing Bu hao dong*]' (Alitto 1986, 289). This had been Liang's consistent view, as he wrote about peasants that 'Their character is deeply conservative. Therefore, to talk to peasants about revolution is banging your head against the wall' (1992 [1933], 176). In 1935, Liang reflected on one of the difficulties of the RRM:

> In the so-called 'village movement', the village does not move [*hao cheng 'xiangcun yundong,' xiangcun Bu dong*] … Villagers were indifferent and it was only outsiders making a big noise …. The most ideal RRM is when villagers move and we cheer them on. Or it at least ought to be like this: they want to move, and we lead them. But it's now completely the opposite, we move, but they don't. Not only do they not move, but also because we move, our relationship with them has become such that it's almost difficult for us to continue our work …. We thought our work beneficial to the villages, but villagers do not welcome it'. (Liang 1989, vol. 2, 575)

Even regarding the modestly successful program of compulsory adult education, Liang remarked in 1935 with disappointment that the peasants were still the object of the transformation, while intellectuals were the subject and it was not uncommon for the RRM activists to compare passive villagers to 'stones' (Thogersen 1998, 152).

With the CCP's experience in peasant mobilization not only for land reform but also for vibrant rural cooperatives in their rural bases, Mao instantly disagreed with Liang's characterization of the peasantry, 'You are wrong! Peasants want action. How can you say that they want to be passive?' (Alitto 1986, 289). After reading Liang's *Theories of rural reconstruction* and listening to Liang expounding his cultural theory and RRM for a week, it became apparent to Mao that their differences lie in whether China need a reform or a revolution and whether class analysis is relevant to Chinese society. Mao summed up their differences:

> Chinese society has its own particularities, its own cultural tradition, and its own ethics, which is not wrong for you to emphasize. Chinese society also has qualities in common with Western societies, which include class opposition, contradiction and struggle. These are its most fundamental attributes, which determine social progress. You overemphasize its special nature and neglect its universal nature.

Liang answered resolutely, 'The reason why China is China lies in these special features. You overemphasize its common qualities and neglect its special qualities' (Allito 1986, 289, Ai 2003, 208).

Many of Liang's admirers and followers today lament the unfortunate disruption of Liang's rural reconstruction by the Japanese invasion or some argue that the merit of the RRM should not be measured in terms of success or failure (Pan 2012). Liang himself was more reflective and forthright about the grave structural difficulties his RRM faced. First was the contradiction between the RRM's social reform position and its dependence on local warlord political power. The local power Liang's RRM depended on was headed by an old warlord Han Fuqu, who murdered CCP members and their supporters, suppressed peasants' armed uprisings, and promoted the KMT's 'new life movement', which was a combination of Chinese traditions and European fascism. The second and, in Liang's view, most painful contradiction was already mentioned above: 'in the so-called "village movement", the village does not move'. Liang was also aware that because of the RRM's inability to respond to peasant problems of taxation and land, it had not been able to address issues that really matter to villagers and 'therefore cannot win them over' (Liang 1989, vol. 2, 581). Although Liang admitted that the problem was unavoidable, he thought it would gradually ease up. Up until 1949, Liang had been confident about his assessment that China lacked a class base for revolution. After witnessing the CCP's success in the land reform, however, Liang acknowledged that he had a static view of Chinese society and had failed to see that disparity could develop into class-based conflicts. He reflected that he made a mistake by overemphasizing the particularity of China's problems (Liang 1989, vol. 6, 866, 950–1, Wang 2004, 157).

3. Contemporary rural advocacy and the debates about cooperatives

The post-Mao rural reform disbanded the rural communes and instituted two-tiered land rights, land ownership right belongs to the village collective while land use right was equally divided and leased to families for family contract farming. Within this land institution, a rural household can further lease part or whole of their land use right to others within or beyond the village. To this day, farmland in China has not been treated as a full commodity.[12] However, the reform over the years has solidified de facto privatization of land use right while at the same time has hollowed the collective ownership right to the extent that the collective is left with little ability to coordinate village public infrastructure.[13]

Three decades of reform and a continuous process of marketization has brought about commodification of agricultural inputs, labor, public goods and technical services, a steady exodus of educated rural youth as migrants to cities, the aging and feminization of rural producers, fragmentation of familial life, estrangement of social relations within villages, growing rural disparity,[14] etc. This complex of interrelated problems has been termed '*san nong* problems', that is, rural sustainability has been threatened in three dimensions, rural livelihood and its reproduction (literally, *nongmin* or peasants), the coherence of rural society (literally *nongcun* or the countryside), and the sustainability of agricultural production (literally *nongye*, agriculture). The new RRM is a response to 'san nong problems' (Day and Hale 2007, Day 2008, 2013). The beginning of the new RRM was marked by the first New Rural Reconstruction conference in Beijing in 2002,[15] the founding of the James Yan Rural Reconstruction Institute in 2003 and the Liang Shuming Rural Reconstruction Center in 2004 by some university-affiliated intellectuals and students and non-governmental or para-governmental rural support organizations. The New RRM includes mobilizing and training student volunteers for rural support, incubating rural advocate student groups on university campuses, policy critique and advocacy at public forums, articulating alternative ideas of rural development, etc. Since then, other intellectuals have rural advocacy each in his or her own way.

---

[12]For a good examination of rural households and farm land distribution, see Unger (2009). There is a considerable force within and beyond China that pushes for land privatization. See China Left Review (2008) for a special issue on this topic, http,//chinaleftreview.org/?page_id=98.

[13]Hu Jing, professor at Huanan Normal University, has a searing critique of the increasing marginalization of the collective right in Chinese central government's rural policies (Hu 2011). Forrest Zhang (F.Q. Zhang 2012) argues that the current land institution still provides some protection for rural producers when they contract with companies.

[14]While this is the general trend, there are also regional differences in rural social connectedness, due to the varying levels of local kinship and clan relations (see X. He 2009).

[15]We thank Matt Hale for this information.

Intellectual advocacy and promotion of rural cooperatives, including training rural cooperative leaders, had been a key component of the new RRM since its beginning,[16] although since the return to family-based farming three decades ago, informal mutual aid practices and cooperative associations have existed in rural China.[17] Following the new RRM, the Chinese central government in 2005 initiated 'constructing a new socialist countryside' in its eleventh five-year plan, which, different from the new RRM, mainly aimed to divert investment to and stimulate demand in the countryside.[18]

The intellectuals' concern for rural sustainability and their promotion of a new RRM and cooperatives place them as critics of the mainstream modernization trope that has been prevalent in the process of reform. With 200 million or so rural households as individuated producers, mainstream policy makers and economists have hoped that agro enterprises will play a leadership role like 'dragon-heads' in streamlining and scaling up production and bringing rural producers into the process. The Chinese government has been supporting 'dragon-head enterprises' [*longtou qiye*] with preferential policies since the mid 1990s. 'Company + households', with company being the integrator through which households are connected with the market, has been promoted as a win-win formula for both companies and rural producers. Since the formal implementation of the Law on Specialized Farmer Cooperatives, government agricultural bureaus and some para-governmental agencies have been active in promoting cooperatives and training cooperative leaders. Cooperatives have mushroomed. However, the nature of cooperatives is highly controversial, which will be discussed below. The remarkable development of cooperatives has not really lessened the above-mentioned agrarian problems. The mainstream vision for dealing with the '*san nong*' problem remains urbanization of much of rural population and capital-led vertical integration of agriculture (*chan ye hua*).[19]

Sharing a normative 'rural support' stance, intellectual proponents of cooperatives not only organize training workshops for farmers and rural support students, but also engage in on-the-ground experiments. They also debate about what kind of cooperation should be promoted and to what end. Some see cooperatives as a welcome revision of the 'company + households' model, because the reformed 'company + cooperative + households' operation supposedly allows farmers more collective power in their relationship with the company. Some others advocate that a displacement of the dragon-head company by cooperatives can offer a more direct and rural-beneficial link to the market. Li Zhonghua, head of the Institute for Cooperatives in Qingdao Agriculture University,

---

[16]As an exception, Wen Tiejun began to advocate rural financial cooperation as early as 1994 (Wen 1994).

[17]In the 1980s such cooperation often existed in the form of associations of agricultural technology, but the content of cooperation often went beyond technologies. For example, according to my field research in 2009 in Hejian county, Hebei, the county had over 100 such associations in 1980s, organized by rural cash crop producers. Over the years, most had dissipated and disappeared. One cotton research association has grown into a prominent share-holding cooperative company (Nongye jingji hezuo zuzhi yanjiu ketizu 2005).

[18]According to Wen Tiejun (2012b), it was economist Justin Yifu Lin who in 1999 first proposed to the central government the idea of construction of a new countryside. In the wake of the Asian financial crisis, Lin proposed the idea as a solution to the problem of the dual surplus that China was already facing, overcapacity and overcapitalization. Lin estimated that many industries in China already had 30 percent overcapacity in 1999 (Y. Lin 1999).

[19]*Chan ye hua* is sometimes translated as agricultural industrialization. It typically refers to a process of integrating agricultural production, processing and marketing in a value chain –which is often led by dragon-head enterprises – in order to create greater economic efficiency and profit.

draws on his experience in Japan to push cooperatives to directly supply supermarkets [*nongchao duijie*] (Z. Li *et al.* 2008). Yuan Peng, a scholar at the Chinese Academy of Social Science (CASS), argues that the cooperative organization is both a product of the market and has an anti-market quality (Yuan 2001). A more critical approach, taken by anti-corporate intellectuals and students, tries to forge an alternative linkage between rural producers and urban consumers, bypassing corporations (Ku *et al.* 2009).

For some intellectual proponents of rural cooperation, the form and purpose of cooperation goes to the heart of the larger issue of China's path of development. Their advocacy is based on re-situating and re-assessing China's national conditions and experiences in a global comparative context. Indeed, their advocacy involves a change of reference. Contrary to the mainstream elite discourse in China, where the US or the West has been the dominant reference for envisioning development, these intellectuals are very clear in their rejection of the US/Western model as relevant for China. Steering away from ideological debate, they base their rejection by frequently emphasizing the conditions of rural China development. Even with urbanization, rural China will continue to host a large population – about 900 million – with dwindling farm land and resources. On this constrained premise, rural China has to achieve the goal of a prosperity and sustainability that are shared by most, not a few (T. Yang 2011, 36–7).

What is becoming shared knowledge among many rural support intellectuals is clearly stated by Yang Tuan, a scholar at CASS. The US/Western model, associated with de-peasantization [*qu nongmin hua*], industrialization, and urbanization,[20] is a model that works for a small number of capitalist farmers and corporations who enjoy big government subsidies, while China needs to find a way to sustain a large rural population (T. Yang 2011, 38). Wen Tiejun, Dean of the school of agriculture and rural development at Renmin University, discusses three types of agriculture around the world: the large-scale plantation agriculture found in America is a product of colonialism; the medium-scale agriculture in EU is mainly run by middle-class, part-time farmers, the small-scale agriculture in Japan and South Korea, operating under a tight population-land resource condition, is the only useful reference for China (Wen 2011, 29). A consensus exists among rural advocates that neither the US nor the European model of agriculture can be useful examples for China. Rural advocates also take warning from the modernization experiences of other developing countries, particularly in South Asia and Latin America (Wen 2004, 8–12, 2007). In what he calls 'the Philippines Path', well-known rural advocate Li Changping sees how the American model of agrarian capitalism has damaged rural livelihood in the Philippines and how the 'company + households' model there has only sped up the bankruptcy of rural households (C. Li 2009, 85–9). Both Yang and Li point out that the application of the 'company + households' model in China has only benefited a few.[21] Li most forthrightly argues that company and households have an unequal relationship that involves exploitation, only resulting in companies squeezing rural producers out of processing, transportation, circulation, rural finance, etc. (C. Li 2009, 76–8, 85). Zhang Xiaoshan, another scholar at CASS, remarks that determining who benefits from cooperatives is a serious question of principle and directionality (X. Zhang 2009).

---

[20]Yang Tuan (2011) used the term 'de-peasantization' [*qu nongmin hua*] in her original text, although this reference is a little odd in the US context.

[21]Wen Tiejun and his team observes that 'company + households' has an unstable relationship in China and 80 percent of such contract relations are undermined by farmers opting out ('Jianshe' 2009, 6).

Going against the thrust of the mainstream that promotes rural cooperation as singularly centered on commodity production, Wen Tiejun, Yang Tuan and Li Changping envision empowerment through a cooperation that encompasses village and farmers and conjoins production and reproduction, economy and culture. They see such a 'comprehensive cooperation' practiced in Japan, South Korea and Taiwan. In the Japan-South Korea-Taiwan (JKT) model, or sometimes referred to as 'the East Asia model', the state supports the national farmers union to coordinate and integrate production, transportation, finance, etc. and provides laws that limit the presence of corporate capital in agriculture. Arguing that the path of 'comprehensive cooperation' fits with China's own six decades' rural development, Yang proposes the township to be the building block of comprehensive farmers' association (T. Yang 2011, 38). Li proposes a community-based collective economy that has finance cooperation at its core and farm land cooperation as the foundation (2009, 75). Contrary to mainstream silence and aversion, Li further lauds the 'new collective economy' embodied by the existing few thousand collective villages that have survived de-collectivization (81). He Huili, a faculty member at Chinese Agricultural University who experiments with rural cooperatives in Henan province, advocates the local state to be a key agent in organizing community-based comprehensive cooperation (H. He 2007, 28).

While agreeing with the rejection of the US-European models of agriculture, He Xuefeng, director of China Rural Governance Research Center at Huazhong University of Science and Technology, questions whether JKT's rural conditions are actually sufficiently comparable to those in China for the JKT model to be relevant (X. He 2012, 125). Similarly, Philip Huang and his co-authors also see a significant divergence between Chinese agriculture and the so-called East Asia model, because unlike that in China, Japan's rural population is less than 10 percent of the national total and its family farming is greatly overshadowed by the expansion of wage-labor-based capitalist agriculture (P. Huang *et al.* 2012, 141). He Xuefeng's discussion provides a different assessment of the major problem faced by rural China and asks what problems cooperation is able to address. Estimating that 70 percent of Chinese farmers have to grow staple grains, He asserts that there is little room for staple grains to increase in price within the existing macro structure. He thus argues that the pressing need for most farmers – hence the purpose of cooperation – is not about their empowerment vis-à-vis the market, but about improving socio-cultural and infrastructural public goods provision in villages. Villagers suffer not so much from the current level of income as from the corrosive effect of modernity penetrating the rural space and subjectivity: atomization, the problem of the old age, the left-behind children, the community deterioration that becomes the push factor for migration, etc. Hence He and his team focus on the potential agency of rural governance in creating such public goods or creating conditions for such public goods. Provision of socio-cultural public good can help rebuild the rural social fabric and re-make the meaning of rural life. Provision of infrastructural public goods can improve conditions of production for fragmented and individuated rural producers.

While other rural advocates mentioned above see in cooperation a way for small farmers to be organized so as to collectively wrestle with the market and capital as external forces, He argues that rural sustainability can be achieved via re-socialization within the village. Rather than looking elsewhere for inspiration, He suggests that the People's Republic of China (PRC)'s own history could serve a useful reference for rural organization. In his proposal, the existing rural governance structure – which was rooted in Mao-era collectivization, but played a notorious role in 1990s predatory taxation (see Day's elaboration of state involution in this collection – can be reformed in the post-taxation era to better mediate between the now-increased state subsidy and the actual needs of villagers and bring about improved socio-cultural and economic public goods provision (X. He 2012, 124–30).

Wen Tiejun, the leading intellectual in the new RRM and the most scathing critic of modernization, interprets China's fin-de-siècle rural decline as an outcome of a relentless drain of rural resources that include money, labor and land. In the past three decades, the city has gained the benefit of marketization and capitalization of these resources, while the countryside is left with the negative social and economic cost of such capitalization, manifested as the *san nong* crisis. The new RRM is necessary both because rural small producers will continue to have a large and long-term presence in China and because its preservation will shelter China from the sharp impact of periodic capitalist crises. Contrary to the mainstream view that attributes the *san nong* crisis to rural backwardness and deficiency in modernization, Wen argues that the crisis is not sui generis, but has its root in the very dynamics of modernization. The rural hinterland, whose resources escape complete monetization and capitalization, rather than being a drag on modernization, is modernization's potential rescuer. It has the practical capacity to buffer the country against the worst impacts of modernization's crisis (Wen 2012a). For example, rural China cushioned the social impact of the large layoff of 20 million labor migrants during the 2008–2009 financial crisis.

Wen's critique of modernization is not limited to the post-Mao period or to China. His deconstruction of modernization as a false universalism exposes the historical and particular kinds of violence and resource expropriation that have made modernization attainable in only a small number of countries (Wen 2004, 3–22, 2007). His deconstruction of modernization in China stretches across the entire twentieth century. The defining characteristic of rural China, in Wen's view, is having a large population of scattered [*fensan de*] small producers who have scarce resources and accumulate little surplus. The constant challenge for modernization in twentieth-century China was how surplus can be extracted from such numerous and meager producers for the sake of industrialization. Thus Wen is interested in explaining how the contradiction between rural China and modernization has played out through various waves and crises of modernization that included the KMT's industrialization in the 1930s, as well as the industrialization blunder in the Mao era (Wen 2012a). Going against the mainstream discourse that continues to push China along the path of modernization, Wen believes deconstruction and disenchantment of modernization is the necessary condition to win the public understanding of the new RRM (Wen 2010a).

In the face of the *san nong* problems, Wen observes both 'market failure' and 'government failure'. As the outflow of rural resources has been through marketization, Wen argues that the conventional market is unable to address the *san nong* problems. He terms this inability 'market failure'. As small producers' farming yields too little surplus itself, a superstructure of conventional modern governance in rural China is beyond rural affordability and therefore has to be supported via large subsidies and fiscal transfers. That only results in reinforcing ponderous and compartmentalized bureaucratic self-interests. This is demonstrated by how government bureaucracies and rural elites in the post-taxation era have intercepted much of the subsidy for rural producers, leading to failed implementation of rural support policies (Wen 2001a, 25, 'Jianshe' 2009). This 'government failure' manifests itself as a negative relationship: the more modern the political-governance superstructure, the less it is able to treat the decline of the rural economic sector (Wen 2012b).

Based on the thesis of deconstructing modernization, Wen and his team propose 'comprehensive cooperation' as a way to organize rural producers into larger subjects that can negotiate and transact with the state and market (S. Yang and Wen 2011). Cultural reconstruction, referring to creating village-based cultural activities, will help rebuild the social capital lost in the fragmentation of village social relationships and facilitate cooperation in other dimensions. Wen and his team envision 'comprehensive cooperation' to cover finance, marketing and production and call on the state to grant new organizations preferential policies and exclusive rights in fields that involves agriculture. He Huili argues that the new RRM is about building a post-capitalist new civilization that respects farming traditions and values ecological agriculture. In her experiment with comprehensive cooperation, villagers pool land resources together and cooperate in production, marketing and building ecological houses (H. He 2009).

Rural advocates differ from the liberal mainstream on the question of rural self-organization. Liberal scholars such as Qin Hui denounce China's Mao-era experience of rural collectivization on the basis that it involved a top-down state coercion. In contrast, rural support intellectuals urge the state to support the cooperative movement and acknowledge that it is very difficult for farmers to self-organize (S. Yang and Wen 2011). Li Changping avers that while it is now fashionable to promote voluntary self-organization in rural China, even well-endowed urban middle-class home owners have not achieved much self-organization against real estate developers, let alone rural producers, who since the 1990s have been exploited by policy-supported dragon-head enterprises in all four fields – finance, processing, circulation and agricultural inputs – and cannot afford the cost of self-organizing. With the relationship between the state and rural producers shifted from extraction through taxation to compensation through subsidy, Li opines that the state should have no cause to fear, but have reason to support rural organization (C. Li 2009, 82). Although not particularly associated with 'san nong' studies, Lao Tian analyses the historical difficulty of peasant self-organization by examining William Hinton's documentation of complex dynamics and tendencies among peasantry in the land reform processes (Lao 2009).

Similarly, Wen and his team argue that the capacity and mechanism for cooperation has been weakened by industrial and commercial capital expansion into the rural sector, which has reduced the benefit – or organizational rent – of cooperation. Hence a vicious cycle, lesser cooperative capacity leads to deficient village governance, which makes it less effective in mobilizing for external subsidy and support (Wen and Dong 2010, 22–3). Wen and his team point to the Japanese state's role in the formation of Japan's national farmers union and argue that 'cooperatives that ensure fairness and protect the disadvantaged … will have to result from strategic support by the state that represents integrative and long-term social interests' (S. Yang and Wen 2011, 45).

If one of Liang's predicaments was 'in the so-called 'village movement', the village does not move', then contemporary support for the cooperative movement is confronted with the predominance of 'fake cooperatives', in which small producers barely participate. Liang lamented that villagers had a very indifferent attitude towards the presence of his RRM institute. Indifference has also been observed today: '"cooperation" becomes an activity that concerns government officials and elites and ordinary farmers are indifferent' (D. Zhang 2011, 59). Among the 272,000 cooperatives formally registered by 2010 in China, it is estimated by many observers that 80–95 percent of them are fake (Liu Laoshi 2010, 54). Such cooperatives are fake in several different and overlapping ways: they are empty shells that exist only in name; they are actually controlled by 'big households' (rich farmers) that rarely involve small producers' cooperation; they are actually dragon-head enterprises or 'company + households' putting on a new hat as cooperatives; they are organized by government departments (D. Zhang 2011). Among the 136 registered cooperatives in a county in Anhui, 125 were set up by big households, four by government departments, five by dragon-head enterprises and two by village committees (X. Zhang 2009, 14). In Hubei province, only 10 percent of the 4375 cooperatives are found to be economically active and 95 percent are dominated respectively by village 'big men' (*neng ren*) who specialize in farming, processing and trading (55 percent), agro-technology associations and government's agro-technology departments, etc. (30 percent), and dragon-head enterprises (10 percent) (K. Zhang and Zhang 2007, 62–3). In these cooperatives, small producers are 'being cooperativized' [*bei hezuo*] (D. Zhang 2011), meaning their 'cooperation' is orchestrated and managed by a dominant minority without their actual participation. With the introduction of the cooperative law in 2007, the situation of 'fake cooperatives' has only worsened ('Jianshe' 2009, 11, Zhao 2010, iii). These pretenders mushroomed in recent years in order to benefit from central and local governments' preferential policies and subsidies for cooperatives (C. Li 2010). In the face of this adverse situation, some concerned scholars have called for a crackdown on the fakes (Liu Laoshi 2010, 54).

How to account for this phenomenon is an issue for debate. Many attributed it at least in part to farmers' deficiency in cooperative consciousness, cooperative culture, or cooperative capacity or to the problem of an insufficient legal framework (Tong and Wen 2009, 21–2, Yuan 2010, 17). Yuan Peng also suggests that the heterogeneity of cooperative membership – referring to varied endowment of resources among members – accounts for a lack of democratic participation within cooperatives. Yuan argues that the problem can eventually be solved when members complete their own transformation into homogenous entrepreneurial commercial farmers (Yuan 2010, 17). Liu Laoshi, the late leader of Liang Shuming Rural Reconstruction Center, who dedicated his life to the rural support cause, argued that the seeming predominance of fake cooperatives is a problem of criteria (L. Liu 2010). He figured that one percent of rural cooperatives in China would meet the seven principles adopted by International Co-operative Alliance in 1995. No more than 10 percent would reach the criteria set out in the 2007 Chinese law on cooperatives, most importantly the criteria of 'one member one vote' and patronage refund (income in excess of expenses generated by the cooperative members' use of their business that is refunded to them). No more than 20 percent institute patronage refund. Yet in Liu's view, the problem is not whether cooperatives are fake, but whether these criteria, particularly the Western criteria, would fit China's complex situation. Liu instead proposed 'one member, one share of power' as a more flexible 'autochthonous' [*bentu zizhi de*] criterion and believed that this can enable farmers' control of cooperatives (L. Liu 2010, 59). He argued that as long as farmers have the right to control cooperatives, they can choose to allow big shareholders to gain more at an early phase of a cooperative development. So long as farmers are in control, then farmers control capital and such capital, regardless of where it is sourced, is people's capital (59–60). The political logic here eerily resembles what is applied to the Chinese state that, by its own self-identity as the helmsman of national economy, is believed to be in control of capital and thus qualifies as socialist.

By Liu's reckoning, cooperatives fall into three groups. Less than 20 percent run smoothly and can be considered 'genuine' cooperatives. Most of these are not strictly local, but have external – non-governmental organizations (NGOs) or universities – involvement. Less than 30 percent are fakes, empty shells without actual members and controlled by big households or companies. More than 40 percent are between genuine and fake and between cooperative and company. They include those initiated by big households, those having 'company + households' arrangements and those initiated by local governments. Cooperatives in the last category have members and rule books, but the actual operations do not follow rules; rather, they follow the decisions of a few. Liu believed that the ambiguous 40 percent are nevertheless local initiatives and their problems are not their own, but derive from the larger context. Liu thus argued that these 40 percent should be regulated and guided, rather than be denounced as fakes. Liu warned that talk of crackdown on the fake would discourage these 40 percent, resulting in stifling and even destroying the development of cooperatives (L. Liu 2010, 62). Pan Jiaen similarly dismisses the categorization of true and fake cooperatives as an 'artificial binary' (Pan 2012, 145) and shares Liu's prioritization of the movement. This logic is resonated in the words of a cadre in Heilongjiang who went all-out to encourage local cooperatives, 'let hair grow first before having a haircut'.[22]

A rare in-depth ethnographic study of one cooperative by Christopher Lammer finds that it lacks active involvement by average members and is dominated by a few (Lammer 2012, 166). Yet it is usually considered 'genuine' within the rural support circle. Like most cooperatives, its management is also dominated by men (166).[23] Lammer argues that the new rural reconstruction effort in this case has actually led to 'the development of capitalist class relations *within* the village' (153, emphasis original). Some villagers who contest the social relations and practices within the cooperative have labeled it 'a fake cooperative'. (160–3). Wary that the term 'fake cooperative' invokes intention on the part of the cooperative founders, Lammer differentiates this kind of cooperative from those that are no more than empty-shells and terms it 'an imagined cooperative' in order to mark the tension, as well as the possible potential of transformation between the reality and the representation: 'they are nominal cooperatives and imagined as cooperatives by those who dominate them and who support them. They are non cooperatives in the sense that patron-client and class relations exist between those involved in the organizations' (169). Transformation of this kind of imagined cooperatives would face challenges that are both internal and external. Matthew Hale's study of four cases of cooperative experiments in China points to a structural contradiction: their success in commercial projects requires a deeper integration with capitalist dynamics and thus the 'internal' social-community principles cannot be neatly separated from 'external' market dynamics (Hale 2013).[24]

Liu's prioritizing the movement and his attribution of fake cooperatives to external factors is not shared by all rural advocates in China. Tong and Wen argue that the fast growth of cooperatives dominated by relatively wealthy households, characterized as 'big farmers exploiting small farmers' [*danong chi xiaonong*], presents only a fake blossoming of cooperatives [*xujia fanrong*] (Tong and Wen 2009). The fast growth of cooperatives is not only due to the need of scattered producers in linking up with the market, as many have hoped to see, but also propelled by the interest of capital and government departments (Tong and Wen 2009, 15). The formation of fake cooperatives, rather than being due solely to exogenous factors, results from three dynamics in the past three decades, the process of rural differentiation into big and small producers produces rural capital and agricultural wage labor, rural capital is joined by urban capital and together they expand from marketing to processing, and now to agriculture, government departments not only play a

---

[22]Field research in Nehe county, Heilongjiang, July 2012.

[23]See He Yufei and Ju Zheng (Y. He and Zheng 2013) and Ku Hok Bun (2013) for notable exceptions.

[24]The existing collective villages in China also face a similar contradiction. The famous Nanjie village articulates a spatial separation between the internal (vis-à-vis collective members) and external (vis-à-vis the market) principles as '*wai yuan nei fang*' (round on the outside, square on the inside). While members of these villages do enjoy much better collective welfare and egalitarian treatment, collective enterprises do build its accumulation based on the exploitation of non-members who are migrant workers and who form the main workforce in these enterprises. See Liu Yongji (2008).

leading role in facilitating capitalization of agriculture, but also seek to increase their own revenue and subsidize their operations by providing fee-based services.

These dynamics feed on each other and produce three further dynamics: government capital is converted to bureaucratic capital and even private capital; rural differentiation, serving the necessary condition of capital expansion in agriculture, becomes intensified; fake cooperatives, dominated by big producers who are favored by capital, government departments, are instrumental in furthering rural differentiation. Such cooperatives play the role of a middle-man between capital and small producers and cannot be expected to empower small producers on the market (Tong and Wen 2009, 13–8). In fact, it is found that 'the key economic goals of co-operatives are also successfully filtered by elites to allow and even reinforce the existence of monopolies and retain inequalities between farmers and business' (Zhao 2010, 168). Tong and Wen propose as an alternative community-based multiple cooperation and call on the state – 'the most authoritative public power' – to intervene. Their proposed interventions include preventing government departments from engaging in for-profit operations, restraining capital, supporting small producers and encouraging mutual benefit and cooperation between big and small producers (Tong and Wen 2009, 23).[25]

4. The past and the present: resonances

The new RRM shares with the 1930s RRM a similar self-positioning as a social reform that explores within the dominant political-economic structure an alternative to the ongoing agrarian change perceived to threaten rural sustainability. Not unlike the early RRM intellectuals, rural advocates today also look for a third way for rural China and attempt to 'go beyond' both the left and the right (Pan 2012, 6). Philip Huang's proposition can more or less represent this shared vision:

> An alternative perspective is that agriculture remains mainly based on the peasant family and not on capitalist farming, and its ideal direction of development should be toward neither capitalism nor socialism, but something different, along the lines of marketized cooperatives, in the manner originally envisioned by Chayanov. (P. Huang et al. 2012, 140)[26]

Deng Xiaoping's market reform, launched in 1978 for what is called the primary stage of socialism, is based on an assumption which he already held in 1956, which is that Chinese society has eliminated class divisions and has only a division of labor (Meisner 1999, 453). Deng's post-revolutionary assumption about Chinese society in a way resonated with Liang Shuming's assumption. In the post-Mao trend of

---

[25]The interventions proposed by Tong and Wen resemble the vision of Sun Yat-Sen (1866–1925), the founding father of KMT and Republic of China. Sun and the progressive force within the KMT were unable to realize his vision.

[26]Most rural support intellectuals do not reference Chayanov. Wen Tiejun in his dissertation-turned-book argues that both Alexander Chayanov's vitality of family farming and Theodor Schultz's rational peasant farmer encounter challenges in contemporary rural China. With labor migration becoming a key economic activity among rural laborers, the emergence of labor price has played the role of a 'shadow wage' for those engaged in agriculture such that households now make labor input decisions in the context of comparative gains. Within China's larger political-institutional context, farm land – not a commodity – serves the triple function of production, guarantee of subsistence and social stability. Thus Schultz's presumed market-based peasant rationality does not quite work either in the Chinese context (Wen 2009, 27–9).

de-radicalization and 'farewell to revolution' (Z. Li and Liu 1995), the social reformist position of the 1930s RRM – including its assumption of rural society as an undifferentiated whole – although a target of political critique by leftist intellectuals then and later in the Mao-period – has again appeared respectable and progressive.[27] In fact, one might say that the political evaluation of RRM is in an inverse relationship with that of revolution in contemporary Chinese discourse. To be sure, the contemporary search for the third way in China is not limited to rural support scholars. Cui Zhiyuan, for example, makes use of the idea of rural producer cooperatives in his well-known manifesto of 'petty bourgeois socialism' that proposes to make the petty bourgeoisie the social foundation for market socialism (Cui 2003).[28]

'Radical' has become a negative signifier in the post-Cultural Revolution liberal discourse, whose signification has been expanded to include the twentieth-century Chinese revolution. In this context, Wen Tiejun makes his own re-signification of 'radical' by calling the twentieth-century efforts at modernization radical, thereby making modernization – particularly the current thrust of urbanization and capitalization – an object for deconstruction and rejection (Wen 2010b). Pan Jiaen, a key member of Wen's team, includes in the category of radical the following, the May Fourth movement, communist revolution, the Cultural Revolution, neoliberalism, as well as fast urbanization, modern governance for rural society and developmentalism (Pan 2012, 11). Wen argues that neither yesterday's Soviet Model nor today's American model should be the reference for Chinese emulation. In Wen's view, the link between the two RRMs lies in the persistent need to disengage from modernization, which is even more urgent with today's more severe resource stress (Wen 2010a, 17). Pan (2012) and Day (2008) frame the historical and current RRM as Polanyi's social protectionism.

Rural advocates today also share with their 1930s forerunners some basic assumptions and conceptualization about rural China. Their rural support position is based on taking rural society as a whole and assuming that the threat to rural sustainability is primarily external. The 'wholeness' is an epistemological assumption embedded in the very term *san nong* – nongmin, nongcun, and nongye. While the policy concept of *san nong* – embraced by government planners, general academia, as well as rural advocates – succeeds in making social subjects, place and production an inter-related set of problems, it makes differentiation within rural society invisible or insignificant.[29] This outlook resonates with the early RRM conceptualization of rural society as largely undifferentiated and fits

---

[27]As it is observed, 'The political evaluation of the RRM was mostly negative before the 1990s, but has turned much more positive since then' (Qiu 2002, 100). In the words of Catherine Lynch, Liang's intention for a non-capitalist industrialization represented 'fresh possibilities of a progressive modernity' (2010, 161).

[28]During the workshop where Cui's paper was presented, Wen Tiejun pointed out that rural households' land entitlement was an outcome of the twentieth-century revolutions, not of petty bourgeoisie sentiment and agency (Cao 2003, 224). Lin Chun reminded Cui that the petty bourgeoisie as a class is not inclined towards socialism (Cao 2003, 225). For a fuller critique, see Cao (2004).

[29]Even in the Mao era's rural commune system, known for its high-level egalitarianism, there was notable differentiation. See Huaiyin Li's summary of research findings on this topic: the top 10 percent of households took 28 percent of the total income, while the bottom 40 percent had 16 percent, and per capita income of the top 25 percent of households was 2–3 times that of the bottom 25 percent. As land holdings were collective and collective labor contributed to 20–30 percent of total household income, Li convincingly made demographic differentiations the primary factor in such differentiation (H. Li 2009, 208).

in with the post-Mao dominant ideology that admits stratification, but disavows class analysis.[30]

To be sure, most recently, some rural support intellectuals, particularly He Xuefeng's team, have begun to incorporate rural stratification in their case studies, by examining diverse compositions of rural household incomes.[31] Having been concerned about troubles in public goods provision, rural social disorder and governability, and particularly worried about the accelerating trend of the rich dominating rural governance (X. He 2012, 290–307),[32] they find 'the middle stratum' – particularly the medium-sized commercial farmers (who they call zhongnong or middle farmers) – to be the guardian of rural social order and the stabilizing force for rural production and governance. This unwittingly mirrors the growing liberal normative assumption about the urban middle class as good for China's future.

Yet whether this rural stabilizing force can maintain its own stability remains an open question. Perhaps recognizing that the middle farmers are not inherently stable, He and his team call for the state to support this particular stratum. In addition to the differentiating tendency inherent in market conditions, the middle farmer stratum also depends on a number of factors for their own reproduction. One of these factors is the uncertainty of keeping the land use rights they have accrued from rural-to-urban migrants, who often constitute about a third of village laborers (e.g. H. Lin 2012, 61) and whose labor employment in the city is contingent on the global and regional economic dynamics. What He and his team termed 'middle farmers' reproduce themselves in the context in which agro-business has been gaining increasing control of means of production (F.Q. Zhang and Donaldson 2008, 26) and has demonstrated a significant tendency of reaping growing gains in the value chain (Wu 2012).

Wen makes a similar but more sweeping generalization about rural society as an autonomous social body. Referencing Liang Shuming, Wen is convinced that a combination of ethics (culture), rural elites and village commons constitutes what he calls 'internality' (Wen 2009, 7) or a form of local institutional arrangement that is conducive to low-cost governance and development (Wen and Dong 2010, 21). Within this institutional arrangement, traditional elites are said to have their interests integrated with those of the rural body as a whole. Similar to Liang Shuming, Wen argues that land ownership was dispersing instead of concentrating in pre-revolutionary China. Thus Wen reinterprets the cause of Chinese revolution: rather than the dual social contradictions created by imperialist oppression and land-based class conflicts, the real cause of revolution was the extortion of agriculture and rural society by industrial and commercial capital, brought on by the KMT's decade-long modernization (1927–1936) (Wen 2001b, 7, 2009, Ch. 4 and 5). In Wen's view, the outcome of both the Chinese revolution and the post-Mao rural reform was the making of rural producers into petty bourgeoisie. In the overall context of excessive extortions during the KMT modernization, benevolent rural gentry became displaced by

---

[30]For disavowals of class in post-Mao ideology and intellectual discourse, see Pun and Chan (2008) and H. Yan (2008, Ch. 5). Philip Huang complains that class categories have largely disappeared from official statistical practices (P. Huang *et al.* 2012, 141).

[31]See X. He (2011). For a collection of their case studies that makes this argument, see the special issue of Kaifang shidai (*Open Times*) in 2012, which includes articles by Chen Baifeng, Lin Huihuang, and Yang Hua. See Day and Hale (2008) for an introduction and collection of essays by scholars in He Xuefeng's team.

[32]He maps different kinds of rich people and views their wealth generation as external to rural place and agriculture (X. He 2012, 290–307).

malevolent ones, exacerbating the tension. Wen sees a similar replay of the displacement of the benevolent 'community elites' by malevolent elites in China's 1990s predatory taxation practices, which resulted in elite interest diverging from the overall community interest. In most rural areas, these elites now intercept and appropriate for themselves most opportunities and government subsidies, described by Wen as 'elite capture' (S. Yang and Wen 2011, 44).

How rural support intellectuals understand the current status of rural society impinges on their advocacy of the rural cooperative movement. Wen's perception of the external threat to the rural social body resonates with one of Philip Huang's characterizations, that the main threat for rural producers in China today is commercial capital and thus it is relations of circulation, rather than relations of production, that matter most (Huang 2012, 95). In conjunction with this, Huang and his co-authors also argue that Chinese agriculture is undergoing capitalization – referring to increased capital input – without proletarianization, with wage-labor calculated to be about 3 percent of the total labor input in agriculture, in contrast to India's 45 percent (P. Huang *et al.* 2012). Due to unevenness in the value addition of agricultural products, Huang and his co-authors also find that in 2009 wage labor was 8.5 percent of the total labor input for vegetable farming, 6.7 percent for cotton farming, 40 percent for apple farming, 28 percent for dairy farming and 27 percent for chicken egg farming.

Yet, while Huang *et al.* argue that China's land institution puts a brake on proletarianization and enhances the viability of family farming, they do not rule out the possibility of Chinese agriculture developing into 'capitalization with proletarinization'. Wu Guanghan's examination of rural producers' shrinking share in the value chain of agriculture products reaches a similar view of the threat of commercial capital to rural households, but suggests that rural producers' subjection to merchants is such that rural producers already experience semi-proletarianization. Rural producers in Wu's conceptualization are both petty bourgeoisie and proletarians, but their apparent autonomy in production elides their subsumption to capital in circulation. Rather than manifesting a viability of family farming, this dual nature in Wu's view is not a stable condition, but a transitional one. For there is a detectable trend that rural producers' subsumption to capital will expand from circulation to production (Wu 2012, 108). On the basis that rural producers as a whole are under an external threat, Huang explicitly (Z. Huang 2010) and Wu implicitly (Wu 2012) advocate support for rural cooperatives to integrate production, processing and marketing in order to counterbalance the power of commercial capital. Yet capital interests are no longer external to the village, but are part of the forces that are remaking the rural society. This is demonstrated in a report by a rural cooperative organizer:

> In today's countryside, businesses that have scale and profit are in the hands of dragon-head enterprises, businesses that have small scale and profit are in the hands of rich villagers, [any] business that have profit are in the hands of all-pervasive small merchants, 70–80% of households in the village make money only by selling primary products or labor ... There are already monopolistic interests in the village that treat some businesses as food on their plate. If a cooperative touches on their interests, the organizers will face harassment and even reprisal. (quoted in Pan 2012, 143)

To conclude, rural support intellectuals increasingly understand China's agrarian problem in a global context that is furnished by a set of references different from those of modernization theory. They form a critical mass that challenges the ongoing dominant neoliberal vision that promotes urbanization and capitalization of agriculture as a viable future for China. While differences and contradictions exist among rural support intellectuals, many have questioned or disagreed that specialized farmer cooperatives would benefit the majority of rural producers. Cooperatives dominated by 'big households' are certainly not unique to China. Interestingly, many rural support intellectuals in China would not hesitate to include such cooperatives among 'fake cooperatives' and denounce them as such. Even as many dismiss the Mao-era rural commune system, this recent past experience has perhaps sharpened their practical – though not always theoretical – egalitarian sensibility on the one hand, and has on the other hand made it easier for them to project a communal coherence that they hope to revive in the future.

It is not difficult to see parallels between the 1930s RRM and the contemporary one. Like their forerunners, rural support intellectuals today politically look for a third way, intellectually adopt the rural support position in their policy/social advocacy and practically use community as the building block for rural reconstruction. For these political, intellectual and practical reasons, they conceive rural society as a largely undifferentiated whole. Class analysis of agrarian change has little place in the study or advocacy of cooperatives or in rural study in general. In that, rural support intellectuals do not depart from the mainstream intellectuals in China who disavow class analysis. Yet, as Tong and Wen (2009) suggest, the dominant presence of rich farmers in rural cooperatives is the very product of rural differentiation, which is an open dynamic that involves local and translocal interests. Also similar to Liang Shuming for whom 'China's problems are externally induced, not internally derived' (Liang 1989, vol. 2, 577), leading rural support intellectuals today apply the same perspective to rural China. The rural support stance which takes rural producers as a whole is logically connected with the perception of the externality of threat.

Since the introduction of the 'company + households' to China in the mid 1990s (Q. Lin 1994), dragon-head enterprises now perhaps already involve about a quarter of rural producers in China (P. Huang *et al.* 2012, 165). Even with the land institution in China that puts a break on land concentration and commodification in the fast advance of agrarian capitalization, rural producers are found to be differentiating into commercial farmers, contract farmers, semi-proletarian producers and proletarian farm workers (F.Q. Zhang and Donaldson 2008, 32, Y. Chen 2013). While Philip Huang argues that the current characteristic of Chinese agriculture is 'capitalization without proletarianization', he acknowledges that producers in contractual relationship with enterprises are in effect 'semi-proletarianized' (P. Huang *et al.* 2012, 166). Among rural support intellectuals, Tong and Wen (2009) provides a rare in-depth understanding of the mutually reinforcing dynamics that are producing rural differentiation, capitalization of agriculture and the large presence of fake cooperatives. Their finding and its implications have yet to be widely discussed and debated.

In addition to taking rural society as a whole, contemporary rural support intellectuals also resemble their forerunner Liang Shuming in having a strategic assumption about the state. Whether rural support intellectuals pin their hopes for rural sustainability on the middle farmers or on community-based comprehensive cooperation, they reason with and appeal to the state to restrain capital as the necessary condition for rural sustainability. The state is assumed – not naively, but strategically – as the public power that rises above various interests. Yet it is clear that the state (both the central and local governments) has provided much stronger support to dragon-head enterprises than to cooperatives (Z. Huang 2010, 25). In 1935, Liang Shuming reflected on two difficulties faced by the RRM. First, the RRM positioned itself as a 'social movement' (Liang 1989, vol. 2, 377), but it depended on political power for its social reform. Second, in the so-called village movement, the village does not move. Liang's candid reflection and the 1930s debate about RRM can still serve as a useful reference for our reflection on the contemporary rural cooperative movement.

References

Ai K. [Allito, G.]. 2003. *Zuihou de ru jia* [*The last Confucian*], translated by Wang Zongyu and Ji Jianzhong. Nanjing: Jiangsu renmin chubanshe.

Ai K. [Allito, G.]. 2006. *Zhege shijie hui hao ma? Liang Shuming wannian koushu* [*Has man a future? Liang Shuming's narration in his late years*]. Shanghai: Dongfang chuban zhongxin.

Allito, G. 1986. *The last Confucian: Liang Shu-ming and Chinese dilemma of modernity*. Berkeley CA: University of California Press.

Bo, G. 1994. Zhongguo sanshiniandai de hezuo yundong ji xiangcun gailiang chao [Co-operative movements and rural reform in China in the 1930s]. *Zhongguo jingji shi yanjiu* [*Researches in Chinese Economic History*], 4, 122–34.

Cao, T., ed. 2003. *Xiandaihua, quanqiuhua yu zhongguo daolu* [*Modernization, globalization and China's path*]. Beijing: Shehui kexue wenxian chubanshe.

Cao, T. 2004. Xiaokang, xiaozi yu shichang shehuizhuyi [Modest affluence, petty bourgeoisie and market socialism]. *Dushu* [*Reading*], 3, 10–15.

Chen, H. 1980 [1939]. *Industrial capital and Chinese peasants: a study of the livelihood of Chinese tobacco cultivators*. New York: Garland Pub.

Chen, Y. 1983. *Zhonghua hezuo shiye fazhan shi* [*The history of co-operatives development in China*]. Taibei: Taiwan shangwu.

Chen, Y. 2013. Zibenzhuyi shi jiating nongchang de xingqi yu nongye jingying zhuti fenhua de zai sikao [The rise of Capitalist Family Farms and a reflection on the differentiation of agricultural subjects]. *Kaifang shidai* [*Open Times*], 4, 137–156.

Chiang, Y. 2001. *Social engineering and the social sciences in China*. Cambridge: Cambridge University Press.

China Review News. 2013. Gongshang zongju: nongmin zhuanye hezuoshe yi da 68.9 wan jia [State Adminstration of Industry and Commerce: specialized farmer cooperatives have reached 689,000] [online]. 10 January. Available from: http://www.chinareviewnews.com/doc/1023/9/6/8/102396817.html?coluid=123&kindid=0&docid=102396817 [Accessed 1 February 2013].

Cui, Z. 2003. Ziyou shehui zhuyi yu zhongguo de weilai [Liberal socialism and the future of China]. *In*: T. Cao, ed. *Xiandaihua, quanqiuhua yu zhongguo daolu* [*Modernization, globalization and China's Path*]. Beijing: Shehui kexue wenxian chubanshe, 192–226.

Day, A. 2008. The end of the peasant? New rural reconstruction in China. *Boundary 2*, 35(2), 49–73.

Day, A. 2013. *The peasant in postsocialist China: history, politics, and capitalism*. Cambridge: Cambridge University Press.

Day, A. and M. Hale, eds. 2007. Special issue on 'new rural reconstruction'. *Chinese Sociology and Anthropology: A Journal of Translations*, 39(4).

Day, A. and M. Hale, eds. 2008. Special issue on 'The Central China School of rural studies'. *Chinese Sociology and Anthropology: A Journal of Translations*, 41(1).

Du, R., *et al.*, eds. 2002. *Dangdai zhongguo de nongye hezuozhi* [*Contemporary China: agricultural cooperative system*]. Beijing: Dangdai zhongguo chubanshe.

Falüjie, 2007. Law of the People's Republic of China on Specialized Farmers Cooperatives [online]. Available from: http://bk.mylegist.com/bilingual/4394.html [Accessed 15 November 2012].

Fei, H. 2002. *Jiangcun jingji: zhongguo nongmin de shenghuo* [*Peasant life in China: a field study of country life in the Yangtze valley*]. Beijing: Shangwu yinshuguan.

Guo, R. 2003. Xinminzhuzhuyi Lilun de Xueli Tanyuan [An exploration of the theory of New Democracy]. *Zhonggong dangshl yanjiu* [*Researches of the history of Communist Party of China*], 3, 50–6.

Hale, M.A. 2013. Tilling sand: contradictions of 'social economy' in a Chinese movement for alternative rural development. *Dialectical Anthropology*, 37(1), 51–82.

Han, X. 2005. *Chinese discourses on the peasant, 1900–1949*. Albany: State University of New York Press.

He, G. 1937. *Zhongguo shehui xingzhi wenti lunzhan* [*Debate on the character of the Chinese society*]. Shanghai: Shenghuo shudian.

He, H. 2007. Nongmin hezuo de jiegou yu guocheng [The structure and process of farmers cooperation]. PhD Dissertation, Beijing University.

He, H. 2009. Chengxiang lianjie yu nongmin hezuo [Urban-rural links and farmers' cooperation]. *Kaifang Shidai* [*Open Times*], 9, 5–6.

He, J. 2007. Liang Shuming de nongye hezuohua sixiang yu shijian [Liang Shuming's thought and practice on rural cooperatives]. *Dongnan xueshu* [*Southeast Academic Research*], 1, 17–24.

He, X. 2009. *Cunzhi moshi [Modes of rural governance]*. Jinan: Shandong renmin chubanshe.

He, X. 2011. Quxiao nongyeshui hou nongcun de jieceng ji fenxi [An analysis of rural stratification after the abolishment of agricultural taxes]. *Shehui kexue [Social Sciences]*, 3, 70–9.

He, X. 2012. *Zuzhi qilai [Get organized]*. Jinan: Shandong renmin chubanshe.

He, Y. and J. Zheng. 2013. Neifaxing shequ fazhan: Shanxi Yongji shehui jingji anli [An endogenous development: a case of social economy in Yongji, Shanxi]. *A Radical Quarter in Social Studies*, 91, 261–80.

Hu, J. 2011. Jiating chengbao zao yi mingcunshiwang [Family-based contract system has long ceased to exist except in name] [online]. Available from: http://www.snsnsn.net/article/article.asp?typeId=17&id=1454 [Accessed 30 November 2012].

Huang, P.C.C., Y. Gao and Y. Peng. 2012. Capitalization without proletarianization. *Modern China*, 38(2), 139–73.

Huang, Z. [Huang, P.C.C.]. 2010. Longtou qiye hai shi hezuo zuzhi? [Dragon-head enterprises or cooperative organizations?] *Zhongguo laoqu jianshe [China's Construction of Old Revolutionary Basic Area]*, 4, 25–6.

Huang, Z. 2012. Xiaononghu yu da shangye ziben de Bo pingdeng jiaoyi: zhongguo xiandai nongye de tese [An unequal transaction between small rural households and big commercial capital: a distinctive feature of China's contemporary agriculture]. *Kaifang shidai [Open Times]*, 3, 88–99.

Jiang, X. and X. Jia. 2008. Minguo xiangcun gongzuo taolunhui pingyi [On symposium of country-side works in the period of republic of China]. *Xuzhou shifan daxue xuebao [Journal of Xuzhou Normal University]*, 34(3), 76–81.

Jianshe shehuizhuyi xin nongcun mubiao, zhongdian yu zhengce yanjiu ketizu ['Target, focus, and policy research of new socialist rural areas construction' group]. 2009. Bumen he ziben xiaxiang he nongmin zhuanye hezuo jingji zuzhi de fazhan [Governments and capital entering rural areas and [the impact on] the development of specialized cooperative economic organizations of farmers]. *Jingji lilun yu jingji guanli [Economic Theory and Business Management]*, 7, 5–12.

Ku, H.B. 2013. Funü, shouyi yu hezuo jingji [Women, skills and cooperative economy]. *A Radical Quarterly in Social Studies*, 91, 241–60.

Ku, H.B., *et al.* 2009. Chengxiang lianjie yu nongmin hezuo [Urban-rural links and farmers' cooperation]. *Kaifang shidai [Open Times]*, 9, 6–8.

Lammer, C. 2012. Imagined cooperatives: an ethnography of cooperative and conflict in new rural reconstruction projects in a Chinese village. Thesis (M.A.). University of Vienna.

Lao, Tian. 2009. Zhongguo xiangcun biange zhong de zi zuzhi kunjing. Renwen yu shehui [online]. Available from: http://wen.org.cn/modules/article/view.article.php/article=838 [Accessed 5 Oct 2012].

Li, C. 2009. *Da qi hou [The great climate]*. Xi'an: Shannxi renmin chubanshe.

Li, C. 2010. You zheyang yige hezuoshe [There is this such a cooperative] [online]. *Cunwei zhuren [The Village Cadre]*, 5. Available from: http://www.caogen.com/blog/Infor_detail.aspx?ID=38&articleId=20515 [Accessed 15 November 2012].

Li, H. 2009. *Village China under socialism and reform: a micro history*. Stanford, CA: Stanford University Press.

Li, Z. 1982. Zhongguo nongcun yundong de lilun yu shiji [Theory and practice of China's rural movement]. *In*: L. Zhong and F. Yang, eds. *Zhongguo xiandai zhexue shi ziliao huibian: cunzhi pai pipan [A collection of works about Chinese modern philosophy: a critique of the village governance school]*. Shenyang: Liaoning daxue chubanshe, pp. 63–4.

Li, Z., X. Gao and C. Cao. 2008. Xiangcun jianshe yu xiangcun hezuo yundong jujiang: Liang Shuming [The great master of rural construction and rural cooperative movement: Liang Shuming]. *Zhongguo hezuo jingji [China Co-operation Economy]*, 5, 46–9.

Li, Z. and Z. Liu. 1995. *Gaobie geming: ershi shiji zhongguo duihua lu [Farewell to revolution: dialogues about twentieth-century China]*. Hong Kong: Cosmos Books Ltd.

Liang, S. 1989. *Liang Shuming quanji [Complete works of Liang Shuming]*, Vols. 2, 5, 6. Jinan: Shandong renmin chubanshe.

Liang, S. 1992 [1933]. *Zhongguo minzu zijiu yundong zhi zuihou juewu: xiangcun jianshe lilun [The final social awareness of the Chinese national salvation movement: theories on rural construction]*. Shanghai: Shanghai shu.

Lin, H. 2012. Jianghan pingyuan de nongmin liudong yu jieceng fenhua: 1981–2010 [Peasant mobility and stratification on the Jiang-Han plain, 1981–2010]. *Kaifang shidai [Open Times]*, 3, 47–70.

Lin, Q. 1994. Gongsi jia nonghu: yindao nongmin zou xiang shichang de youxiao xingshi [Company plus household: an effective way to direct farmers towards the market]. *Nongye jingji wenti [Issues of Agricultural Economy]*, 5, 6–9.

Lin, Y. [Justin]. 1999. Xin nongun yundong yu qidong neixu [New rural movement and boosting internal demand]. *Zhongguo wuzi liutong [Materials Circulation in China]*, 9, 8–12.

Liu, L. 2010. Hezuoshe shijian yu bentu pingjia biaozhun [Social practice of rural cooperatives and local evaluation criterion]. *Kaifang shidai [Open Times]*, 12, 53–67.

Liu, Q. 2010. Kangzhan shiqi Jin-cha-ji bianqu de hezuoshe shulun [Policies on cooperatives in Jin-Cha-Ji border region during the Anti-Japanese Period]. *Dangshi wenyuan [Culture and History of Chinese Communist Party]*, 2, 13–5.

Liu, Y. 2008. Yi Nanjiecun wei li, dui reng baochi de jitizhi jinxing hezuozhi gaige [Nanjie Village as a case: transforming the existing collective system into a cooperative one] [online]. *China Left Review*, 1. Available from: http://chinaleftreview.org/?p=17 [Accessed 20 July 2013].

Lynch, C. 1989. Liang Shuming and the populist alternative in China. PhD Dissertation, University of Wisconsin-Madison.

Lynch, C. 2010. The country, the city, and vision of modernity in 1930s China. *Rural History*, 21(2), 151–63.

Ma, Y. 1992. *Liang Shuming pizhuan [Biography and commentary on Liang Shuming]*. Hefei: Anhui remin chubanshe.

Mao, Z. 1943. Lun hezuoshe [On cooperatives] [online]. Available from: http://gxs.changyang.gov.cn/art/2012/5/11/art_3172_79829.html [Accesse 18 July 2013]

Mao, Z. 1982. *Mao Zedong nongcun diaocha wenji [Collected works of Mao Zedong on rural investigation]*. Beijing: Renmin chubanshe.

Mao, Z. 1991. *Mao Zedong xuanji [Collected works of Mao Zedong]*. Beijing: Remin chubanshe.

Mei, D. 2004a. Gongheguo chengli qian geming genjudi huzhu hezuo zuzhi bianqian de shili kaocha [Case investigation of the transformation of cooperative organizations in the revolutionary base before the foundation of PRC]. *Zhongguo nongshi [Agricultural History of China]*, 2, 102–7.

Mei, D. 2004b. Guomindang zhengfu shiqi nongcun hezuoshe zuzhi bianqian de zhidu fenxi [An institutional analysis of rural cooperative organization during the rule of Kuomintang]. *Minguo dangan [Republican Archives]*, 2, 84–7.

Meisner, M. 1999. *Mao's China and after*. New York: the Free Press.

Nongye jingji hezuo zuzhi yanjiu ketizu ['Research on the organizations for agricultural economic cooperation' Group]. 2005. *Hezuo sheng jin: guoxin nongyanhui ershi nian fazhan yu sikao (1984–2004) [Co-operation makes miracles: reflection on the 20-year development of Guoxin agriculture association (1984–2004)]*. Beijing: zhongguo nongye chubanshe.

Pan, J. 2012. Shuangxiang yundong shiye xia de zhongguo xiangcun jianshe [China's rural construction seen as a double movement]. PhD Dissertation, Lingnan University.

Pun, N. and C. Chan. 2008. The subsumption of class discourse in China. *Boundary 2*, 35(2), 75–91.

Qian, J. and Z. Li. 1982. 'Zhongguo xiangcun jianshe pipan' bianzhe xu [Editors' introduction to 'A critique of China's rural construction']. *In*: L. Zhong and F. Yang, eds. *Zhongguo xiandai zhexue shi ziliao huibian: cunzhi pai pipan [A collection of works about Chinese modern philosophy: a critique of the village governance school]*. Shenyang: Liaoning daxue chubanshe, pp. 63–4.

Qiu, Z. 2002. Dui Liang Shuming xiangcun hezuo yundong de fansi [Reflections on the rural co-operative movements led by Liang Shuming]. *Zhongguo shehui jingjishi yanjiu [The Journal of Chinese Social and Economic History]*, 2, 100–4.

Selden, M. 1971. *The Yenan way in revolutionary China*. Cambridge: Harvard University Press.

Shi, J. 1957. *Zhongguo nongye hezuohua yundong shi liao [Historical documents on China's agricultural cooperative movement]*. Beijing: Sanlian shudian.

Thogersen, S. 1998. Reconstructing society: Liang Shuming and the rural reconstruction movement in Shandong. *In*: K.E. Brodsgaard and D. Strand, eds. *Reconstructing twentieth-century China: state control, civil society, and national identity*. Oxford: Claredon Press, pp. 139–62.

Tong, Z. and T. Wen, 2009. Ziben yu bumen xiaxiang yu xiaononghu jingji de zuzhi hua daolu [The transference of capital and departmental capital to rural areas and the organization of the farming household economy]. *Kaifang shidai [Open Times]*, 4, 5–26.

Unger, J. 2009. Families and farmland in Chinese villages: unexpected findings, *In*: M. Farquhar, ed. *Twenty-first century China: views from the South*. Cambridge: Cambridge Scholarly Publications, pp. 138–55.

Wang, D. 2004. *Liang Shuming wen da lu* [*Dialogue with Liang Shuming*]. Wuhan: Hubei remin chubanshe.

Webb, A.K. 2008. The countermodern movement: a world-historical perspective on the thought of Rabindranath Tagore, Muhammad Iqbal, and Liang Shuming. *Journal of World History*, 19(2), 189–212.

Wei, B. and Y. Zeng. 2010. Minjian huzhu, hezuo yundong, geming celue [Folk mutual aid, cooperative movement and revolutionary strategies]. *Journal of Gannan Normal University*, 2, 32–8.

Wen, T. 1994. Nongcun hezuo jinrong yanjiu yu fazhan de jiben silu [The basic idea for research and development of rural cooperative finance]. *Nongcun hezuo jingji he jingying guanli* [*Management and Administration on Rural Cooperative Economy*], 1, 27.

Wen, T. 2001a. Shichang shiling + zhengfu shiling: shuangchong kunjing xia de 'sannong' wenti [Market failure + government failure: the 'sannong' problem under double predicaments]. *Dushu* [*Reading*], 10, 22–9.

Wen, T. 2001b. Bainian zhongguo, yi bo si zhe [Twists and turns in twentieth century China]. *Dushu* [*Reading*], 3, 3–11.

Wen, T. 2004. *Jiegou xiandaihua* [*Deconstructing modernization*]. Guangzhou: Guangdong renmin chubanshe.

Wen, T. 2007. Deconstructing modernization. *Chinese Sociology and Anthropology*, 39(4), 10–25.

Wen, T. 2009. *Sannong wenti yu zhidu bianqian* [*Rural issues and institutional evolution*]. Beijing: zhongguo jingji chubanshe.

Wen, T. 2010a. Weishenme women hai xuyao xiangcun jianshe [Why we still need rural construction]. *Zhongguo laoqu jianshe* [*Construction of China's Old Liberated Areas*], (3), 17–8.

Wen, T. 2010b. Gaobie bainian jijin [Farewell to hundred years' radicalization] [online]. Renmin luantan wang. Available from: http://www.rmlt.com.cn/News/201012/201012311103335611_6.html [Accessed 15 November 2012].

Wen, T. 2011. Zonghexing hezuo jingji zuzhi shi fazhan qushi [Comprehensive cooperative economic organization is the development trend]. *Zhongguo hezuo jingji* [*China Co-operation Economy*], 1, 29–30.

Wen, T. 2012a. Ba ci weiji yu ruan zhuoluo [Eight crises and soft landing] [online]. Renwen yu shehui. Available from: http://wen.org.cn/modules/article/view.article.php/3485 [Accessed 15 November 2012].

Wen, T. 2012b. Xiandaihua weiji yu zhongguo xin nongcun jianshe [Crisis of modernization and new countryside construction in China] [online]. Zhongguo xiangcun faxian wang. Available from: http://www.zgxcfx.com/Article/51702.html [Accessed 30 November 2012].

Wen, T. and X. Dong. 2010. Cunshe lixing: pojie 'sannong' yu 'sanzhi' kunjing de yige xin shijiao [Village-community rationality: a new perspective to solve the predicaments of 'sannong' and 'sanzhi']. *Zhonggong zhongyang dangxiao xuebao* [*Journal of the Party School of the Central Committee of the CPC*], 14(4), 20–3.

Wu, G. 2012. 'Zhongjianshang + nongmin' moshi yu nongmin de ban wuchanhua [The model of 'intermediary merchants plus peasants' and the semi-proletarianization of the peasants]. *Kaifang shidai* [*Open Times*], 3, 100–11.

Xu, L. and J. Zhao. 2011. Liang Shuming de tudi lilun [Land theories of Liang Shuming] [online]. Lianhe shibao, 5 July. Available from: http://shszx.eastday.com/node2/node4810/node4851/node4864/userobject1ai47610.html [Accessed 20 November 2012].

Yan, H. 2008. *New masters, new servants: migration, development and women workers in China*. Durham, NC: Duke University Press.

Yan, Y. 1989. Xiangcun yundong chenggong de jiben tiaojian [Basic conditions for the success of rural movements]. *Yan Yangchu quanji* [*Complete works of Yan Yangchu*], Vol. 1. Changsha: Hunan jiaoyu chubanshe.

Yang, F. 1999. Liang Shuming yu shehuizhuyi [Liang Shuming and socialism]. *Shehuizhuyi yanjiu* [*Socialism Studies*], 5, 37–9.

Yang, F. 2001. *Liang Shuming hezuo lilun yu Zouping hezuo yundong* [*Liang Shuming's cooperative theory and Zouping's cooperative movement*]. Chongqing: Chongqing chubanshe.

Yang, S. and T. Wen. 2011. Nongmin zuzhihua de kunjing yu pojie [The predicament of and solution to farmers' organization]. *Renmin luntan* [*People's Forum*], 29, 44–5.

Yang, T. 2011. Zonghe nongxie: xin nongcun jianshe lujing xuanze [Comprehensive agricultural cooperatives: the path for new rural construction]. *Jinri guancha* [*Observation Today*], 10, 36–9.

Yu, S. 1929. *Hezuoshe zhi lilun yu jingyin* [*Theory and practice of cooperatives*]. Shanghai: Zhonghua shuju.

Yuan, P. 2001. Zhongguo nongcun shichanghua guocheng zhong de nongmin hezuo zuzhi yanjiu [A study of farmer cooperative organization in the process of rural China's marketization]. *Zhongguo shehui kexue* [*Social Sciences in China*], 6, 63–73.

Yuan, P. 2010. Hezuoshe minzhu guanli de yiyu he mianling de tiaozhan [The significance of democratic management for cooperatives and the challenge they face]. *Zhongguo nongmin hezuoshe* [*China Farmers' Cooperatives*], 6, 16–7.

Zhang, D. 2011. 'Pibao hezuoshe' zheshe chu lai de guanmin guanxi [The government-people relationship reflected by 'brief-case' cooperatives]. *Renmin luantan* [*People's Tribune*], 25, 58–9.

Zhang, F.Q. 2012. The political economy of contract farming in China's agrarian transition. *Journal of Agrarian Change*, 12(4), 460–83.

Zhang, F.Q. and J.A. Donaldson. 2008. The rise of agrarian capitalism with Chinese characteristics: agricultural modernization, agribusiness and collective land rights. *The China Journal*, 60, 25–47.

Zhang, K. and Q. Zhang. 2007. Nongmin zhuanye hezuoshe chengzhang de kunhuo he sikao [Puzzles and thoughts about the growth of farmers' professional cooperatives]. *Nongye jingji wenti* [*Agricultural Economic Issues*], 5, 62–70.

Zhang, X. 2009. Nongmin zhuanye hezuoshe yin chao shenme fangxiang fazhan [To what direction should peasant specialized cooperatives develop]. *Zhongguo laoqu jianshe* [*China's Construction of Old Revolutionary Base Area*], 2, 13–4.

Zhao, J. 2010. The political economy of farmer cooperative development in China. PhD Dissertation, University of Saskatchewan.

Zhu, H. 1996. *Liang Shuming xiangcun jianshe yanjiu* [*A study of Liang Shuming's rural construction*]. Taiyuan: Shanxi jiaoyu chubanshe.

Yan Hairong teaches in the Department of Applied Social Sciences at Hong Kong Polytechnic University. She is the author of *New masters, new servants: migration, development and women workers in China* (Duke University Press, 2008). Her current interests include China-Africa links, agrarian changes and the food sovereignty movement. She is conducting research on rural China in globalization, focusing on China's soybean crisis as a case study.

Chen Yiyuan is a PhD student in the Department of Applied Social Sciences at the Hong Kong Polytechnic University. She received her MA in sociology at Renmin University of China in 2011 and BA in social work at Huazhong University of Science and Technology in 2009. Her research interests include agrarian change and food security in China.

Chinese discourses on rurality, gender and development: a feminist critique

Tamara Jacka

During the 1980s and 1990s, peasants, especially peasant women, were mostly ignored in elite Chinese discourse on development, or portrayed as a 'backward', 'low quality' group, who put a drag on modernization. But since then, a number of elite discourses have emerged, which try to address the disadvantages suffered by the peasantry. In this paper I critique two of these recent discourses, relating to 'participatory development' and 'new rural reconstruction'. Drawing on Nancy Fraser's conceptualisation of 'injustice' and her analysis of 'affirmative' and 'transformative' strategies for overcoming it, I argue that these discourses make important, but limited, contributions to efforts to overcome injustice. The main focus of the paper is on new rural reconstruction discourse, because it promises a more radically transformative approach to injustice. However, advocates of new rural reconstruction elide gender inequalities in rural society. Far from being incidental, I argue, this elision is an integral component of an essentially affirmative approach, which reproduces injustice rather than providing the theoretical tools and language with which to address it. Comparing new rural reconstruction discourse with that of participatory development helps illuminate the limitations and strengths of each.

Introduction

Between the second half of the twentieth century and the first several years of the twenty-first, shifts in China's political economy have been accompanied by dramatic changes in elite discourse[1] about rural women and men. During the Maoist era from the 1950s to the late 1970s, 'peasants' were accorded a relatively high status, being seen to stand with industrial workers at the vanguard of the revolution, and 'peasant women' were hailed as 'holding up half the sky'. In contrast, in the 1980s and 1990s, peasants, especially peasant women, were portrayed mostly as a 'backward', 'low-quality' group, who put a drag on modernization. Since the late 1990s, however, concern has grown that rural-

The author wishes to thank Yang Lichao, Ye Jingzhong and two anonymous reviewers for their helpful comments and advice.

[1]In this paper, I use the term 'discourse' to refer to bodies of knowledge, the concepts, norms and values that underpin them, and the language that frames and communicates them. The term 'elite discourse' refers to discourse generated by scholars, state policy makers and advisers, and social activists, the vast majority of whom come from the urban, educated elite. The connections and overlaps between these groups are strong. Together, they play a dominant role in shaping both state policy and public attitudes toward national development.

urban inequalities and social conflict have reached crisis levels. In response, a number of elite discourses have emerged in which the disadvantage and lack of well-being suffered by rural women and men is foregrounded.

In this paper I discuss and critique two of the most influential of this last group of discourses, framing research and activism in, firstly, 'participatory development' and, secondly, 'new rural reconstruction'. Both discourses have been influenced by and can be seen as belonging to a broad range of 'alternative development' discourses that have emerged over the twentieth and twenty-first centuries. Reacting against dominant capitalist, and in some cases communist, discourses of development, these have sought to protect 'Third World' rural communities, and within such communities, promoted the principles of participation and empowerment for the poor, anti-consumerism, environmental sustainability, and respect for local, indigenous knowledge and cultures.[2]

My discussion and critique of participatory development and new rural reconstruction discourses is aimed primarily at evaluating their potential to help overcome injustice in Chinese society. In this critique I draw on the work of feminist critical theorist Nancy Fraser. Two aspects of Fraser's work are particularly important: her conceptualisation of 'injustice', and the distinction she draws between 'affirmative' and 'transformative' strategies for overcoming injustice.

With Fraser, I am concerned about two kinds of injustice: economic and cultural. Economic injustice involves the unequal distribution of material resources and can come in the form of exploitation (the appropriation of the fruits of one person's labour for the benefit of another), economic marginalization or discrimination, or deprivation. Cultural injustice involves the institutionalized unequal recognition of human worth and status. It includes cultural domination (being subjected to forms of culture that are alien or hostile to one's own), disrespect and disparagement, cultural marginalization and subordination, and non-recognition, that is, the rendering of a person or persons as invisible (Fraser 1997, 14). Following Fraser, I argue that, in order to progress toward a truly just world, we need to find effective means to achieve both forms of redistribution that will overcome economic injustice and institutions of social recognition that will overcome cultural injustice. Both economic and cultural injustices are problems in their own right, but they are commonly also intertwined. As Fraser puts it,

> economic injustice and cultural injustice are usually interimbricated so as to reinforce each other dialectically. Cultural norms that are unfairly biased against some are institutionalized in the state and the economy; meanwhile, economic disadvantage impedes equal participation in the making of culture, in public spheres and in everyday life. The result is often a vicious circle of cultural and economic subordination. (Fraser 1997, 15)

In contemporary China, as in much of the world, rural citizens, especially rural women, suffer the brunt of just such a vicious circle of cultural and economic subordination. In the

---

[2]For discussion and critique of alternative development discourses, see Pieterse 1998. In this paper, I have focused on participatory development and new rural reconstruction discourses as being two of the most progressive and influential of those recent elite Chinese discourses that are aimed primarily at enhancing rural socio-economic development and overcoming the economic and cultural disadvantages suffered by rural citizens. Aside from these discourses, another progressive and influential elite discourse on rural development relates to rights protection (*weiquan*). However, rights protection discourse, propounded by the liberal legal scholar Yu Jianrong among others, differs from participatory development and new rural reconstruction discourses in being less directly concerned with socio-economic disadvantage, focusing instead on protecting rural citizens' legal rights and improving their political representation. For further discussion of rights protection discourse, see Day (2007, 312–9) and Yu (2006).

1980s and 1990s, dominant elite discourses relating to development contributed to this vicious circle by failing to recognise or else demeaning rural citizens and women, as well as by failing to suggest effective strategies for overcoming inequalities in the distribution of material resources between urban and rural citizens and between men and women. More recent discourses about participatory development and new rural reconstruction are different. Neither discourse draws on the language of 'injustice', but criticism of and strategies for overcoming forms of injustice very similar to those identified by Fraser are central to both. In this paper, I argue that these discourses potentially help to address economic and cultural injustice by shifting understandings of 'development' and how it is achieved, and by changing perceptions of rural citizens and rural culture. This is particularly true of the discourse on new rural reconstruction and, for this reason, this discourse is the main focus of this paper.

At the same time, however, the political potential of both participatory development and new rural reconstruction discourses is undermined by a failure to develop effective strategies for overcoming gender injustice; specifically, the economic and cultural injustices that rural women suffer as women. In the case of participatory development, I draw on a large body of feminist critique to argue that this failure stems from the political difficulties of proposing and implementing a strategy that will truly transform basic social institutions.[3] In the case of new rural reconstruction discourse, the problem is even more fundamental, because proponents fail to give adequate attention to the gendered nature of basic social institutions and to the major gender inequalities and differences characterising rural Chinese society. In this respect, new rural reconstruction discourse contributes to the reproduction of injustice, rather than providing the theoretical lens and language with which to challenge it.[4] A comparison of new rural reconstruction discourse with that relating to participatory development helps to illuminate both the contribution that new rural reconstruction discourse makes to overcoming the injustices suffered by rural Chinese citizens, as well as its limitations in this regard.

Further insights into the political implications of participatory development and new rural reconstruction discourses can be gained by analysing them in terms of the distinction that Fraser draws between 'affirmative' and 'transformative' approaches to justice. Put simply, 'affirmative' approaches 'aim to correct inequitable outcomes of social arrangements without disturbing the underlying social structures [i.e. institutions] that generate them,' while 'transformative' strategies 'aim to correct unjust outcomes precisely by restructuring the underlying generative framework' (Fraser 2003, 74).

---

[3]Throughout this paper, the term 'social institutions' refers to the sets of laws, rules, customs and norms that structure social roles and interactions.

[4]In making this argument, I draw on an original analysis of the publications of the two leading scholarly advocates of new rural reconstruction, Wen Tiejun and He Xuefeng. To my knowledge, there has previously been no published critique of Wen and He's failure to address gender inequalities and differences. Some Chinese colleagues working on gender and development have privately commented on the sexism of Wen Tiejun and He Xuefeng, but they have not published these comments. The reason for this may relate to hierarchies of power. Wen and He are very influential figures, and it is politically difficult to challenge them openly. This is particularly so in Wen's case, because he is seen as having initiated the focus on the 'three rural problems' that has characterised recent state development policy. In contrast, outside of a relatively small network of scholars and activists working in gender and development, most senior Chinese policy makers, social activists and scholars do not recognise gender inequality as a social problem, or regard the problem as trivial. Among gender and development scholars and activists, only a tiny handful have anything like the kind of scholarly and political clout wielded by Wen and He.

Transformative approaches, Fraser suggests, are more likely to overcome social and economic injustices. In this paper, I argue, however, that participatory development discourse in China in recent years has involved an approach to injustice that is best characterised as 'weakly affirmative'. In comparison, some aspects of new rural reconstruction discourse suggest a more transformative approach to injustice. But the transformative potential of this discourse is undermined by its elision of gender inequalities and differences within rural society. Drawing on Fraser's analytical framework, I argue that this elision is by no means incidental. Rather, it should be understood as an integral, and profoundly problematic, component of a characteristically affirmative approach to injustice.

In the remainder of this paper, I first briefly outline the history of changing discourses relating to rurality, gender and development in China since the mid-1980s. This part of the paper provides the political context for the critique of discourses on participatory development and new rural reconstruction undertaken in the second part of the paper. The focus in this critique is on the concepts, ideals and strategies propounded by advocates of participatory development and new rural reconstruction, and on the extent to which they address key aspects of rural Chinese society. Some examples of the practical measures that proponents of participatory development and new rural reconstruction have taken to enact their strategies for change are also discussed. However, a thorough investigation into the politics of how and to what extent participatory development and new rural reconstruction ideals have been put into practice is beyond the scope of this analysis. The final part of the paper summarises key lines of critique, framing these in terms of the distinction that Fraser draws between 'affirmative' and 'transformative' approaches to justice, and an evaluation of the transformative potential of participatory development and new rural reconstruction discourses.

From the 'spectralization' of the rural to the 'three rural problems'

From the mid-1980s to the mid-1990s, industrialization and urbanization dominated Chinese elite discourse on development. During this period, national policy advisers paid relatively little attention to the rural sector, assuming that the rural economic reforms of the early 1980s, involving decollectivization and marketization, had been successful in improving agricultural productivity, and generating the capital and labour 'surplus' necessary for rural industrialization. With this success, policy makers could turn their attention to reforming the urban industrial sector (Day 2007, 46).

At the same time, this period saw members of the urban educated elite seeking to reclaim a positive status and future for both themselves and the Chinese nation in the aftermath of late Maoist zealotry, in part by emphasising the 'backwardness' of the peasantry. Whereas Maoist discourse had cast the peasantry as a revolutionary, or at least potentially revolutionary, force, post-Mao intellectuals increasingly saw peasants and 'peasant consciousness' as an obstacle to modernization.[5] This perception was only strengthened by the one-child policy and resistance to it among rural citizens. The one-child policy saw the emergence of a discourse about population 'quality' (*suzhi*). Aimed at improving the 'quality' of the national population, as well as reducing its quantity, this core element in the state's efforts to improve economic growth and achieve modernization was being

---

[5]For further discussion of shifts in elite Chinese discourse on the peasantry, see Kelliher (1994), Flower (2002), and Day (2007).

stymied, it was felt, by an enormous 'low-quality' peasantry (Greenhalgh and Winckler 2005).

Through the 1980s and 1990s, concerns about the need to raise people's 'quality' infused public sentiment and state rhetoric and policy across the board. This was no more apparent than in the work of the state-affiliated All China Women's Federation (henceforth Women's Federation).[6] In 1986, leading Women's Federation officials at a rural work conference stressed that

> Organizing and mobilizing women to take part in commodity production and helping women to speedily overcome poverty and get rich is the task bestowed on rural women's work in the new period. How must the Women's Federation grasp this task, what should the focus of our work be, and how will we achieve breakthroughs? The key lies in raising women's quality. (Zhang 1986, cited in Jacka 2006, 588)

To raise rural women's quality, these officials claimed, it was necessary to improve the quality of their thinking so that women could 'liberate themselves from the fetters of traditional, small-scale production and egalitarianism', and to raise their capabilities in commodity production (Jacka 2006, 588). Subsequently, in order to fulfil these aims, the Women's Federation launched the 'double study, double compete' (*shuangxue, shuangbi*) campaign. The main aim of the campaign, which has continued to the present, is to help and encourage rural women to improve their skills and productivity in commercial agricultural production. Leaders of the Women's Federation explain this focus on agriculture as both a reflection of its importance in the national economy and of the fact that, as men have moved into non-agricultural employment, women have increasingly dominated the agricultural labour force (Jacka 2006, 587).

Meanwhile, at the national level through the 1990s, an urbanist teleology grew and with it what Yan Hairong has termed, following Gayatri Spivak, the 'spectralization' of agriculture and the rural (Yan Hairong 2003). This decade saw an unprecedented growth in the number of rural-registered people migrating out of the countryside in search of work in urban centres and in the export-oriented processing zones of the southeastern coastal regions, and increases in the length of rural migrant workers' sojourns away from home. In the early part of the decade, official policy was to encourage rural residents to 'leave the land but not the countryside' (*li tu bu li xiang*) and seek employment in local township enterprises. However, declines in the competitiveness and subsequent collapse of a large proportion of township enterprises across central and western China led to the realization that rural-based industrialization would not be sufficient to absorb the 'surplus labour' coming out of agriculture, and a shift occurred in national policy toward encouragement of rural residents to 'leave the land *and* the countryside' (*li tu you li xiang*). In both elite and popular discourse by this time, the countryside had become a wasteland and the place of the past. As Yan Hairong put it,

> In the 1990s, production land has become welfare land absorbing ill, injured, and unemployed bodies and enabling a cheap reproduction of the next generation of migrant workers. ... The process of spectralization is a process of violence that appropriates economic, cultural, and ideological values from the countryside. The problem for rural youth is that they cannot find a path to the future in the withering countryside. (Yan Hairong 2003, 2)

---

[6]The Women's Federation is a 'mass organization'. As such, it is officially separate from the state, but its policies and leadership are determined by the state, as is its funding.

For most Chinese scholars and policy advisers, as well as rural citizens, the 'path to the future' led out of the countryside. There was considerable elite interest in the youth escaping the 'withering countryside' but much less attention paid to those 'left behind', despite the fact that migrants, though very numerous, continued to be vastly outnumbered by the people remaining in the village.

From the mid-1990s, however, this began to change, with elite anxiety increasingly expressed about widening rural-urban inequalities and rural poverty, and more attention directed toward improving the circumstances of the rural population. Much of the impetus for this came from two overlapping sources: overseas development agencies promoting participatory development, and Chinese scholars teaching and conducting research on rural issues.

Overseas development agencies had begun to fund research, training and projects in participatory development in the late 1980s. One of the earliest initiatives was the Centre for Integrated Agricultural Development (CIAD), established at the China Agricultural University in 1988, with support from the German Agency for Technical Cooperation (GTZ). Through CIAD, numerous scholars were exposed for the first time to foreign ideas about 'integrated' and 'participatory' development, and to interdisciplinary approaches to the study of rural issues, and many went on to undertake further studies and training in Europe and the United States. These scholars became a core cohort of consultants for overseas-funded development projects and influential proponents of participatory development (Ye 2010).[7]

In the 1990s, the Ford Foundation also began supporting a range of participatory development training and research programmes and development projects, especially in Yunnan. These too played a crucial role in supporting the emergence of a cohort of non-governmental researchers and activists, within and outside the academy, working in the area of rural development, and in promoting ideas and strategies for achieving 'sustainable' development, and 'empowerment' and 'participation' in rural communities, especially for poor women.

Meanwhile, in 1998, the Ministry of Education agreed to a recommendation from the China Agricultural University that 'rural regional development' (*nongcun quyu fazhan*) be established as an accredited academic specialisation. In the same year, China Agricultural University established the country's first College for Rural Development, a move subsequently copied by numerous tertiary institutions and other research bodies. Subsequently, that college was merged with the College of Humanities and Social Sciences to form the present-day College of Humanities and Development Studies, which includes CIAD and several other centres (Ye 2010). These various moves reflected a broad shift in scholarly approaches to rural issues, away from a predominant focus on achieving increases in agricultural productivity toward a broader, more wholistic conceptualization of rural social and economic development.

Aside from those at China Agricultural University, another extremely influential scholar in this broad shift was Wen Tiejun, an economist and Dean of the School of Agricultural and Rural Development at Renmin University. In the early 1990s, some reference had been made in elite discourse to the 'three rural problems' (*san nong wenti*) of 'agriculture,

---

[7]For example, in the early 2000s, Li Xiaoyun, the first Director of CIAD, was the leader of a team piloting a participatory approach to rural poverty reduction, subsequently implemented through the state's new County Poverty Alleviation Planning processes. For discussion and (self) critique, see Li and Remenyi (2004) and Li and Liu (2010).

rural society and the peasantry'. By the turn of the century, Wen succeeded in re-ordering dominant elite understandings of the 'three rural problems', bringing peasants to the fore and putting agriculture last, in the process re-centering debates about rural China on the plight of the peasantry (Day 2007, 196).

In 2000, public attention was drawn to the three rural problems with the publication of an open letter to Premier Zhu Rongji, written by Li Changping, a rural cadre from central Hubei province. Li's much cited claim that 'the peasants' lot is really bitter, the countryside is really poor, and agriculture is in crisis' helped crystallise elite anxieties about the rural situation and heightened pressure on the state leadership to find an effective response. This opened up political space for a range of rural development discourses and projects, including in participatory development and also in 'new rural reconstruction' (*xin xiangcun jianshe*) initiated by Wen Tiejun at about this time.

In 2006, the state launched a major new program for the 'construction of a new socialist countryside' (*shehuizhuyi xin nongcun jianshe*, subsequently abbreviated to *xin nongcun jianshe*). Despite the similarity in its title and the shared aim of addressing rural disadvantage, the main political thrust of the new program was different from new rural reconstruction. Indeed, the state's construction of a new socialist countryside has been shaped far more by former Chief Economist and Senior Vice President of the World Bank Justin Yifu Lin, who advocates the privatization of land, increased urbanization and greater state investment in rural infrastructure as a step toward increasing domestic consumption as a driver of market-led economic growth, objectives explicitly rejected in new rural reconstruction discourse (Day and Hale 2007, 7, note 2). Nevertheless, proponents of both new rural reconstruction and participatory development have been supportive of the state's program and have sought to identify themselves with it. This identification has in turn lent further legitimacy to these discourses, enabling proponents to put their ideas into practice.[8]

Participatory development

Discourse and practice in participatory development have been influenced by two main trends in global alternative development discourse: Gender and Development (GAD), and the use of participatory action research methodologies. In China, the promotion of GAD and of participatory methodologies has been led by several individuals, organizations and networks.[9] Here I focus on the work of West Women and its founder and director, Gao Xiaoxian. West Women, which is based in Xi'an, Shaanxi province, is one of China's longest running women's non-governmental organizations (NGOs).[10] Like other NGOs, it is funded almost solely by overseas donor agencies. Since the mid-1990s, it has run

---

[8]This is particularly obvious in the case of new rural reconstruction discourse, advocates of which have increasingly moved away from the term *xin xiangcun jianshe*, instead identifying their ideas and activities with the label *xin nongcun jianshe*. This is identical to the term used to refer to the state's program for the 'construction of a new socialist countryside' (*shehuizhuyi xin nongcun jianshe*), except that it omits the word 'socialist'. Some activists have successfully implemented strategies for new rural reconstruction through cooperatives and other initiatives launched under the auspices of the state's 'construction of a new socialist countryside' program (Day and Hale 2007, note 2).
[9]English-language translations of papers written by several participatory development activists are included in Plummer and Taylor (2004).
[10]The official name of this NGO is The Shaanxi Research Association for Women and Family. For convenience's sake, I use their shorter web name, West Women. West Women was founded in 1986 as a research group affiliated with the Shaanxi provincial Women's Federation, of which Gao

numerous participatory development projects across Shaanxi, focusing primarily on rural women's health, poverty alleviation, post-disaster reconstruction and sustainable ecological development.

Intellectually, Gao Xiaoxian and West Women have been influenced by trends in global feminism, with which Gao came into contact as a result of her involvement in the Ford Foundation's Reproductive Health program, begun in 1992, and through interactions with participants in the Fourth World Conference on Women and its NGO Forum, held in Beijing in 1995 (Gao 2001, 198). Her exposure to overseas feminism led Gao to critique the Women's Federation for its focus on rural women's 'low quality' and for the approach they took in the 'double study, double compete' campaign. There are, she wrote, deficiencies in the way in which the campaign was conceptualized:

> The theoretical assumption underlying the campaign is that rural women's lesser participation in rural modernization can be explained by their cultural and technical backwardness, and, therefore, that attempts to improve it should focus on literacy and technical training. This closely resembles the first stage in international approaches to women and development, that is 'Women in Development' (WID). But this approach has been criticized because it does not challenge the underlying reasons for this state of affairs [i.e. women's cultural and technical backwardness], and has been supplanted by … 'Gender and Development' (GAD) approaches. (Gao, cited in Jacka 2006, 598. See also Gao 2008, 21–2)

The WID approach that Gao identifies here was first developed in the 1970s by liberal feminists who sought to improve women's economic competitiveness with men and the benefits they gained from modernization by increasing their access to education, technology and other resources. By the 1980s, the WID approach had been taken up by almost all major international development agencies. However, as Gao suggests, it was strongly criticised by feminists, especially socialist feminists, on the grounds that it remained within the framework of neo-classical modernization theory and did not challenge its most basic premises. Reacting against WID, socialist feminists advocated a Gender and Development (GAD) approach, which paid more attention to the role played by social institutions, especially the patriarchal family and the gender division of labour, in simultaneously boosting 'development' and reproducing gender inequalities. GAD theorists also focused on the interactions between gender, class and race in shaping social and economic disadvantage, and tried to overcome the colonial mentality of the WID approach, by encouraging alliances between women across intra- and inter-national divides and by emphasising grassroots women's agency and 'empowerment' in the local community (Beneria and Sen 1981, Jacka 2006, 594–5). Because they challenged basic social institutions and power relations, GAD insights were harder than WID to integrate into dominant approaches to development. All the same, in the 1990s, most development agencies, including United Nations (UN) bodies and the World Bank, made a policy commitment to GAD (Jacka 2006, 594–6). The key to this absorption (or co-option) was the marginalization of challenges to social institutions such as the gender division of labour, and greater relative emphasis on the 'participation' and 'empowerment' of the disadvantaged within communities.

This turn toward participation and empowerment was heavily influenced by the work of Robert Chambers, author of *Rural development: putting the last first* (1983). Chambers argued for a number of 'reversals', including in the control over development and the

---

Xiaoxian was (and continues to be) a member. The Association was registered as an NGO (*shehui tuanti*) in 1999. For more details see West Women. n.d. and Jacka (2010).

flow of benefits to be gained from it, from urban populations and the rural rich, especially men, to the rural poor, including women and children. He also called for a 'reversal of learning' in which local rather than professional outsiders' knowledge would guide development (Chambers 1983, 147). To effect these aims, Chambers recommended that outside development workers employ a package of action research techniques termed 'Rapid Rural Appraisal', so as to enable them to learn from the rural poor, and the rural poor to express their needs and be involved in shaping projects that would meet those needs. Later, Chambers and his associates replaced Rapid Rural Appraisal with 'Participatory Rural Appraisal' (PRA) and other tools.[11]

In China, West Women has been one of the main recipients of overseas aid for participatory development projects. Rather than trying to mould women into higher-quality contributors to development as in the 'double study, double compete' campaign, the main aim of these projects has been to empower rural women, overcome the socio-economic disadvantages and discrimination they face as women and improve their well-being. Central to the projects have been four core strategies for increasing women's opportunities, involvement and power in village decision-making processes and economic activities, and for improving their health, well-being, self-confidence and social recognition and status. The first is the use of PRA and other tools to establish the needs of villagers, especially poor women. The second is the provision of training sessions in gender awareness as well as technical skills. These are provided to local cadres, peasant women and occasionally peasant men. The third strategy is the funding of micro-credit schemes, with loans given mainly to women. And the fourth is the establishment of rural women's groups, headed by local female community leaders (*funü gugan*), who organize village women into community activities, such as health checks, cultural activities, including dancing and singing performances, and village clean-up campaigns (Jacka 2010, 104–9).

West Women presents most of its work in rural areas in positive terms, as enhancing development, rather than as seeking to overcome gender (and other) 'injustices' or 'inequalities'. This should be seen as a political strategy: it is much easier to obtain international funding and local, government approval for projects that will contribute to community and national development, than by drawing attention to injustices for which the state might be held accountable. Yet it is clear from their public statements and the work they do that Gao Xiaoxian and her colleagues are deeply concerned about gender injustices, and are committed to overcoming them in order to 'build a society in which resources are shared, opportunities are equal and there is no gender discrimination'.[12]

For all their dedication, though, West Women have had only limited success in overcoming the economic and cultural injustices facing rural women. The group claims that one of their most successful projects is a participatory environmental reconstruction and community development project undertaken in three villages in Pingxi County (a pseudonym), with funding from Oxfam Hong Kong (Yang 2011a, 48). Yang Lichao argues, however, that this project 'neither empowered women within households nor at the village level' (Yang 2011a, 229). Oxfam Hong Kong, she suggests, did not provide adequate guidance and support to Pingxi project workers, and failed to commit enough

---

[11]Research methods commonly employed in PRA include semi-structured interviewing, village mapping exercises for the purposes of resource identification, and participatory needs assessment and ranking exercises, carried out in villager sub-groups.
[12]This aim is stated in a banner written across the top of West Women's website (West Women n.d.).

resources to village-level gender awareness and other training to ensure that poor women were genuinely involved and benefited from the project (Yang 2011a, 230).

Aside from this, Yang argues that the Pingxi project's effectiveness in empowering women was severely limited by its inability to address prevailing gender divisions of labour, and its failure to transform existing structures of power. *Funü gugan* in all three villages felt that women's domestic responsibilities were a major obstacle to both their own involvement in the project and their efforts to involve other village women in project activities. Women's heavy workloads in agriculture as well as domestic work and the care of children and the frail elderly meant that they simply did not have time for project activities (Yang 2011a, 230–1).

In two villages, *funü gugan* reported that villagers' disapproval of women involved in the public sphere and their skepticism about women's abilities as leaders also impeded their work. This problem was compounded by lack of support and in some cases abuse and violence from within the family, with husbands and parents-in-law sometimes highly resistant to the idea of women working outside the home. Such resistance was particularly strong in one village with a very high rate of male outmigration. Men's absence, Yang suggests, increased their own and other family members' sense of insecurity and anxiety about the marital relationship, further compounding husbands' and in-laws' wishes to control young wives' behavior and ensure they conform to conventional norms (Yang 2011a, 231–2). In some cases, women wanted to participate in project activities, but withdrew or declined to be involved in the face of resistance from family members. In others, women themselves either did not feel it right that they should engage in project activities, or were not confident in their abilities to do so. To some extent, cultural activities and performances organized by *funü gugan* helped to overcome these feelings. They increased women's self-confidence in the public sphere and this sometimes enabled them to be more assertive at home (Yang 2011a, 196–9). Not all women were willing or had the time to join in such activities, however, and others who did nevertheless had to withdraw after a short time because of resistance from family members (Yang 2011a, 196–9).

The project's failure to transform village power relations was particularly noteworthy in one village, where project workers selected Mrs Shang, the sister-in-law of the village Party Secretary, to head the project's *funü gugan* group, the village Women's Management Committee. According to Yang, Mrs Shang's 'access to project funding also became a means through which the political patronage system of her family was strengthened' (Yang 2011a, 145). Most of the other members of the Management Committee were also senior women of the most powerful families in the village. Younger members from poorer, less powerful families were less involved in the committee's decision-making processes and had less control over project resources (Yang 2011a, 146). According to Yang, their involvement in the Pingxi project did foster greater confidence in the few women selected to be *funü gugan*, enabled them to develop skills, for example in coordinating community activities and writing project reports, and gained them increased recognition in the village. However, this did not translate into real decision-making power in village affairs, which was dominated by the male members of the village branch of the Communist Party and the Village Committee. In any case, 'there was no evidence that other village women were "empowered" or that the project had brought broader changes in gender relations in the other two project villages' (Yang 2011a, 147).[13]

---

[13]For other critiques of participatory development in China, see Li and Liu (2010), Liu (2010), Yang (2011b) and Zhao (2011). Several non-China related critiques are included in the collections edited by Cooke and Kothari (2001) and Parpart *et al.* (2002).

Yang's critique of West Women's Pingxi project suggests that its failure to overcome gender injustices stems partially from inadequate implementation of the participatory approach. Had project workers received more guidance and resources, Yang implies, they might have been more effective in enabling women in Pingxi to participate in and benefit from the project.

At the same time, Yang's critique also resonates with a growing body of critical literature from activists and scholars around the world, which attributes the failures of participatory development not just to poor implementation, but to failures of conceptualization inherent in the very discourse of participatory development. This discourse, it is claimed, has become 'the new tyranny' – an obligatory framework and language which, however, cannot effect meaningful, change in the lives of poor rural people or the power hierarchies in which they are trapped (Cooke and Kothari 2001). 'Participation' and 'empowerment', it is argued, have become 'motherhood' terms, readily agreed to by development agencies because they have positive connotations, but their meanings are vague. Often, they mask strategies and behaviours that fail to promote, and may even detract from, the actual realization of empowerment among poor villagers, especially women. As feminist scholar Jane Parpart argues, the main drawbacks of PRA and other participatory approaches to empowerment derive from a failure to recognize that local hierarchies of power, usually dominated by elite men and their families, are deeply embedded in local, regional, national and even international institutions, and cannot be overcome merely through encouraging and giving voice to poor women and other subaltern groups at the local level (Parpart 2002, 170–2; Jacka 2006, 596).

Taking the feminist critique yet further, Wendy Harcourt argues that the women's agenda has been subverted by the UN and other powerful development agencies. Even while protesting its disempowering effects, Harcourt suggests, women working within such structures have been complicit in a discourse that

> essentially continued to create a colonized poor and marginalized woman who needed to be managed and educated; whose capacity for work and local decision making needed building; and who needed to be controlled reproductively and sexually through a series of development interventions designed for 'women's empowerment'. (Harcourt 2005, 43)

Today, disillusion and cynicism about the participatory development policies and projects of the UN and other agencies is widespread. For Harcourt and many others, they no longer offer the most promising approach to overcoming the economic and cultural injustices faced by poor women and other subaltern groups. Instead, Harcourt writes, 'women's movements are much more readily finding a space in the alternative globalization movement' (Harcourt 2005, 46).

It is important to realize, however, that the alternative globalization movement encompasses a wide range of groups, with diverse discourses and ideals, some more conducive to the women's movement than others. Among the Zapatistas in Mexico, women's activism has played a major role and overcoming gender injustices has been a core objective (Speed *et al.* 2006). Gender equality is also part of the agenda for social change espoused by the Landless Workers' Movement (MST) in Brazil, although Rute Caldeira suggests that in MST practice, women's activism and concerns about gender injustices have been marginalized (Caldeira 2009). In contrast, in China's new rural reconstruction discourse, which also contributes to the alternative globalization movement, there is no significant recognition of or concern about gender injustice. This is despite evidence that these scholars' understanding of rural problems in China has been influenced by their observations of the situation in Mexico and Brazil, and strong resonances between the strategies for change that they advocate, and other aspects of the discourse and practice of the Zapatistas and the MST (Day 2007, 320; Wen 2007, 17). As we will see in the next section, new rural reconstruction discourse offers very little space for those concerned to overcome the injustices faced by rural women.

New rural reconstruction

New rural reconstruction discourse takes its name from the Rural Reconstruction Movement, a Chinese populist movement of the 1920s–1930s, led by the scholars Liang Shuming and Yan Yangchu. In seeking a way out of national crisis and an approach to development that would overcome the problems of western-style urban-centred industrialization, Liang and Yan eschewed the Communist Party's efforts at rural mobilization and instead urged intellectuals to lead a rural cultural revival, a reconstruction of rural society and economy, and experimentation in building voluntary rural cooperatives (Day 2008, 59–60). Since the early 2000s, Wen Tiejun and others have sought to rebuild a similar movement, with activism, experimentation and training organized at the Yan Yangchu Institute for Rural Reconstruction, near the site of Yan Yangchu's 1930s Rural Reconstruction Centre in Ding County, Hebei and the Liang Shuming Centre for Rural Reconstruction on the outskirts of Beijing. Funding and academic direction for these centres has come from the Rural Reconstruction Centre at Renmin University, directed by Wen Tiejun.

Another academic centre for new rural reconstruction has been the Centre for Rural Governance Studies at Huazhong University of Science and Technology in Wuhan, directed by the sociologist He Xuefeng. He and his associates, referred to by some as the Central China *xiangtu* (earthbound) school of rural sociology, draw inspiration from China's most famous sociologist, Fei Xiaotong, and the ethnographic studies of rural society that he conducted in the 1930s and 1940s.[14]

The following critique of new rural reconstruction draws primarily, though not exclusively, on the work of Wen Tiejun and He Xuefeng. Many others have devoted more effort to practical experiments in new rural reconstruction, but Wen and He have had the most influence over the ideas and conceptualizations guiding those practical efforts. There are significant differences between the ideas and strategies propounded by Wen and He and their respective followers. However, both scholars emphasize the distinctiveness of Chinese peasant economy and society, and simultaneously elide gender and other differences and inequalities within rural communities.[15] I will argue that these two aspects of their scholarship are closely connected, and this connection is one of the most troubling features of new rural reconstruction discourse.

The intellectual foundations of that discourse lie in a rewriting of the history of national development undertaken by Wen Tiejun. Wen claims that China's earliest efforts to industrialize were undertaken in a context radically different from that in the West. Lacking the

---

[14]The name *xiangtu* comes from the title of Fei Xiaotong's book, *Xiangtu Zhongguo* [*Earthbound China*], published in 1947. He Xuefeng's first book (He Xuefeng 2003) was entitled *Xin Xiangtu Zhongguo* [*New Earthbound China*].

[15]He Huili, another leading figure in new rural reconstruction, devotes more effort than Wen, He and other proponents to furthering the well-being of rural women, primarily by organizing them into cultural performance troupes. Despite this, however, her writing is similar to that of Wen and He in evincing next to no recognition of the significance of gender injustices or gender-based differences in the lives of rural women and men. See, for example, He Huili (2007).

resources that Western nations acquired through their colonial ventures, China succeeded in industrializing only in the 1950s, by de-linking from the world economy and transferring resources out of the rural sector into urban industry, through a process of primitive accumulation that Wen refers to as 'an unprecedented self-exploitation led by a highly centralized government' (Wen 2001, 292; Day 2007, 207–8). This process was necessary for national development during the Maoist period. However, it also 'divided the urban and the rural into antagonistic positions' (Wen 2001, 293), producing the present-day 'three rural problems'. Today, Wen argues, industrialization no longer needs to suck the countryside dry – the process of primitive accumulation has been completed – so attention can be turned to solving the three rural problems.

Liberal policy advisers have argued that the way to overcome rural problems is to re-link China with the global, capitalist economy as fast as possible and let market forces do the work of 'liberating' rural labour. For Wen, in contrast, the structures of global capitalism combined with the legacies of Maoism, especially the hierarchical divide between urban and rural sectors, mean that it is highly inadvisable for China to fully re-link with global capitalism and adopt the American model of development, based on privatization, marketization and urbanization. A whole-scale adoption of such a model, Wen fears, would lead to unpalatable consequences similar to those that can be seen in Mexico, Bangladesh and elsewhere in the developing world: an exploitation of rural migrant labour that further entrenches rather than overcomes rural poverty, the creation of vast slums of poor rural migrants on the outskirts of urban centres, and growing rural protest (Wen 2004; Day 2007, 215–6).

Reacting against this potential scenario, proponents of new rural reconstruction argue for the social protection and maintenance of semi-autonomy for the 'small peasant economy' (*xiaonong jingji*) which, following Fei Xiaotong, they claim is naturally anti-market and operates on a different logic from the modern market economy (Day 2007, 222–3). This small peasant economy can and should, they argue, form the basis for a new 'ecological civilisation', which respects local 'indigenous culture' (Wen *et al.* 2012) in China's interior, agricultural heartland.[16] To protect the small peasant economy, they emphasise, two approaches are needed: the maintenance of a system that distributes land equally to all peasant households, rather than privatizing it and concentrating it in the hands of a few, and the organization of the peasantry, reconstruction of peasant culture and redevelopment of cooperative relations among peasants.

With regard to the first approach, Wen Tiejun, He Xuefeng and others argue that urban migration alone cannot solve the problem of surplus rural labour, so a way must be found to protect the livelihoods of those remaining in the countryside. The maintenance of small holdings of farmland, distributed equally among peasants, is crucial to this, for it ensures the subsistence of each household, providing the sense of security that peasants need to embark on more risky non-agricultural ventures (Wen 2012, 29). Further to this, it enables peasants to implement a family livelihood strategy entailing an inter-generational division of labour between agriculture and non-agricultural activities. In this division of labour, middle-aged and older family members farm the land, growing their own grain

---

[16]New rural reconstruction discourse is based primarily on observations of the economies and cultures of Han-dominated villages in crop-growing regions, especially in central and western China. One of the discourse's limitations is that many of these observations do not apply to other ethnic populations, especially in Xinjiang, Inner Mongolia, Tibet, Guangxi and Yunnan. Many also do not match the lives of rural-registered people in highly developed regions, for example in Guangdong, and in recently rural but now peri-urban districts.

and other food, and generating enough income to cover their own basic living costs and that of dependent children. This makes it possible for younger family members to leave home in search of waged work, and to use their earnings for purposes other than the basic subsistence of family dependants (He Xuefeng 2012).

Added to this, the second leg of new rural reconstruction is the organization of the peasantry into voluntary cooperatives and the rebuilding of rural culture and cooperative social relations. 'Culture' and cooperative cultural endeavours, such as performance troupes, are central to new rural reconstruction discourse and practice, but different proponents attach to them different meanings (Day 2008, 61).

As Wen Tiejun and his colleagues see it, rural dwellers are a 'weak' or 'vulnerable' group (*ruoshi qunti*) who suffer economic disadvantage and marginalization because they lack capital. Economic cooperatives can counter this disadvantage, but rural dwellers' very weakness poses problems for self-organization into economic cooperatives. Organizing rural dwellers first of all into cultural cooperatives can help to ameliorate these problems. Cultural cooperatives are, in other words, a means toward overcoming economic injustices. In an article explaining this approach, Wen Tiejun and Yang Shuai write:

> As elderly people, women and other vulnerable groups left behind in the countryside have rather low human, material and social capital, it is generally very difficult for them to set up cooperatives themselves. At the same time, following the atomisation of rural households, village cultural values and interpersonal networks have been lost, so there is also no social climate within the community for the building of a cooperative culture. (Yang and Wen 2011)

Wen and Yang's response to this predicament is to ease the process of organizing by first of all engaging 'vulnerable groups' in women's song and dance troupes, elderly people's associations and the like, and using these as a basis from which to build economic cooperatives (Yang and Wen 2011).[17]

He Xuefeng similarly emphasises the need for cultural cooperatives. However, he sees the nurturing of culture through such cooperatives as an end in itself, rather than a step toward improving peasants' economic standing. In fact, he argues explicitly against the use of cooperatives as a means of increasing peasant incomes. Large increases in income are unlikely, he suggests, because 'in a market economy, peasant cooperatives can only be organized at the periphery of the highly competitive market for very meagre profits, but at a very high cost of organization' (He Xuefeng 2007, 32). In any case, though, for He, the cultural problems faced by peasants are even more important than the economic ones:

> The fundamental problem in the countryside at present is not that there is a lack of increase in peasant incomes, but that the speed of increase in expenditures is much greater. The problem is not one of food and clothing but one of consumer culture telling them that lack of spending money is a shameful thing. Peasants are marginalized in this marketized society, and they also look down on themselves … . [T]he core of new rural construction from [peasants'] perspective is to reconstruct (*chongjian*) their way of life in order to give them meaning … It is about constructing a way of life with 'low consumption and high benefit', one that is different from consumer culture. A way of life thus constructed could help improve peasant satisfaction without money being the major criterion for value of life. (He Xuefeng 2007, 34)

---

[17]An example of this two-step process is discussed in He Huili (2007), 52–3.

This quotation suggests that, although he does not refer explicitly to 'cultural injustice', He Xuefeng's major concern is precisely with the type of cultural marginalisation and subordination that Fraser terms 'cultural injustice'.

Along with 'culture', another key concern for proponents of new rural reconstruction, especially He Xuefeng and the *xiangtu* school, is rural 'governance' (*xiangcun zhili* or *cunzhi*). This term frames a series of 16 books edited by He Xuefeng, discussing a broad range of topics, including changing family relations and values, as well as official and unofficial village decision-making institutions, villager cooperation and the supply of public goods and services. The core premise of the series is that while capitalism has weakened traditional kinship-based ties and forms of governance, these still exist and can and should be nurtured. However, attention must be given to the varying models of kinship-based governance that characterize different parts of the country. In his contribution to the book series, He identifies six models, each of which he associates very loosely with one of three different regions of the country – North, South and Central.[18] One model centres on the (patri)lineage (*zongzu*). The book identifies this as the 'southern' model, though it recognizes that lineages have everywhere been greatly weakened and only continue to exist as a strong social force in a small minority of villages, even in the south. Other models of governance, characteristic of the north, are built around smaller, sub-lineage kinship groupings. In the central region, the atomization of rural relations is particularly marked, with kinship groupings no longer forming the basis for governance or a sense of community in many areas. However, in some places, villagers do identify to varying degrees with forms of community created during the Maoist period, namely the production team (*xiaozu*) and the production brigade (now administrative village) (He Xuefeng 2009).

Central to He's political stance toward culture and governance are conceptualizations that he draws from Fei Xiaotong of traditional peasant consciousness as being centred around a dichotomy between 'self' (*si*) and 'public' (*gong*). In Chinese culture, he writes, the smallest unit of selfhood is not the individual, as in the West, but the family, and the dichotomy between self and public 'is not one between the individual and the collective, but between one's own family and other people's' (He Xuefeng 2009, 70). Within the family, there inevitably are conflicts. But these do not arise from a clash between 'rational' individuals seeking to further their own interests. Inside the family, such a concept of individual rationality does not pertain; it does, though, come into play in the relationship between 'our' family and others who belong to the public.

Traditionally, according to He, notions of selfhood extended well beyond the nuclear family, and the interests of the nuclear family were subsumed to varying degrees within the interests of larger family groupings, and the lineage. The 'self' of the lineage further coincided with that of the village, with generations of men in the same patrilineage forming families, and continuing to live and work the land in the same place. This formed an 'acquaintance society' (*shuren shehui*) in which everyone knew each other and a strong sense of morality based on human sentiment (*renqing*) contributed to social cohesion and the maintenance of order (He Xuefeng 2010, 240–3).

He Xuefeng portrays traditional ideas about self and public as the core of a 'logic of peasant action' (*nongmin xingdong luoji*) that continues to be vital for the peasant economy and society today. The problem, as he sees it, is that the size of the kinship groupings that people identify as 'self' has shrunk, and the intensity of identification with them

---

[18]See the author's preface to He Xuefeng (2009) for a summary and explanation of the division between the three regions.

has been eroded first through forced collectivization and then, even more sharply, as a result of consumer-capitalism since the 1990s. These trends have resulted in a destructive shift toward individualism and rational calculation (*lixinghua*), which weakens the bonds forged through *renqing*, increases conflict between people and makes it difficult to organize for the provision of public goods. Consequently, he argues, a sense of rural community and 'self' needs to be rebuilt (He Xuefeng 2011).

The strategies for doing this, He writes, and the potential effectiveness of these strategies for building cooperative relations and overcoming conflict within villages, will vary, depending on the size of the group that villagers currently identify as 'ours' and the intensity of identification with the group among its members (He Xuefeng 2009, 35). The suggestion is that villages in which kinship groupings are still important in social relations are likely to find it easier to rebuild effective cooperative relations and village governance. In comparison, in places where kinship ties are weaker and do not extend as far, there are likely to be more conflicts (between extended families and other sub-lineage kinship groups), which will be harder to overcome in the interests of harmonious, cooperative village relations.

In addition to the consequences for rural intra-household governance, He also discusses the problems associated with a growing individualism and trend toward 'rational calculation' within families, especially among young people. The impact of this on the elderly is particularly serious, he claims, with declines in filiality and respect for elderly parents resulting in their neglect, consequent declines in their physical and mental health and self-esteem, and high rates of elderly suicide (He Xuefeng 2010, 135–42).[19] For He and other members of the *xiangtu* school, village associations and cooperatives are an important forum for the cultivation of a cooperative spirit that can counter these problems. Wang Ximing claims, for example, that seniors' associations set up by He Xuefeng and He Huili in Hubei and Henan enhance seniors' pleasure in life, relieve their loneliness, and protect them by providing the collective strength and confidence to speak up against family members' neglect (Wang 2009, 142).

If proponents of participatory development might be accused of failing to come up with an effective means of empowering rural women, they nevertheless have given serious consideration to gender as a cause of differentiation and inequality within rural communities. The same cannot be said of the main proponents of new rural reconstruction, who, like so many scholars of peasant studies, implicitly equate 'the peasant' with a man (Razavi 2009, 200). Their neglect of women and gender has two aspects. On the one hand, both Wen Tiejun and He Xuefeng mention women only rarely and, when they do, it is usually to portray them as a problem, rather than as key agents and subjects of rural (re)construction. On the other hand, neither scholar recognises the gender-specific injustices and difficulties faced by women.

He Xuefeng claims that the status of women has risen dramatically since the Communist Party came to power in 1949, and today, regional variations notwithstanding, men and women are more or less equal: within the family, both women and men participate in decision-making, and in public affairs, women participate less than men only because they are less interested (He Xuefeng 2009, 9–12). These claims are only partially valid. My own fieldwork observations suggest that women do indeed have a high degree of involvement in routine family decision-making, sometimes higher than their husbands care to admit. They are, however, much less involved in village governance. As Yang's critique

---

[19]For similar observations, see Yan Yunxiang (2003). He Xuefeng refers to Yan's research.

of the Pingxi project suggests, the reasons for this relate to women's subordinate position in the family and to gender divisions of labour; they cannot be summed up in terms of individual preference, though it is also true that many women feel village governance to be irrelevant to their lives. Aside from this, as I will discuss shortly, He's statements elide a number of other institutionalized gender inequalities that continue to impede women's well-being in rural China today.

Both the neglect of women's agency and the neglect of gender-specific difficulties faced by women diminish the value of new rural reconstruction as a way of overcoming economic and cultural injustice. In order to examine this issue further, let us look again at what Wen Tiejun and He Xuefeng say (and do not say) about two institutions that they recognise as crucial to rural economy, culture and governance: divisions of labour and kinship.

Above, I cited Wen and his colleague talking about the difficulties associated with the fact that most of those remaining in the village today are women and the elderly. This is one of few signs that Wen appreciates that across much of China, the village-based economy is now dominated by women, especially older women.[20] However, he sees this only as an impediment, rather than in terms of women's contribution to rural life. His portrayal of women and the elderly involves a form of cultural injustice – a misrecognition – that is very similar to that characterising discourse about the 'low quality' of peasants and women. Similar to the Women's Federation, Wen attributes 'deficiencies' to women themselves, rather than examining the underlying institutional causes of those deficiencies. Furthermore, he stereotypes all rural women and elderly people as deficient, failing to recognise the variations in capital and capabilities that exist among rural dwellers and that cut across gender and age differences.

He Xuefeng is more positive about the capabilities of the left-behind. Countering common concerns about the ageing of the agricultural labour force as resulting in declines in agricultural productivity, he argues that the left-behind elderly are, in fact, highly productive farmers. A large proportion are actually not so old, being in the age bracket 40 to 65, and they are generally physically fit and have plenty of experience and technical know-how under their belts. They can add meaning to their lives by working on the land and, at the same time, provide food and income for their families (He Xuefeng 2012, 9–10). Aside from the paternalism and romanticism underpinning these observations, it is striking that He does not 'see' women in the left-behind population at all, acknowledging neither that most of the left-behind are older women, nor that a further sizable fraction are young women with pre-school aged children. Nor does he see the large amount of work that the left-behind, especially women, do outside of agriculture – in raising domestic livestock, cooking and cleaning, and tending to the needs of the frail elderly, as well as small children and grandchildren left in their care by their outmigrating parents.[21] Perhaps because of this, He portrays the workloads of the left-behind as much lighter than do other scholars, who have expressed grave concern about the heavy burdens borne by older women in particular (Pang *et al.* 2004; Mu and van de Walle 2009).

The work performed by the left-behind forms one side of a gender division of labour in China that builds on a distinction between a sphere 'inside' the family and one 'outside' the

---

[20]If Wen had noted that, due to male-female differences in longevity, most of those termed 'left-behind elderly' are also women, this would have been made even clearer.

[21]The invisibility of reproductive and 'domestic' labour, critiqued by feminists for decades, remains a common feature of studies of political economies the world over. For a critical review, see Razavi (2009, 205–8).

family. As a basic social institution, this division of labour varies from one place to another and has changed radically in some respects over time, but in other respects has remained fairly constant for centuries. Since the 1980s, the most obvious change that has occurred in the inside/outside division in rural areas has been to the content of 'inside' and 'outside' work. While trading and village governance continue to be considered 'outside' work, with industrialization, and in particular labour outmigration, agriculture, once considered 'outside work', has joined the 'inside sphere', along with domestic work and the care of dependants. Meanwhile, the gendering of the division between 'inside' and 'outside' has not changed, and nor have the basic social meanings attached to each. 'Inside work' is considered the responsibility of women, especially older women, and, partly because of this and because it commonly earns less cash income than industrial employment, it is considered inferior to men's outside work; in fact, it often is not recognised as 'work' at all (Jacka 1997).

Aside from the cultural injustice to women that results from the devaluation and lack of recognition of their inside work, one of the consequences of the inside/outside gender division of labour is that village governance and public affairs tend to be dominated by men. As Yang's critique of the Pingxi project shows, this is in part because women are too busy working in the 'inside sphere' to feel like taking on more work outside. And in part it is because women are not seen as having the abilities for outside work, and it is not regarded as appropriate for them to interact with others, especially men, in the outside sphere. During the Maoist era, the state tried to change these views, but they have proved remarkably resilient. Given that involvement of women in public cultural and governance-related associations is a core component of new rural reconstruction, just as it is in participatory development projects, the neglect of the gender division of labour poses limitations, not just to the ability of new rural reconstruction to address gender injustice, but to its very sustainability.

A failure to appreciate the significance of the gender division of labour also undermines Wen Teijun's and He Xuefeng's analysis of capitalist exploitation of the rural population. Huge capitalist profits and economic growth in China's metropolises and export-oriented processing zones over the last two decades have depended on the gross exploitation of young rural migrant labourers. That exploitation in turn has only been possible because the reproduction of migrant labour is achieved by means of the under-recognised and under-remunerated labour of older women back on the farm. Wen and He show little concern about the latter exploitation. In one piece, He writes critically about the exploitation involved in rural parents' subsidisation of migrant workers' life in the city. His concern, though, is with the financial problems this causes the elderly, rather than the exploitation of older women's labour (He Xuefeng 2010, 38–41).

Wen and He's blindness to the injustices suffered by older rural women raises important questions about the value of their strategies for 'protecting' and 'reconstructing' the countryside. It is hard to see, in fact, how 'new rural reconstruction' does not just mean more of the same: a continued spectralization of older rural women, whose 'inside' work – unrecognised as it is – forms the foundation for both rural survival and national development.

He Xuefeng's promotion of kinship as the basis for reconstructing village culture and governance raises yet further questions about the value of new rural reconstruction as an approach to justice. It is important to realize that 'the family' in which He invests so much cultural meaning is patriarchal, patrilineal and virilocal. Virilocal marriage, which He takes as the norm, involves the bride leaving her natal family and going to live with her husband's family in another village. One consequence of this is that women are relatively socially isolated after marriage. Being an outsider to her husband's village, it can take many years for a virilocally-married woman to be accepted as part of the village 'self' and to establish the kind of connections necessary to be elected into village government or any other form of local leadership, to run a business or, indeed, to seek help in everyday work-life. Social isolation also increases a woman's vulnerability to domestic violence. This is a particularly serious problem in villages dominated by patrilineal male kin, who tend to stick together.

Virilocal marriage also underpins gender inequalities in rural property rights. Legally, women and men have equal rights to land usage-rights. In practice, married women's rights are invested in their husband's household, and are lost in the case of divorce. In most instances of divorce, women lose all access to land in their husband's village and are unable to gain land in their natal village or anywhere else. The consequent risks involved in divorce weaken married women's power in the family, making them even more vulnerable to abuse. This is a particularly serious problem for older women whose prospects of finding non-agricultural employment are poor, and who therefore rely on land to survive (Jacka 2012, 17–8).[22]

Apart from ignoring these problems, perhaps the most disturbing aspect of He's promotion of 'the family' is the way in which it closes off possibilities for alternative, more just social relations and forms of governance. Putting questions about cultural distinctiveness to one side, the family is indisputably central to rural Chinese culture. However, He's understanding of rural Chinese culture and of the family is too essentialist and homogenizing. Aside from ignoring the diversity of non-patrilineal, non-virilocal family forms that prevail among the Han as well as other ethnic groups, He's model of the family is male-centric. He does not acknowledge that for many rural women, ties with their natal family, continuing long after marriage, constitute a 'self' that is sometimes as important as that constituted by their ties with their husband and his family. Yet, such ties play an important role in maintaining married women's well-being and in strengthening the rural economy. When they suffer domestic violence in their marital home, for example, women commonly seek help from their siblings and parents. In addition, women often borrow money from natal kin, and in some places mutual aid between women and their natal families in neighbouring villages is crucial during the busy agricultural season (Jacka 2012, 14–5. See also Judd 1989, Judd 2010; Zhang 2009). He Xuefeng and his associates give only negative examples of the problems that women's ties with their natal families can cause their husband's family.[23] They do not recognise the considerable social benefits to be derived from such ties.

Conclusions: affirmative vs transformative approaches to justice

As I noted above, one particularly useful aspect of Nancy Fraser's discussion of injustice and how to overcome it is the distinction she draws between two approaches: transformative strategies, which aim to overcome injustices by transforming the social institutions that generate them, and affirmative strategies, which seek to correct unjust outcomes of social arrangements without significantly altering underlying social institutions. The distinction,

---

[22]For discussion of other gender injustices associated with virilocal customs and unequal property rights in rural China, see Sargeson and Song (2010) and Bossen (2012). For a brief review of non-China related feminist studies of gender and land, see Razavi (2009, 211–4).

[23]For example, He's former student, Chen Baifeng, cites one example of a woman calling on her brother and his neighbours in their natal village to beat up the woman's elderly mother-in-law (Chen 2008, 37).

she suggests, can be applied to approaches to both economic and cultural injustice. With respect to the former, an affirmative strategy redresses inequalities in resource distribution by transferring additional resources to the disadvantaged, without, however, disturbing the underlying economic structure. In contrast, a transformative approach changes the division of labour, forms of ownership and other economic structures and institutions (Fraser 2003, 74). With respect to cultural injustice, an affirmative approach revalues group identities, according greater respect to one relative to another, without, however, trying to change the content of either identity or the institutionalized differentiation that underlies them. In contrast, a transformative approach redresses subordination by deconstructing and destabilising the symbolic oppositions that underlie institutionalized patterns of cultural value (Fraser 2003, 75).

Considered in the abstract, Fraser argues, a combination of transformative approaches to economic and cultural injustice is most likely to overcome both. In comparison, affirmative approaches have two major drawbacks. First of all, when used to address economic injustices, they can provoke a backlash of misrecognition. In contrast, transformative approaches to economic injustice tend to seek to restructure the general conditions of labour and cast entitlements in universalist terms, and in this way they address inequalities without compounding misrecognition by creating 'stigmatized classes of vulnerable people perceived as beneficiaries of special largesse' (Fraser 2003, 76–7). Secondly, when used to address cultural injustices,

> affirmative remedies tend to reify collective identities ... [They] tend to pressure individuals to conform to a group type, discouraging dissidence and experimentation. Suppressing exploration of intragroup divisions, they mask the power of dominant fractions and reinforce cross-cutting axes of subordination. (Fraser 2003, 76)

In contrast, transformative approaches to cultural injustice, when successful, foster interaction across differences, rather than promoting separatism or repressive communitarianism, as affirmative approaches tend to do (Fraser 2003, 77).

While transformative approaches to injustice might appear ideal when considered in the abstract, in the real world in which we live today, they are rarely feasible. However, 'reforms that appear to be affirmative in the abstract can have transformative effects in some contexts, provided they are radically and consistently pursued' (Fraser 2003, 78). Given this, Fraser argues for consideration of 'transitional' strategies or what she calls 'nonreformist reforms'. These 'engage people's identities and satisfy some of their needs as interpreted within existing frameworks of recognition and distribution' (Fraser 2003, 79), while simultaneously involving those people and others in a movement that works toward a transformation of existing institutions and structures, such that affirmation of particular identities is no longer required:

> When successful, nonreformist reforms ... alter the terrain upon which later struggles will be waged. By changing incentive structures and political opportunity structures, they expand the set of feasible options for future reform. Over time their cumulative effect could be to transform the underlying structures that generate injustice. (Fraser 2003, 79–80)

China's participatory development and new rural reconstruction discourses do not conform entirely to the characteristics of either the affirmative or transformative approach to injustice delineated by Fraser. Nevertheless, I suggest, the affirmative/transformative distinction can usefully serve as an heuristic tool for evaluating the political implications and potential of these discourses.

With respect to participatory development discourse, early GAD theorists argued for an approach that was radically transformative. However, in the participatory development discourse that has come to dominate the policies and practice of development agencies today, that transformative potential has been lost. Largely, I argue, this has come about through a shift in focus away from tackling institutionalized gender inequalities toward participation and empowerment for women and the poor within local communities. In China, Gao Xiaoxian and other women activists have made efforts to challenge institutionalized gender inequalities, for example through gender awareness training and by nurturing women's involvement in the 'outside' sphere of governance. However, their efforts have been hampered by a lack of support from overseas donors and a lack of recognition from the state of the significance of the gender-unequal nature of basic social institutions. In a political environment unconducive to efforts at gender transformation, and lacking resources, these activists have been confined to a rather weak affirmative approach to injustice, which seeks to add poor women to development projects, increase their respect and self-respect and give them a little more say in decision-making processes, without fundamentally altering underlying social institutions and structures. In the example of the Pingxi project critiqued above, there is some evidence that this approach has exacerbated inequalities between women in the village. However, the project does not appear to have engendered a backlash of misrecognition. This is no doubt because the few resources that were transferred came from outside the community and were easily appropriated by local elites.

Evaluating China's new rural reconstruction discourse in terms of affirmative and transformative approaches to injustice is more complicated. In some respects, the discourse has considerable transformative potential. Proponents' historical analysis of the relationship between the rural sector and industrialization is particularly important in this regard. So too is their critique of capitalist-consumerist values. However, in other respects, their approach is more clearly affirmative. It does not try to tackle global capitalism head on, and nor does it seek to overcome China's rural-urban divide. Instead, it tries to overcome the economic and cultural injustices suffered by the peasantry by highlighting the distinctiveness of Chinese peasant economy, culture and social relations, and organising, protecting and strengthening peasant society against the vagaries of global capitalism.

We should, perhaps, see this as a 'transitional strategy', with the potential to develop into a more transformative approach toward injustice in the future. However, some features of new rural reconstruction discourse are likely to weaken its transformative potential. For example, a view of the peasantry as inherently separate from the capitalist economy and value system provides poor grounding for efforts to transform that economy and value system into something more equitable. Furthermore, highlighting the distinctiveness of peasants may undermine efforts to develop solidarity among the disadvantaged across the rural-urban divide. Yet such a solidarity, one might argue, is crucial if a truly transformative approach to justice for all Chinese citizens is to be developed.

The way in which a distinction between peasant Chinese society and others in new rural reconstruction discourse maps onto a further dichotomy between the evil capitalist, modern imperialist west and the virtuous, anti-capitalist, non-modern, non-imperialist Third World is also cause for concern. This dichotomy, common among many alternative and post-development discourses, may limit the possibilities for creative, transformative approaches to injustice by reifying the 'anti-west' (as well as the west). Following Meera Nanda, I suggest that the main danger of such a reification is that it minimizes the theoretical space for a critical assessment of non-western cultures (Nanda 2002, 215).

In India, Nanda argues, the reification of cultural differences with the west in post-development discourses, including Vandana Shiva's ecofeminism, has led to a 'strengthening [of] the already formidable power of upper-caste, rich rural males' (Nanda 2002, 222). In China, also, there is a risk that new rural reconstruction, with its essentializing approach to 'traditional' rural culture and elision of gender (and other) differences and inequalities, could strengthen the power of male village elites, while compounding the injustices suffered by poor rural women.[24]

However, if new rural reconstruction was to incorporate an understanding of rural culture and society that was more sensitive to differences and divisions within communities, as well as between regions, this might provide the foundation for a strong 'transitional strategy' with transformative potential. Such a strategy might combine measures already taken in participatory development as well as new rural reconstruction projects. It could also introduce new 'affirmative' measures, which would help move toward a gradual deconstruction, rather than reconstruction, of problematic social institutions, and which would lead to improvements in the socio-economic circumstances and well-being of the whole rural population, including men, women and children. These might include, for example, state support for and promotion of women's ties with their natal family, the state-subsidised provision of free childcare for the left-behind, and pressure from the state for proper recognition and remuneration from local government, business leaders and villagers for 'inside work', including care-work and agricultural work. In the last several years, state leaders have directed very little effort toward overcoming gender injustices. But they have been highly responsive to suggestions on how to overcome rural-urban inequalities, and have begun substantially to increase investment in the countryside in the form of government welfare as well as infrastructural projects. This creates a political environment quite conducive to the development of a transitional strategy for justice of this kind.

References

Beneria, L. and G. Sen. 1981. Accumulation, reproduction and 'women's role in economic development': Boserup revisited. *Signs*, 7(2), 279–98.

Bossen, L. 2012. Reproduction and real property in rural China: three decades of development and discrimination. *In:* T. Jacka and S. Sargeson, eds. *Women, gender and rural development in china*. Cheltenham, UK and Northampton, MA: Edward Elgar, pp. 97–123.

Caldeira, R. 2009. The failed marriage between women and the landless people's movement (MST) in Brazil. *Journal of International Women's Studies*, 10(4), 237–58.

Chambers, R. 1983. *Rural development: putting the last first*. Essex, UK: Longman Group Ltd.

Chen Baifeng. 2008. The influence of changing peasant values on familial relations. Translated in *Chinese Sociology and Anthropology*, 41(1), 30–42.

Cooke, B. and U. Kothari, eds. 2001. *Participation: the new tyranny?*. London and New York: Zed Books.

Day, A. 2007. Return of the peasant: history, politics, and the peasantry in postsocialist China. PhD dissertation, University of California, Santa Cruz.

Day, A. 2008. The end of the peasant? New rural reconstruction in China. *boundary 2*, 35(2), 49–73.

Day, A. and M.A. Hale. 2007. Guest editors' introduction. *Chinese Sociology and Anthropology*, 39 (4), 3–9.

Flower, J. 2002. Peasant consciousness. *In:* P. Leonard and D. Kaneff, eds. *Post-socialist peasant? Rural and urban constructions of identity in Eastern Europe, East Asia and the former Soviet Union*. Houndmills, Basingstoke and New York: Palgrave, pp. 44–72.

---

[24]For further critiques of the inequality-strengthening tendencies of social movements and development discourses that focus on 'traditional' and 'local' cultures and social institutions, see Razavi (2009, 217–221) and Mohan and Stokke (2000). As these authors observe, such tendencies are common among participatory development projects as well as alternative globalization projects and movements.

Fraser, N. 1997. *Justice interruptus: critical reflections on the 'postsocialist' condition*. New York and London: Routledge.

Fraser, N. 2003. Social justice in the age of identity politics: redistribution, recognition, and participation. *In*: Nancy Fraser and Axel Honneth, eds., *Redistribution or recognition? A political-philosophical exchange*. London: Verso, pp. 7–109.

Gao Xiaoxian. 2001. Strategies and space: a case study. *In*: Ping-Chun Hsiung, Maria Jaschok and Cecilia Milwertz, with Red Chan, eds. *Chinese women organizing: cadres, feminists, muslims, queers*. Oxford and New York: Berg, pp. 193–208.

Gao Xiaoxian. 2008. Women and development in China: an analysis and reappraisal of practice. *Chinese Sociology and Anthropology*, 40(4), 13–26.

Greenhalgh, S. and E.A. Winckler. 2005. *Governing China's population: from Leninist to neoliberal biopolitics*. Stanford, CA: Stanford University Press.

Harcourt, W. 2005. The body politic in global development discourse: a woman and the politics of place perspective. *In*: Wendy Harcourt and Arturo Escobar, eds. *Women and the politics of place*. Bloomfield, CT: Kumarian Press, Inc., pp. 32–47.

He Huili. 2007. Experiments of new rural reconstruction in Lankao. Translated in *Chinese Sociology and Anthropology*, 39(4), 50–79.

He Xuefeng. 2003. *Xin xiangtu Zhongguo: Zhuanxingqi xiangcun shehui diaocha biji* [New earthbound China: notes on investigations into rural society during the transition period]. Guilin: Guangxi Shifan Daxue Chubanshe.

He Xuefeng. 2007. New rural construction and the Chinese path. Translated in *Chinese Sociology and Anthropology*, 39(4), 26–38.

He Xuefeng. 2009. *Cunzhi moshi* [Patterns of village governance]. Jinan: Shandong Renmin Chubanshe.

He Xuefeng. 2010. *Xiangcun shehui guanxi: Jinru 21 shiji de Zhongguo xiangcun sumiao* [Keywords of rural society: a sketch of China's rural society in the early twenty-first century]. Jinan: Shandong Renmin Chubanshe.

He Xuefeng. 2011. Lun shuren shehui de renqing [A discussion of human sentiment in an acquaintance society]. *Nanjing Shifan Daxue Bao (Shehui Kexue Bao) [Journal of Nanjing Normal University (Social Science)]*, 4, 20–7.

He Xuefeng. 2012. Tudi wenti de shishi yu renshi [Understanding and recognizing the land problem]. *Zhongguo Nongye Daxue Xue Bao (Shehui Kexue Ban) [Journal of China Agricultural University (Social Science)]*, 29(2), 5–19.

Jacka, T. 1997. *Women's work in rural China: change and continuity in an era of reform*. Cambridge: Cambridge University Press.

Jacka, T. 2006. Approaches to women and development in rural China. *Journal of Contemporary China*, 15(49), 585–602.

Jacka, T. 2010. Women's activism, overseas funded participatory development, and governance: a case study from China. *Women's Studies International Forum*, 33, 99–112.

Jacka, T. 2012. Migration, householding and the well-being of left-behind women in rural Ningxia. *The China Journal*, 67, 1–21.

Judd, E.R. 1989. *Niangjia*: Chinese women and their natal families. *Journal of Asian Studies*, 48(3), 524–44.

Judd, E.R. 2010. Family strategies: fluidities of gender, community and mobility in rural West China. *The China Quarterly*, 204, 921–38.

Kelliher, D. 1994. Chinese Communist political theory and the rediscovery of the peasantry. *Modern China*, 20(4), 387–415.

Li Xiaoyun and Remenyi, J. 2004. Toward sustainable village poverty reduction: the development of the county poverty alleviation planning (CPAP) approach. *In*: Janelle Plummer and John G. Taylor, eds. *Community participation in China: issues and processes for capacity building*. London: Earthscan, pp. 269–303.

Li Xiaoyun and Liu Xiaoqian. 2010. Stalemate of participation: participatory village development planning for poverty alleviation in China. *In*: N. Long, Ye Jingzhong and Wang Yihuan, eds. *Rural transformations and development – China in context: the everyday lives of policies and people*. Cheltenham, UK and Northampton, MA: Edward Elgar Publishing Ltd, pp. 312–26.

Liu Jinlong. 2010. How local politics shape intervention practices in the Xiaolongshan forest region of Gansu, NW China. *In*: N. Long, Ye Jingzhong and Wang Yihuan, eds. *Rural transformations and development – China in context: the everyday lives of policies and people*. Cheltenham, UK and Northampton, MA: Edward Elgar Publishing, Ltd., pp. 327–56.

Mohan, G. and K. Stokke. 2000. Participatory development and empowerment: the dangers of localism. *Third World Quarterly*, 21(2), 247–68.

Mu, R. and D. van de Walle. 2009. Left behind to farm? Women's labor re-allocation in rural China [online]. Policy Research Working Paper WPS5107, The World Bank. Available from: http://elibrary.worldbank.org/content/workingpaper/10.1596/1813-9450-5107 [Accessed 1 September 2012].

Nanda, M. 2002. Do the marginalized valorize the margins? Exploring the dangers of difference. *In*: Kriemild Saunders, ed. *Feminist post-development thought: rethinking modernity, postcolonialism and representation*. London and New York: Zed Books, pp. 212–23.

Pang, L., A. de Brauw and S. Rozelle. 2004. Working until you drop: the elderly of rural China. *The China Journal*, 52, 73–94.

Parpart, J.L. 2002. Rethinking em(power)ment, gender and development: the PRA approach. *In*: Jane L. Parpart, Shirin M. Rai and Kathleen Staudt, eds. *Rethinking empowerment: gender and development in a global/local world*. London and New York: Routledge, pp. 165–81.

Parpart, J.L., S.M. Rai and K. Staudt. 2002. *Rethinking empowerment: gender and development in a global/local world*. London and New York: Routledge.

Pieterse, J.N. 1998. My paradigm or yours? Alternative development, post-development, reflexive development. *Development and Change*, 29, 343–73.

Plummer, J. and J.G. Taylor. 2004. *Community participation in China: issues and processes for capacity building*. London and Sterling, VA: Earthscan.

Razavi, S. 2009. Engendering the political economy of agrarian change. *Journal of Peasant Studies*, 36(1), 197–226.

Sargeson, S. and Song Yu 2010. Land expropriation and the gender politics of citizenship in the urban frontier. *The China Journal*, 64, 19–46.

Speed, S., R.A.H. Castillo and L.M. Stephen, eds. 2006. *Dissident women: gender and cultural politics in Chiapas*. Austin, TX: University of Texas Press.

Wang Ximing. 2009. Seniors' organizations in China's new rural reconstruction: experiments in Hubei and Henan. Translated by Matthew A. Hale in *Inter-Asia Cultural Studies*, 10(1), 138–53.

Wen Tiejun. 2001. Centenary reflections on the 'three dimensional problem' of rural China. Translated by Petrus Liu in *Inter-Asia Cultural Studies*, 2(2), 287–95.

Wen Tiejun. 2004. Jiegou xiandaihua [Structural modernization]. *In: Jiegou xiandaihua – Wen Tiejun yanjiang lu* [Structural modernization – the lectures of Wen Tiejun]. Guangzhou: Guangdong Renmin Chubanshe, pp. 8–22.

Wen Tiejun. 2007. Deconstructing modernization. *Chinese Sociology and Anthropology*, 39(4), 10–25.

Wen Tiejun. 2012. Weiwen daju yu 'sannong' xinjie [A new understanding of the maintenance of stability and the 'three rurals']. *Zhongguo Hezuo Jingji* [*Chinese Cooperative Economy*], 3, 29–32.

Wen Tiejun, *et al.* 2012. Ecological civilization, indigenous culture, and rural reconstruction in China [online]. *Monthly Review*. Available from: http://monthlyreview.org/2012/02/01/ecological-civilization-indigenous-culture-and-rural-reconstruction-in-china [Accessed 1 September 2012].

West Women. n.d. www.westwomen.org. [Accessed 30 October 2013].

Yan Hairong. 2003. Spectralization of the rural: reinterpreting the labor mobility of rural young women in post-Mao China. *American Ethnologist*, 30(4), 1–19.

Yan Yunxiang. 2003. *Private life under socialism: love, intimacy and family change in a Chinese village 1949–1999*. Stanford, CA: Stanford University Press.

Yang Lichao. 2011a. Development interventions, gender and social change in rural China: a case study of three villages in Shaanxi. PhD dissertation, the Australian National University.

Yang Lichao. 2011b. Myths and realities: gender and participation in a donor-aided project in northern China. *In*: Tamara Jacka and Sally Sargeson, eds. *Women, gender and rural development in China*. Cheltenham, UK and Northampton, MA: Edward Elgar, pp. 208–28.

Yang Shuai and Wen Tiejun. 2011. Nongmin zuzhi de kunjing yu pojie – hou nongye shui shidai de xiangcun zhili yu nongcun fazhan [Predicaments and breakthroughs in peasant organization – rural governance and development in the post-tax era] [online]. *Zhongguo Nongcun Yanjiu Wang* [*China Rural Research Net*]. Available from: http://www.snzg.cn/article/2011/1119/article_26308.html [Accessed 15 August 2012].

Ye Jingzhong. 2010. Zai lun 'canyushi fazhan' yu 'fazhan yanjiu' ['Participatory development' and 'development studies' revisited. Foreword]. *In*: Li Ou, ed. *Canyushi fazhan yanjiu yu shijian fangfa* [*Methodologies of participatory development studies and practices*]. Beijing: Shehui Kexue Chubanshe, pp. 1–8.

Yu Jianrong. 2006. Conflict in the countryside: the emerging political awareness of the peasants. *Social Research*, 73(1), 141–158.

Zhang Weiguo. 2009. 'A married out daughter is like spilt water'? Women's increasing contacts and enhanced ties with their natal families in post-reform rural North China. *Modern China*, 35, 256–283.

Zhao Jie. 2011. Developing Yunnan's rural and ethnic minority women: a development practitioner's self-reflections. *In*: Tamara Jacka and Sally Sargeson, eds. *Women, gender and rural development in China*. Cheltenham, UK and Northampton, MA: Edward Elgar, pp. 171–189.

Tamara Jacka is a Senior Fellow in the Department of Political and Social Change, College of Asia and the Pacific, the Australian National University. Her publications include *Women's work in rural China: change and continuity in an era of reform* (Cambridge University Press, 1997); *On the move: women and rural-to-urban migration in contemporary China* (co-edited with Arianne Gaetano, Columbia University Press, 2004); *Rural women in urban China: gender, migration and social change* (M.E. Sharpe, 2006; winner of the American Anthropological Association's Francis Hsu award for best book in East Asian Anthropology, 2007); and *Women, gender and development in rural China* (co-edited with Sally Sargeson, Edward Elgar, 2011).

Finance and rural governance: centralization and local challenges

John James Kennedy*

The ability of local governments to raise revenues has been dramatically altered over the last several decades in rural China, and this has had a significant influence on local governance. The fiscal reforms in 1994 and after 2002 significantly strengthened central government capacity to collect revenues at the provincial, county and township levels. The centralization of tax policies has also fundamentally changed the relationship between central and local administrative levels. This paper examines the cost and benefits of strengthening central government capacity and the increasing reach of the state in rural China. One of the challenges facing local officials is to provide social services and at the same time deal with underfunded mandates from higher authorities. Local officials and villagers have adjusted to changes in finance and tax policies in diverse ways. Moreover, when analyzing fiscal reforms and local governance it is important to take into account the differing perspectives from the village, township, county and center.

Introduction

Financial reform after 1994 is a story of recentralization and national policies designed to recapture local revenues and create a standardized tax system as well as increased local service provision. The story begins with initial market reforms in the 1980s and decentralization of resource allocation at the lower levels to allow for local economic development decisions. This created the impetus for economic growth in the 1980s and increasing local revenues. However, tax restructuring lagged behind market reforms and the central government continued to lose revenues relative to provincial governments. By 1993, central revenues were about a third of total provincial revenues. Moreover, the central government ran a deficit while the combined provincial government budgets had a surplus (see Table 1). In order to regain fiscal control at the macro level, the central government launched the 1994 tax reform that established a tax sharing system with the provinces. This reform strengthened the capacity of the national government to collect revenues from the provincial governments, but the center also shifted responsibility for a number of expenditures to local governments. The 1994 reform provided greater autonomy to sub-provincial governments at the county and township level to levy local taxes and fees in order to make up the balance between central remittance and local expenditure needs. Soon after, the reform officials at the lowest administrative level shifted their key duties from service provision to revenue

We would like to acknowledge Kevin O'Brien, Emily Ting Yeh and Saturnino Borras for their insightful comments as well as the Northwest Socioeconomic Development Research Center (NSDRC) in Xian China and the Ford Foundation, Beijing, for their support.

Table 1. Central and provincial revenues and expenditures[1] from 1980–2010.

Year	Central		Provincial	
	Revenues	Expenditures	Revenues	Expenditures
1980	284	667	875	562
1985	770	795	1235	1209
1990	992	1004	1945	2079
1991	938	1091	2211	2296
1992	980	1170	2504	2572
1993	958	1312	3391	3330
1994	2907	1754	2312	4038
1995	3257	1995	2986	4828
1996	3661	2151	3747	5786
1997	4227	2533	4424	6701
1998	4892	3126	4984	7673
1999	5849	4152	5595	9035
2000	6989	5520	6406	10,367
2001	8583	5768	7803	13,135
2002	10,389	6772	8515	15,281
2003	11,865	7420	9850	17,230
2004	14,503	7894	11,893	20,593
2005	16,549	8776	15,101	25,154
2006	20,457	9991	18,304	30,431
2007	27,749	11,442	23,573	38,339
2008	32,681	13,344	28,650	49,248
2009	35,916	15,256	32,603	61,044
2010	42,488	15,990	40,613	73,884

Notes: [1]100 million yuan
Source: China Statistical Abstract (2011).

collection. The change in funding expenditures created uneven tax burdens on villagers throughout rural China. In order to further centralize the tax system down to the micro level and reduce villager burdens, the central government initiated the tax-for-fee reform that eliminated most local fees in 2002, and then the abolition of basic agricultural taxes in 2006. This is a remarkable achievement. In fact, it is the first time in over 2000years that the rural population has not been subject to an agriculture-based tax or fee. The 2002 tax reform successfully reduced villagers' burdens and extended the reach of the state.[1] The main function of township officials shifted from tax and fee collection back to service provision, but the new reforms also weakened township governments and increased government debts at the township and county levels. Thus, tax reform in China is a story of recentralization with both costs and benefits for local finance and governance.

The fiscal reforms dramatically altered the ability of county and township governments to raise revenues, and this has a significant influence on governance, especially at the township level. The town (*zhen*) or township (*xiang*) is the lowest level of official administrative authority in China, and in the 1990s rural residents blamed township officials for excessive taxation in the countryside. Indeed, township officials had a difficult job as they attempted

---

[1]This refers to Shue (1988) and the ability of the central government to influence local officials. Also see Oi *et al.* (2012) for a similar discussion

to provide social services and at the same time deal with underfunded mandates from higher authorities. These officials adjusted to changes in finance and tax policies in various ways. For example, soon after the 1994 tax reform, township governments in the inland poorer regions had to impose more informal fees and levies on individual agricultural households, while wealthier townships in the coastal areas relied on local industry for resources, and villager households experienced little change in their agricultural tax and fee rates (Lin and Liu 2007). In some areas, township leaders took out loans to meet county revenues targets (Zhao 2006). After the implementation of the tax-for-fee reform and the abolition of the agricultural tax in 2006, township officials near urban centers adjusted to the loss of funds through land lease revenues due to urban expansion and real estate development (Takeuchi 2013). The diversity of ways in which township governments adjusted to central governments poses challenges in observing and understanding how fiscal policies influence rural governance across China.

The research on the 1994 to 2002 tax reforms is diverse and there is a rich variation in the methods scholars use and geographic areas where they have conducted studies. For example, one of the most influential and damning investigations of rural conditions and excessive taxation was a 2001–2002 qualitative report that incorporated in-depth interviews with villagers in Anhui province (G. Chen and Wu 2004). There were also a number of quantitative studies that tend to examine national or regionally representative surveys as well as government statistical data (Tao and Liu 2005, Lin and Liu 2007). These statistical studies present a different picture of villager burdens. Some scholars also use mixed methods. One example is Zhao Shukai, a leading expert on rural governance from the China Development Research Center, who incorporates qualitative interviews and statistical data to study township governments. Zhao (2006) finds that township leaders face a complex array of challenges and they are not simply unruly rent-seeking officials.

This paper offers a state-of-the-field review of rural financial reform and local governance in China starting in the 1980s, focusing on different approaches, methods and perspectives used to evaluate local tax and fiscal policies. It is divided into three sections. The first section examines rural fiscal policies from the 1980s to 1994, examining the conditions that led to the 1994 reform. While many scholars agree on the fiscal challenges that central and local governments faced adjusting the tax system, these studies tend to approach the issues from difference perspectives as well as various definitions of central-local relations.

The second part examines the causes and consequences of increasing villager burdens in the 1990s from the villager, institutional and township government perspectives. These different views or approaches have a significant influence on how scholars evaluate local governance and the fiscal reform. From the central government perspective, the 1994 tax reform strengthened fiscal relations with the provinces and provided greater fiscal autonomy for sub-provincial governments to fulfill expenditures obligations. However, a major consequence of that reform was a significant increase in villagers' burdens (taxes and fees). The reasons for these burdens (and the amount) varied based on level of regional economic development as well as central and local perspectives. From the perspective of a villager in a poorer region, the increase in burdens was due to uncontrolled rent seeking by township and county officials. The 1994 reform created a system whereby the central and even provincial governments could not effectively monitor local officials to avoid over-taxation and collection of illegal fees (So 1997, T.B. Bernstein and Lü 2003, Chen and Wu 2006). The institutional perspective suggests that the root problem is more systematic and focuses on the regressive tax system, income disparity among the rural population, and uneven implementation of tax regulations and laws (Zhou 2006, Lin and Liu 2007, Tao and

Qin 2007, X. Chen 2009). The township government view holds that officials unduly received the brunt of the blame for villager burdens (Zhao 2005).

The third section examines the fate of township governments and local governance in the post-2002 (2006) Tax-for-Fee Reform era. A number of studies find that the 2002 reform dramatically reduced villager tax and fee burdens (X. Chen 2009, L.C. Li 2012, Liu *et al.* 2012). However, there are still unresolved issues such as the role of the township government and public services provision. Some scholars suggest the township government has lost much of its authority and has become a service provider for the county (Ho 2001, Smith 2009, 2010). Some have even proposed a new definition of 'local' with the elimination of township governments and establishing the county as the lowest administrative level (He 2000, Xu 2004). However, other scholars suggest that the township still plays an important role as the lowest level of government. They suggest that it should have greater budget autonomy (Takeuchi 2013) and that the introduction of direct elections for township government leaders can increase local accountability (governance) and resolve the supervision problem (Lin and Liu 2007, Tao and Qin 2007, Meng and Zhang 2011).

Central-local relations and rural fiscal policies from 1980 to 1994

From the end of the Maoist period (1949–1976) until the early 1990s, the rural financial system went through significant changes at the local and national levels. In 1979, central government revenues were dependent on state-owned enterprises (SOEs) in the urban areas and indirect taxes through grain quotas in the countryside. However, this changed with the introduction of economic reforms in the 1980s, such as the breakup of the commune system and the introduction of private enterprises in the rural and urban areas. Indeed, the central tax system did not adequately adjust to the rise in new private businesses and the central revenues fell throughout the 1980s (Oksenberg and Tong 1991, Hershler 1995, Wang 1997, Wong 2000). Moreover, provincial governments were attempting to protect local revenues from 'predatory' practices of the national government and even avoiding tax payments to central authorities (Wong *et al.* 1995). This gave the impression of a relatively weak central government that was unable to adjust the tax system and collect revenues. However, the 1994 tax reform reflects a re-assertive central government with the capacity to reshape provincial and central fiscal relations (Hershler 1995). The reform improved the revenue stream for the central government, and it altered the relationship between provincial and central government, but it also created a regressive tax system and a new set of problems for sub-provincial-level governments. County and township governments were faced with reduced budgets and increased responsibilities for local expenditures.

In 1979, the fiscal and political system was highly centralized. The central government controlled local expenditures and revenues and the tax system was redistributive. The central government collected taxes from wealthier industrial coastal provinces with SOEs and redistributed revenues to poorer inland provinces. Rural taxes were mostly based on grain quotas collected from over 54,000 communes in the countryside. In 1979, 81 percent of the population lived in the countryside on communes (*gongshe*) which were rural units that consisted of brigades (*dadui*) and production teams (*shengchandui*). The production teams contained 10–20 households and were the basic unit of production. Grain was sold to state grain stations and villagers were paid in work points. Moreover, all grain quotes and taxes were collected at the production team level (Rozelle 1994, Unger 2002). In 1981, there were 54,000 communes, 720,000 brigades and over 6 million production teams (China Statistical Abstract 1982).

While farmers were not directly taxed during this period, they were indirectly taxed through 'price scissors' where the central government sold communes agricultural inputs such as tractors and farming tools for high prices while buying production team grain at depressed prices (Lin and Liu 2007). However, in the early 1980s, the central and local governments broke up the communes in favor of the household responsibility system (HRS), and the new basic unit of agricultural production became the household. Thus, in 1985 there were 83,000 townships, 940,000 administrative villages and over 190 million households (China Statistical Abstract 1986). While the commune grain quota system ended, the price scissors effect continued. Rural households were given plots of land based on family size, but the land remained collectively owned and villagers had land use contracts. Moreover, villagers were subject to grain procurement quotas where farmers had to sell a specific amount of grain at below-market prices to state (township) grain stores. Local cadres supplied farm inputs, such as chemical fertilizer, pesticides and hybrid seeds, through state stores and deliveries (Rozelle 1994). In addition, township officials and village cadres collected taxes and fees through the grain procurement system. The township government deducted taxes and fees from the procurement grain prices and then paid the villagers the remaining balance (Lin and Liu 2007).

After the breakup of the communes, the official administrative hierarchy was established in the 1982 state constitution, and it lays out the three official local administrative levels: province (*sheng*), county (*xian*) and township (*xiang*). However, according to the *China Statistical Abstract*, most provinces have several administrative levels and one lower level: province (*sheng*), prefecture (*diji*), prefecture municipality (*dijishi*), county (*xian*), county seat (*xianshi*), town (*zhen*), township (*xiang*) and village (*cun*). The lowest administrative level is the town or township. The difference between a town and a township is the proportion of residents who are registered as non-agricultural (urban) population. A town has over 10 percent of the population registered as non-agricultural, whereas a township has over 90 percent of the population registered as agricultural (Zhong 2003). Most studies that examine the town and township level of government tend to combine these two into one, referring to townships as both town and township governments. There are also administrative villages (*xingzhengcun*) that have a villager committee with 5–7 members, a village leader and a party secretary, and within an administrative village there are natural villages (*zirancun*) and small groups (*xiaozu*) (see also Ho 2013).

Given the change in the local administrative divisions in the early 1980s, it is also important to clearly define 'local' government. Studies in the 1990s and after 2000 that discuss the rural fiscal system often refer to central and local government relations or higher authorities and local cadres. The usage of the term 'local' depends on the context and how scholars define local government. Some studies refer to local as the province. For example, when discussing central-local relations, Wong (2000) examines the financial policies between Beijing and the provinces. The term 'fiscal federalism' refers to the centralized political system with a decentralized fiscal system between the central and provincial governments (Montinola *et al*. 1995). Others use local to refer to sub-provincial levels such as municipality, county and township. Li (2006) and Kung *et al*. (2009) evaluate the tax system and government relations between county and township levels where the township is local. Often, village studies on taxes and fees refer to the village as local and the township as a higher authority (Tsai 2002, Kennedy 2007).

In the 1980s, the main function of township governments and village committees was to provide local services and implement central policies. However, the additional functions varied across rural China. In the poorer provinces, central transfers provided most of the local (sub-provincial) revenues. Services included local road and irrigation repair, maintenance of local schools, running water and availability of farming needs such as seeds, fertilizers and pesticides. There were also central mandates and unpopular policies that had to be implemented such as family planning and land management. In wealthier provinces, township leaders were also establishing and managing township and village enterprises (TVEs). Indeed, rural economic development in the coastal provinces took off during this time (Oi 1999).

Between 1980 and 1996 the gross domestic product (GDP) grew on average by nearly 10 percent (Wang 1997). This was astonishing growth and it was in part due to the decentralized allocation of resources and local economic decision making at the provincial and sub-provincial levels. However, the central government was also losing revenues at an astounding rate. Central revenues were dependent on provincial and sub-provincial governments to collect taxes and make upward transfers, but entrenched local interests and outdated tax codes resulted in shrinking and reduced central revenues (Wang 1997, A. Chen 2008). This included tax evasion and unauthorized tax breaks that municipal and county governments offered to attract industrial development. As a result, the percentage of total revenues to national GDP dropped from 31 percent in 1978 to 12 percent in 1993 (China Statistical Abstract 1995). Moreover, the central revenues actually declined between 1990 and 1993 while provincial revenues nearly doubled (see Table 1). Provincial and sub-provincial governments were responsible for delivering services, and local expenditures came from the central government through revenue sharing to finance local mandates (Wong 2000). This was important for poorer agricultural provinces because revenue sharing was meant to fill in the gap between central mandates and local revenues. Thus, the redistributive revenue-sharing system benefited the inland provinces, but the wealthier provinces gave more in revenues than they received, and provincial leaders attempted to protect their revenues. As a result, the center had to negotiate the percentage or amount of locally-collected revenues to be remitted, and these contracts varied among provinces (Oksenberg and Tong 1991, Wang 1997). Central government deficit was climbing and the center made three attempts to change or institutionalize 'fiscal contracts' between central and provincial governments, but central leaders were unable to reverse the trend (Hershler 1995, Wong 2000). As a result, the central government scrapped the old system of contracts and introduced a new tax reform in 1994 that created a shared tax system and strengthened central control over the fiscal relationship between the national and provincial governments.

The 1994 tax reform recentralized the financial system, and the central government introduced a rule-based tax system rather than negotiated percentages of remittance from the provinces (Wang 1997, Wong 1997, A. Chen 2008). The new tax assignment system (*fenshuizhi*) divided the taxes into three distinct categories: national, local and shared. National taxes were based on a unified tax code, and shared taxes were given to the central government and then portions were returned to the provinces as remittances. Local taxes were decentralized with various types of taxes determined at the provincial and county levels. As a result, the central government no longer relied on the provincial and sub-provincial governments to collect all taxes. Instead, the task of collecting central taxes now fell on the national tax service, and the central government set up its own revenue collection agencies in the provinces and counties (Wang 1997). This increased the reach of the state and established a parallel tax administration: one for the national and the other for provincial and sub-provincial governments. The central taxes included value added tax (VAT) on industrial products, banks and SOEs. Provincial taxes include local enterprise income tax, animal slaughter tax, agricultural tax, specialty (non-grain products) agricultural tax and land lease revenues (A. Chen 2008, Wang 1997, Oi 1999). The central revenue agencies provided incentives for sub-provincial governments to collect state taxes by allowing the county governments to retain 25 percent of the VAT taxes collected. The provincial and county governments were able to keep local taxes (Zhong 2003). The new system was also known as 'eating in separate kitchens' (*fenzaochi-fan*). Within a few years, the tax assignment system was successful in increasing central revenues and reversing the trend (Hershler 1995, Wang 1997). Table 1 shows the dramatic and immediate turnaround for central revenues after 1993. Indeed, central revenues tripled from 1993 to 1994, while total provincial revenues decreased.

Despite the central government success, there were five systemic problems with the 1994 tax reform. First, the new system eroded the redistributive aspect of the previous tax system. Central transfers were now based on remittances from provincial tax collection and subsidies in the form of special transfer payments. Thus, wealthier provinces retained more of their revenues and poorer provinces had to rely more on local taxes and fees (Wang 1997, A. Chen 2008). The transfer payments were mostly earmark grants for provincial projects such as capital construction and economic development (Lee 2000). However, allocations of these grants were largely political and poorer provinces were underrepresented in the central decision-making process (A. Chen 2008). This put the poorer inland provinces at a disadvantage for some of the larger transfer payments. While poorer provinces did receive transfer payments, one scholar predicted that 'new taxes may counter the redistributive aims of the reform by increasing some of the poorer regions' tax burdens' (Hershler 1995). Second, the central government (Ministry of Finance) introduced revenue growth targets for each province. The central government mandated that each year's revenue growth had to match or surpass that of the previous year (Wang 1997, Wong 2000). In turn, the provinces imposed revenue growth targets on sub-provincial governments with punishments and rewards based on achieving targets. Third, the tax reform provided 75 percent of the VAT from local enterprises to the central government and counties could keep the local enterprise income tax. The problem was that the VAT could be collected even if the state enterprise was unprofitable (Zhong 2003). This left county and township governments with non-profitable enterprises or no local industries at all searching for new revenue sources.

The fourth systemic problem is that the 1994 tax reform did not address extra-budgetary funds because it provided sub-provincial governments the opportunity to make up for lost revenues (Wang 1997, Zhou 2006, A. Chen 2008). These are off-budget funds that consist of informal levies and fees, and they are used to fulfill local fiscal commitments and upper-level mandates. The official budget relies on designated taxes, such as the agricultural tax and local enterprise income tax, but if the local tax base is not enough to cover expenditures then county or township government as well as the village committee may impose unofficial fees and levies to make up for the shortfall. Local governments have relied on extra-budgetary funds since the 1950s and they increased during the 1980s (Wong 2000). Thus while the central government gained greater fiscal control over the collection of revenues from the provinces, county and township governments enjoyed a level of autonomy in levying fees, fines and apportionments on villagers. Indeed, by the late 1990s, township and village authorities were dependent on extra-budgetary funds to provide services and over half the local revenues collected consisted of these informal fees and levies (Wang 1997, Eckaus 2003, A. Chen 2008).

Finally, the central government shifted expenditures downward to the county and township governments (Wang 1997, Zhao 2006, A. Chen 2008). Although the counties and townships had increased local revenues due to greater autonomy in raising

Table 2.   Revenue and expenditures[1] for three counties in Shaanxi Province, 1996–2010.

Year	Hu County		Jia County		Shenmu County	
	Revenues	Expenditures	Revenues	Expenditures	Revenues	Expenditures
1996	0.7	1.0	0.1	0.4	0.7	0.9
1997	0.8	1.1	0.1	0.4	0.8	1.0
1998	0.9	1.1	0.1	0.5	0.9	1.2
1999	1.0	1.3	0.1	0.5	1.0	1.3
2000	1.1	1.4	0.1	0.7	1.3	1.6
2001	1.2	1.8	0.1	0.9	1.8	2.2
2002	1.1	2.1	0.1	1.1	2.2	2.8
2003	1.2	2.2	0.1	1.3	3.4	4.0
2004	1.3	2.6	0.1	1.4	4.4	5.2
2005	1.0	2.7	0.1	1.9	6.7	7.9
2006	1.4	3.8	0.0	2.2	6.8	11.3
2007	1.7	4.8	0.1	3.5	10.0	15.6
2008	2.1	7.2	0.2	4.8	16.5	23.6
2009	2.6	9.3	0.2	8.2	21.6	27.7
2010	3.4	13.2	0.3	9.7	28.4	28.6

Notes: [1]100 million yuan
Source: China Statistical Abstract (2011).

extra-budgetary funds, these local governments had little authority in deciding expenditures. Indeed, expenditures increased faster than revenues after 1994. Between 1994 and 1999, total fiscal revenues at the county level increased by 51 percent, but the total expenditures grew over 100 percent (A. Chen 2008). The situation was worse for poorer counties. Table 2 displays the revenue and expenditures for three counties in Shaanxi province. Jia County is one of the poorest counties, with no natural resources or industry. In 1996, total revenues covered one fourth of the county expenditures. By 2002, total revenues comprised only one tenth of the county expenditures. Many provincial governments did not fare much better. Table 1 shows that total provincial revenues were greater than total expenditures in 1993, but by 1999 revenues only covered 62 percent of total expenditures. This also contributed to the growing local debt; county and township governments had the heaviest debt burdens.

Local perspectives and fiscal policies leading up to the 2002 tax-for-fee reform

The 1994 tax reform influenced villagers and township and county officials differently. As a result, the positive or negative consequences of the tax reforms depend on the level of analysis and perspective taken. Scholars who take on a villager perspective of local burdens have a very different interpretation of tax reforms than scholars who examine the reforms from a township official's viewpoint or from a broader institutional perspective. In addition, not all villagers and township officials were affected the same way. Villagers in the wealthier coastal areas experienced minimal household tax and fee increases, while villagers in the inland agricultural regions were subjected to increasing burdens. Nevertheless, the national leadership continued to centralize fiscal policies. In the 1990s, the central government made several attempts to adjust the tax and fee policies and reduce villager burdens, but these peripheral regulations did little to change the system. As a result, the central government scrapped the old system of local fiscal autonomy and introduced a new tax reform in 2002 that abolished most local fees and eventually the agricultural tax in favor of increased central transfers to pay for local services (X. Chen 2009). This expanded central fiscal control down to the grassroots level.

The villager perspective, especially in the poorer regions, tends to paint a bleak picture of rural China in the 1990s, and township officials are viewed as the key culprits for rising local taxes and fees. For example, G. Chen and Wu (2004), who conducted their 2001–2002 fieldwork in Anhui province, report a very oppressive environment where self-interested township officials use coercive and even brutal measures to collect fees. Despite growing local repression, villagers still resisted when they could, and they viewed the central government as a benevolent leadership unaware of the corruption at the local level. Li (2004) observes the same dynamic between villagers' view of local officials and the central government. Indeed, qualitative studies conducted in the 1990s and early 2000 offer a similar negative depiction of township authorities (O'Brien and Li 1996, Lü 1997, So 1997). Personal interviews provide in-depth understanding of village life and hardships, especially observations of local abuses at the hands of some corrupt township officials. Interviews also offer insight on informal and illegal revenue collection, as well as villagers' reactions and resistance. During my own research in the late 1990s, I found many villagers referring to the 'three wanting cadres' who want your money, your grain and your unborn children. Although the general feeling among villagers was that the township officials were out to enrich themselves, villagers I interviewed also admitted that they were willing to pay for services, but few had faith that fees and levies would go directly towards village services. Many of these observations reflect the problem scholars pointed out earlier with the loopholes in the1994 tax reform and the fact that reliance on extra-budgetary funds could potentially give rise to wasteful spending and corruption.

Official fees were a significant source of villager burdens (nongmin fudan). These were known as the 'five township and three village fees' (wu tongchou, san tiliu). Township budgetary funds that cover the basic local public services are called the 'five township fees' (tongchou), for education, militia training, family planning, road repair and construction, and public health care and epidemic prevention. The 'three village fees' or tiliu are the public accumulation fund, public welfare and administrative fees. While development and social welfare needs determined the amount of money collected annually, village cadres typically decided the actual amount of the tiliu (Lü 1997, Kennedy 2007). In addition, villagers were required to provide corvée labor that ranged from 10–20 days a year of 'voluntary' labor for public projects. Outside of the official fees, township governments and village committees could also collect apportionments (tanpai) or 'self raised' funds for specific investment projects (O'Brien and Li 1996, Kennedy 2007, So 2007). These are unofficial levies charged to the villagers without explicit government regulations or laws. These include basic services such as water and electricity fees. In fact, the list of fees both official and unofficial could be quite long, ranging from a few dozen to more than a hundred items (Wong 1997). Villagers have often referred to these fees as the 'three unrulies' (sanluan): unruly fundraising, unruly fines and unruly apportionments (T.B. Bernstein and Lü 2003). With the combination of budget and extra-budgetary collections, total burdens could reach over 40 percent of villagers' annual income and the majority of payments were typically informal fees (ibid.).

In addition to the fees, villagers were also required to pay agricultural taxes such as the specialty tax. The specialty tax (techan shui) was levied on non-grain agricultural products such as tobacco, fruits and vegetables. Villagers who relied on these cash crops were usually wealthier and could afford the tax, but for villagers who lived mainly on grain products and grew cash crops on the side, the tax was a heavy burden. Indeed, one scholar pointed out that 'the special product tax on those who lived mainly on ordinary products was not just unfair but drove them into an impasse' (A. Chen 2008, 327). Some villagers were so frustrated that they gave up cash crops as a side business. During a village visit in 1998, I met a farmer who decided to plant a few apple trees in 1995 because the market price for apples was relatively high. However, by 1998 the price of apples had plummeted and he still had to pay the specialty tax. He decided not to harvest the apples and let them rot on the ground rather than give the bulk of his sales to the township government.

In order to reduce villager burdens, but not make any significant changes to the tax system, the central government attempted to introduce several regulations and policies to limit local fees. One early example is the 1993 Law on Agriculture that stipulated villager fees could not exceed five percent of their annual income. However, the regulation was vague on how to calculate average household annual income at the village level. Thus, from the onset of the five percent rule, township governments were able to work around the regulation to collect fees. Throughout the 1990s there were several central documents and policies addressing excessive rural taxation, but villager burdens as a proportion of their income continued to increase. These unsuccessful national attempts to reduce local burdens contributed to the perception of a decentralized fiscal system and lack of central oversight. T.B. Bernstein and Lü (2000) state that 'inadequate central fiscal capacity and inadequate controls over bureaucracy were the underlying causes for excessive burdens' (752). While the center was sympathetic to villagers' plight, the central government was also enjoying increased revenues and there seemed to be little incentive to fully address the systemic problems at the county and township levels. Even if the national leadership had the will, some scholars suggest it is extremely difficult for the central government to rein in the systematic corruption at the county and township levels (Wedeman 2000). In an effort to show parallels between imperial (late Qing) and post-Mao revenue collection, Kaiser and Tong (1997) show that the excessive taxes farmers paid during the late Qing did not reflect strong extractive capabilities of the central imperial court, but rather a weak central authority unable to control corrupt county magistrates. This inability to control local officials and rising rural discontent and protests contributed to the fall of the Qing dynasty. T.B. Bernstein and Lü (2000) echo this sentiment for the post-Mao period, stating that 'discontent with burdens was widespread and chronic' (756). Some of the evidence in the 1990s and early 2000s tend to reflect this problem. The number of reported social disturbances increased from 10,000 in 1994 to 24,500 in 1998, to over 40,000 in 2000 (Tanner 2005). Although most protests were relatively small, ranging from 50 to 1000 rural residents, some were massive, involving 60 to 70 townships and 20,000 to 200,000 villagers (ibid.). Moreover, these protests and social disturbances were atomized and dispersed, and directed at the township and county officials rather than the central government (O'Brien and Li 1996, 2006). Another important aspect of the rural protests at the time is that there were no rural and urban linkages. That is, most urban intellectuals did not show any interest in villager plight or protests (T.B. Bernstein and Lü 2000).

While the villager perspective tends to focus the blame for excessive taxes and fees on the unruly township government officials, the institutional viewpoint examines the structural bias against villagers that allowed for an increase in underfunded mandates and excessive burdens (Tao et al. 2004, Yep 2004, Lin and Liu 2007, Tao and Qin 2007, Liu et al. 2012). After the 1994 tax reform, the central government shifted expenditure responsibilities downward and turned a blind eye towards township officials' informal revenue collection. However, the institutional explanations vary as to how and why villager burdens increased after 1994. Some scholars suggest that market liberalization, income disparity and a regressive tax system are the key factors for the rise in villager burdens (Lin and

Liu 2007, Tao and Qin 2007). Other scholars find that villager burdens represent a structural bias against the rural population (Yep 2004).

Lin and Liu (2007) and Tao and Qin (2007) suggest that regional income inequality contributes to variation in villager burdens. Using provincial-level data from the Ministry of Agriculture to evaluate villager burdens across regions, they find that excessive taxation in the countryside was not uniform across China. Indeed, for some of the coastal provinces, such as Guangdong province, burdens measured as reported fees actually decreased in the 1990s. The hardest-hit regions were inland provinces. The combination of income disparity within rural China and a regressive tax system put agricultural regions at a greater disadvantage. Most rural taxes were levied on land and agricultural products, but by the early 1990s an increasing share of rural incomes was coming from non-agricultural sources, such as salaries from TVEs and rural migrant remittances. Moreover, these were not subject to village or township tax administration. However, poorer rural communities have a lower proportion of non-agricultural income and are more vulnerable to direct taxes and fees. Still, both Lin and Liu (2007) and Tao and Qin (2007) argue that increased villager burdens were also due to central and provincial governments granting county and township governments greater informal tax and fee autonomy in order to fulfill central mandates.

Yep (2004) finds that the reduced fiscal transfers as a result of the 1994 tax reform contributed to the underfunded mandates and that the tax-for-fee reform in 2002 may not resolve this structural problem. The lack of local investment created a cycle where the township governments collect fees from villagers because there is no local industry, but when local resources are extracted through informal fees the tax base shrinks (i.e. losing potential investment). This also contributed to overstaffed township governments (Zheng 2000, Zhao 2006). Extra-budgetary funds were the only way to fill the revenue gap, but these informal funds are available to township governments through the loophole from the 1994 reform. In addition, duties for township officials and village cadres shifted from strictly service provision to revenue collection and villager burdens dramatically increased; at the same time villagers were incorporating various ways to resist rising burdens. This began a vicious cycle where township governments needed to employ more staff to collect revenues, but a surge in staff also meant an increase in expenditures for salaries. In turn, greater villager resistance resulted in the need for more staff to counter local resistance (Yep 2004).

Township officials are in a precarious position as agents in the lowest administrative level of government. They are caught between villager dissatisfaction with taxes and fees and the county government's demands to collect revenues. From the township government perspective, collecting taxes and fees was a survival tactic. Township officials had to collect official taxes and fees as well as extra-budgetary funds to meet expenditures. However, township expenditures and revenue sources varied across rural China (A. Chen 2008). For example, a wealthy township in Zhejiang province with a population of 45,000 collected 47 million *yuan* in 2002, of which 17 million were budgetary funds and the remaining 30 million were extra-budgetary. A much poorer township in Ningxia province with a population of 12,000 had 130,000 *yuan* in total revenues. This township submitted all its tax revenues to the county government and then received 630,000 *yuan* in subsidies (Zhao 2006).

Township officials also had burdens, and they needed to fulfill tax and fee quotas set by the county government. While collecting extra-budgetary funds was a common way to meet expenditures and tax/fee quotas, some township officials had to defer salaries or take out loans to meet county revenue targets. The township officials would borrow money from the credit cooperative to fulfill the target that was submitted to the county, then pay back the loan with the county remittance. In his sample of 20 townships, Zhao (2006) found four townships that admitted to using this method, called 'empty running'. In the 1990s, it was not uncommon for township officials to not be paid for months at a time. During a 1997 village visit in northern Shaanxi province, we heard that there was a general tax strike in the township. Villagers had recently refused to pay excessive fees and taxes. When we visited the village leader's home, his wife came out the doorway and sternly marched across the courtyard with a heavy pot in each hand threatening to beat us if we did not leave the courtyard. After we explained that we were not with the township government, she invited us in and served tea. We asked her why she thought a foreigner was with the township government. She believed that the township government was so desperate to end the tax strike that they would even hire a foreigner to help collect the revenues from the villagers. Later that evening we met with two township officials. In order to meet the county revenue targets, they had to forgo getting paid until the end of the tax strike. Although they had not been paid in six months, they were sympathetic to the villagers' demands. This case shows that it is important to understand both the villagers' and township officials' perspectives.

The interaction between provincial leaders and county and township officials also contributed to the rise in excessive fees (L.C. Li 2006, 2012). One of the key factors that define the relationship between county and township officials is the cadre management system (see also Smith 2013). The provincial organization department (*zuzhibu*) sets the promotion standards for county leading cadres and the county party organization department oversees the promotion of township leading cadres.[2] During the 1990s and early 2000s, fulfilling tax and fee quotas (or revenue targets) was a key factor for promotion. Thus, local officials conform to measures of performance (O'Brien and Li 1999). Indeed, according to Zhao (2006), 'Failure to meet quotas not only precludes the possibility of commendations and promotions, but also entails "yellow card warnings" and even summary "dismissals from office"' (44). Central documents and notices were disseminated to reflect central concern with the growing villager burden problem, but the lack of immediate action to reduce burdens did not necessarily signal a weak central government. Instead, the central and provincial leadership strategically allowed the system to continue shifting the blame for increasing villager burdens onto the township governments (O'Brien and Li 2006, L.C. Li 2012).

In response to the social and governance problems arising from the 1994 tax reform and increasing rural taxes and fees, some provincial and county governments introduced local experiments in tax and fee reforms to relieve rural burdens. One reform experiment in Anhui province in 2000 caught the attention of the central leadership. This was the tax-for-fee reform where all fees were eliminated in favor of a single agricultural tax. The move represented a complete break with the previous system, and reduced the autonomy of township and county governments. It also continued fiscal centralization down to the township level. After 2002, county and township governments became more dependent on central and provincial subsidies to pay for basic public services. Once the central leadership adopted tax-for-fee as a national policy in 2002, it was fully implemented down to the township and village level through the cadre management system. For township leaders, the new measure for promotion was implementing the tax-for-fee reform and fulfilling public service duties.

---

[2]Leading cadres are the official leaders at each administrative level. They are either the mayor (government head), vice mayor, party secretary or vice party secretary. Unlike civil servants, leading cadres are transferred every three or four years and they are subject to strict promotion standards from the administrative level above (see Edin 2003).

Local governance and the fate of the township in the post-2002 tax-for-fee reform era

As with the 1994 reform, whether or not the 2002 tax-for-fee reform was a success depends on the perspective taken. The central view is that the reform was very successful in expanding its fiscal authority down to the township and village level (Lin and Liu 2007, Tao and Qin 2007, Liu *et al.* 2012). Unlike the 1994 reform, the central government has taken greater responsibility for funding expenditures through transfers and subsidies (X. Chen 2009, Oi 20102, L.C. Li 2012). From the villager perspective, the reform increased household income after the elimination of most fees and the agricultural tax in 2006. For the township government, official duties have shifted again from revenue collection to service provision. The township governments have also adjusted and found new sources of revenue, especially in the area of land leases (Takeuchi 2010). Yet the township government has also lost a level of autonomy (Smith 2009).

The ongoing debate is over the fate and function of the township government after 2006. One argument suggests the township government should be eliminated as a level of government and serve as a county office (He 2000, Xu 2004). Some scholars believe this has already happened (Smith 2009). An alternative is that the township government should be allowed greater autonomy and that direct elections for township leaders need to be introduced (Lin and Liu 2007, T.B. Bernstein and Lü 2003). Each has distinct implications for local governance. The tax-for-fee reform would weaken the authority and ability of the township government to extract local resources, but maintain and even strengthen county authority in the township. Direct elections would make township leaders more accountable to villagers, but it would also strengthen the autonomy of the township governments.

One of the ways township governments have responded to the 2002 tax reform is to search for alternative sources of revenue. The most abundant resource available (at this time) for this is land. Takeuchi (2013) and H. Li and Kung (2012) suggest that township and county officials are turning to land lease revenues to make up for the loss in direct fees and taxes. Indeed, the number of reported land disputes has increased dramatically since 2006. Sun Liping, a professor at Tsinghua University, suggested that over 180,000 disturbances occurred in 2010 and the majority of these incidents were over land disputes (Orlik 2011). In the early stages of the 2002 tax reform, Ho and Lin (2003) also found that township leaders were tapping into land resources as a way to make up for the tax and fee shortfalls. In his case studies, Takeuchi (2013) finds that township governments are using the land lease revenues to maintain overstaffed offices, even under pressure to reduce personnel. However, the increase in land lease sales to real estate developers is also due to a systematic change in promotion measures from economic (industrial) development to urbanization (H. Li and Kung 2012). Indeed, these new measures have not only changed the behavior of county and township officials, but also the landscape. In 2000, about 37 percent of the population was considered urban. In January 2012, the central government reported that the urban population made up 51 percent of the population. According to these figures, China is now a majority urban nation.

Despite rapid urbanization, township governments still serve about half the population. Yet in the wake of the 2002 tax reform, the function of the township government remains in flux. He Kaiyin (2000), who introduced the tax-for-fee reform innovation in the late 1990s and 2000, suggests eliminating the township altogether. He advocates a new three-tier government system where the county is the lowest level of government. Xu Yong (2004), the director for the Center for Chinese Rural Studies at Central Normal University, contends that the township should become a branch of the county government. The assumption is that the elimination of the overstaffed township offices should reduce local budgets and streamline services. Indeed, township offices are split between local (*kuaikuai*) and county (*tiaotiao*) responsibilities (Mertha 2005). After townships lost the authority to raise revenues, the township offices had little to do and many services were covered by the county offices at the township level.

Smith (2010) suggests that a number of townships have already become county offices by default. The authority to collect revenues defines a local government, and without tax and fiscal autonomy the township becomes a hollow administration unit. He finds that the township offices have fallen under the control of the county bureaus. While this may coordinate service provision and policy implementation between the township and county governments, it has also placed greater responsibility for service provision onto the township officials. Ironically, Smith argues that the township governments are now overwhelmed and understaffed. Counties have a greater interest in wealthier townships with local industry and thus industrial towns have less autonomy than poorer townships. However, greater autonomy for townships also requires improved public oversight.

Direct elections for township leaders were viewed as one solution for the accountability problem as well as reducing villager burdens (So 1997, T.B. Bernstein and Lü 2003, Yep 2004, Lin and Liu 2007). Indeed, the central leadership provided political openings for both tax and local election reforms. Zhu Rongji, at the 9th National People's Congress in 1998, announced the importance of tax and fee reform and encouraged local experiments (Yep 2004). County leading cadres who introduced direct elections for township government heads were responding to central leadership cues such as Jiang Zemin's pledge to expand grassroots democracy at the 15th Party Congress, and Zhu Rongji's hint at supporting elections beyond the village level (L. Li 2002, L.C. Li 2007). County innovators also viewed township elections as a way to resolve the problem of unruly township leaders. Thus, in the early 2000s, there were three possible directions for reducing villager burdens: tax and fee reform, direct elections for township leaders, or both. Direct elections for township government heads look good from the villager perspective because they should provide more accountable leaders, but for the county government they would be a loss of authority to select key personnel at the township level. Still, if the township governments are becoming county branch offices by default, then township elections may become a moot point.

Although the national government increased central transfers and subsidies after 2002, one of the remaining problems for county and township governments is rising deficits (Yep 2004). The gap between revenues and expenditures was a problem for county and township governments throughout the 1990s, but the problem became more acute after 2002 and especially after 2006. In the early 2000s, annual township deficits were in the millions of *yuan*, a significant problem for poorer townships with annual revenues under one million *yuan* (Zhao 2006). The sources of township debts included infrastructure investments, administrative expenditures and unpaid salaries (ibid.). At the same time, many counties were incurring debts due to the growing deficit between annual revenues and expenditures. After 2002, county and township governments were unable to use extra-budgetary funds to make up the difference, so they had to rely more on higher-level transfers and subsidies (X. Chen 2009). Of course, deficit levels vary across townships and counties. Table 2 shows annual revenues and expenditures for three counties in Shaanxi province. Hu County had an agricultural, wheat-based economy, especially in the 1990s. From 1996–2001, revenues covered about 70 percent of county expenditures. By 2006, total revenues covered only 25 percent of expenditures. Jia County is a poor, remote county in the mountainous region of northern Shaanxi. In 1996, revenues covered 25 percent of expenditures, but by 2006 the proportion had dropped to 3 percent. Shenmu County has natural resources, especially coal, and the 2002 tax reform has had little influence on county deficit there.

Thus, the 2002 tax-for-fee reform and the elimination of the agricultural tax in 2006 have reduced villager burdens as well as the authority of the township government. Before 2002, rising deficits were managed at the township and county levels, but after 2002 townships and counties are more reliant on central transfers and subsidies.

Conclusion

The introduction of financial reforms in 1994 and 2002 is a story of centralization and national policies aimed to capture local revenues and increase local service provision in rural China. The central government began in 1994 with macro policies that reshaped the fiscal relationship between central and provincial governments. However, the 1994 tax system allowed for greater local autonomy at the county and township levels to collect taxes and fees. The 2002 tax reforms continued the centralization process by eliminating not only local fees, but also local fiscal autonomy. Success of these tax reforms depends in part on whether the central or local perspective is taken.

A number of questions remain regarding deficits, public service provision and the function of township governments. For example, how long can local governments depend on land leases to help reduce deficits? Will the elimination of the township government in favor of county-level offices improve service provision for the rural population? Will elections for more autonomous township leaders enhance service provision and local accountability in the countryside? These are only some of the questions that will contribute to the rich literature on rural taxation and local governance.

References

Aubert, C. and X. Li. 2002. 'Peasant burden': taxes and levies imposed on Chinese farmers. *In*: Wilfred Legg, *et al*., eds. *Agricultural policies in China after WTO accession*. Paris: OECD, pp. 160–76.

Bernhardt, K. 1997. *Rents, taxes, and peasant resistance: the Lower Yangzi Region, 1840–1950*. Stanford, CA: Stanford University Press.

Bernstein, T.B. and X. Lü. 2003. *Taxation without representation in contemporary rural China*. Cambridge and New York: Cambridge University Press.

Bernstein, T.P. 2004. Unrest in rural China: a 2003 assessment [online]. Center for the Study of Democracy, UC Irvine, Working Paper. Available from: http://repositories.cdlib.org/csd/04-13 [Accessed 12 July 2012].

Chen, A. 2008. The 1994 tax reform and its impact on China's rural fiscal structure. *Modern China*, 34(3), 303–43.

Chen, G. and C. Wu. 2004. *Zhongguo Nongcun Diaocha* [China Rural Survey]. Wuhan, China: Wuhan Publishing House.

Chen, X. 2004. *Zhongguo Nongye Shuizhi Xinzheng Rang Jiu Yi Nongmin Jin Kaiyan* [New agricultural tax policy makes 900 million peasants smile]. *China Rural Science & Technology*, 4, 20–3.

Chen, X. 2009. Review of China's agricultural and rural development: policy changes and current issues. *China Agricultural Economic Review*, 1(2), 121–35.

Edin, M. 2003. State capacity and local agent control in China: CCP cadre management from a township perspective. *The China Quarterly*, 173, 35–52.

Ge, R. 2005. Zhongguo Nongye Jinru 'Wu Shui Shidai' [Chinese countryside has entered 'age of no taxes']. *Outlook Weekly*, 5, 14–6.

He, K. 2001. What way should go the reform of the rural taxes *and* fees? *World of Surveys and Research* [*Diaoyan Shijie*], 8, 14–8.

He, K. and L. Sun. 2000. *Zhongguo nongcun shuifei gaige chutan [The early exploration of the rural tax-for-fee reform]*. Beijing: Zi Gong chubanshe [Zi Gong Publishing].

Herschler, S.B. 1995. The 1994 tax reforms: the center strikes back. *China Economic Review*, 6(2), 239–45.

Ho, S.P. and G.S. Lin. 2003. Emerging land markets in rural and urban China: policies and practices. *The China Quarterly*, 175, 681–707.

Huang, B. and C. Kang. 2012. Are intergovernmental transfers in China equalizing? *China Economic Review*, 23(3), 534–51.

Huang, C. 2005. Premier vows to spread the wealth: agricultural taxes will be abolished next year, two years ahead of schedule, promises Wen Jiabao. *South China Morning Post*, 6 March, 6.

Kennedy, J.J. 2007a. From the fee-for-tax reform to the abolition of agricultural taxes: the impact on township governments in northwest China. *The China Quarterly*, 189,

Kennedy, J.J. 2007b. The implementation of village elections and tax-for-fee reform in rural northwest China. *In*: Elizabeth J. Perry and Merle Goldman, eds. *Grassroots political reform in contemporary China*. Cambridge, MA: Harvard University Press.

Kung, J. 2000. Common rights and land reallocations in rural China: evidence from a village survey. *World Development*, 28(4), 701–19.

Lee, P. 2000. Into the trap of strengthening state capacity: China's tax-assignment reform. *China Quarterly*, 164, 1007–24.

Li, H. and J. Kung. 2012. *Fiscal incentives and policy choice of local governments: evidence from China*. Working Paper, Hong Kong University of Science and Technology.

Li, L. 2002. The politics of introducing direct township elections in China. *The China Quarterly*, 171, 704–23.

Li, L. C. 2007. Working for the peasants? Strategic interactions and unintended consequences in the Chinese rural tax reform. *The China Journal*, 57, 89–106.

Li, L.C. 2012. *Rural tax reform in China: policy processes and institutional change*. Routledge.

Li, X. 2003. Rethinking the peasant burden: evidence from a Chinese village. *Journal of Peasant Studies*, 30, 46–74.

Lin, J.Y., *et al.* 2002. *Urban and rural household taxation in China: measurement and stylized facts*. Working Paper, China Center for Economic Research, Peking University.

Lü, X. 1997. The politics of peasant burden in reform China. *The Journal of Peasant Studies*, 25, 113–38.

Luehrmann, L.M. 2003. Facing citizen complaints in China, 1951–1996. *Asian Survey*, 43, 845–66.

Luo, R., *et al.* 2007. Elections, fiscal reform and public goods provision in rural China. *Journal of Comparative Economics*, 35, 583–611.

Meng, X. and L. Zhang. 2011. Democratic participation, fiscal reform and local governance: empirical evidence on Chinese villages. *China Economic Review*, 22(1), 88–97.

Mertha, A.C. 2005. China's 'soft' centralization: shifting *Tiao/Kuai* authority relations since 1998. *The China Quarterly*, 184, 792–810.

Montinola, G., Y. Qian and B. Weingast. 1995. Federalism, Chinese style: the political basis for economic success in China. *World Politics*, 48(1), 50–81.

O'Brien, K. and L. Li. 1999. Selective policy implementation in rural China. *Comparative Politics*, 31(2), 167–86.

O'Brien, K. and L. Li. 2006. *Rightful resistance in rural China*. New York: Cambridge University Press.

Oi, J. 1999. *Rural China takes off: institutional foundations of economic reform*. University of California Press.)

Oi, J., *et al.* 2012. Shifting fiscal control to limit cadre power in China's townships and villages. *The China Quarterly*, 211, 649–75.

Oksenberg, M. and J. Tong. 1991. The evolution of central-provincial relations in China, 1953–1983: the formal system. *China Quarterly*, 125, 1–32.

Ong, L. 2006. The political economy of township government debt, township enterprises, rural financial institutions in China. *China Quarterly*, 186, 377–400.

Orlik, T. 2011. Unrest grows as economy booms [online]. *Wall Street Journal*, 26 September. Available from: http://online.wsj.com/article/SB10001424053111903703604576587070600504108.html

People's Daily. 2005. Ba Nongcun Shuifei Gaige Quanmian Yinxiang Shenru [An all-round deepening of the rural tax reforms]. *People's Daily*, 8 June, 1.

Qu, C. 2005. Mianzheng Nongyeshui, Xianzhen Zenme Ban? [Agricultural taxes rescinded, what will township governments do?] *People's Daily*, 24 March, 6.

Rozelle, S. 1994. Decision-making in China's rural economy: the linkages between village leaders and farm households. *The China Quarterly*, 137, 99–124.

Rozelle, S. and G. Li. 1998. Village leaders and land rights formation in China. *American Economic Review*, 88(2), 433–8.

Shue, V. 1988. *The reach of the state: sketches of the Chinese body politic*. Stanford, CA: Stanford University Press.

So, A. 2007. Peasant conflict and the local predatory state in the Chinese countryside. *Journal of Peasant Studies*, 34(3), 560–81.

Takeuchi, H. 2013. Survival strategies of township governments in rural China: from predatory taxation to land trade. *Journal of Contemporary China*, 22(83), forthcoming.

Tang, Y.Y. 2005. When peasants sue en masse: large-scale collective ALL suits in rural China. *China: An International* Journal, 3, 24–49.

Tanner, M.S. 2005. Chinese government responses to rising social unrest. Testimony presented to the US-China Economic and Security Review Commission on April 14, 2005 [online]. Available from: http://www.rand.org/pubs/testimonies/2005/RAND_CT240.pdf [Accessed 12 July 2012].

Tao, R. and M. Liu. 2005. Urban and rural household taxation in China: measurement, comparison and policy implications. *Journal of the Asia Pacific Economy*, 10.

Tao, R. and P. Qin. 2007. How has rural tax reform affected farmers and local governance in China? *China and World Economy*, 15(3), 19–32.

Tian, J.Q. 2009. Reorganizing rural public finance: reforms and consequences. *Journal of Current Chinese Affairs*, April, 145–71.

Tsai, L.L. 2002. Cadres, temple and lineage institutions, and governance in rural China. *The China Journal*, 48, 1–27.

Unger, J. 2002. *The transformation of rural China*. Armonk, M.E: Sharpe.

Wang, S. 1997. China's 1994 fiscal reform: an initial assessment. *Asian Survey*, 37(9), 801–17.

Wedeman, A. 2000. Budgets, extra-budgets, and small treasuries: illegal monies and local autonomy in China. *Journal of Contemporary China*, 9, 489–511.

Wong, C.P. 1995. Fiscal reform in 1994. *In*: Kuan Hsin-chi, ed. *China Review 1995*. Hong Kong: Chinese University Press, p. 20.

Wong, C.P. 1997. Rural public finance. *In*: Christine P.W. Wong, ed. *Financing local government in the People's Republic of China*. Hong Kong and New York: Oxford University Press, pp. 167–212.

Wong, C.P. 2000. Central-local relations revisited: the 1994 tax sharing reform and public expenditure management in China. World Bank Office in China, April.

Wong, C.P., C. Heady and W.T. Woo. 1995. *Fiscal management and economic reform in the People's Republic of China*. New York: Oxford University Press.

Xu, Y. 2004. Xianzheng, xiangpai, cunzhi: xiangcun zhilide jiegouxing zhuanhuan [County government, township branch and village management: the structural transformation of rural management]. *Zhongguo Shijie [China Century]*, 771.

Yep, R. 2004. Can 'tax-for-fee' reform reduce rural tension in China? The process, progress and limitations. *The China Quarterly*, 177, 42–70.

Yep, R. 2009. Land conflicts, rural finance and capacity of the Chinese State. *Public Administration and Development*, 29(1), 69–78.

Zhang, X., *et al.* 2004. Local governance and public goods provision in rural China. *Journal of Public Economics*, 88, 2857–71.

Zhao, S. 2006a. Hard pressed township finances. *Chinese Sociology and Anthropology*, 39(2), 45–54.

Zhao, S. 2006b. The debt chaos of township governments. *Chinese Sociology and Anthropology*, 39(2), 36–44.

Zheng, Y. 2000. Xiangzhen zhengquan zuzhi yu zhidu bianqiande tedian, jigou pengzhangde chengyin jiduici tantao [Exploring the causes of overextended staffing in township governments]. *Zhongguo Nongcun Guancha [China Rural Survey]*, 4, 31–42.

Zhong, Yang. 2004. *Local government and politics in China: challenges from below*. Armok, NY: M. E. Sharpe.

Zhou, F. 2006. Shuifei gaige dui guojia yu nongmin guanxi zhi yingxiang [The impact of tax-fee reform on state-peasant relations]. *Shehuixue yanjiu [Sociology Studies]*, 3, 1–38.

John James Kennedy received his PhD at the University of California, Davis in 2002. He is an associate professor in the department of political science at the University of Kansas (KU), and director of the Center for Global and international Studies at KU. He is also affiliated with the Northwest Socio-economic Development Research Center (NSDRC) at Northwest University, Xian. His research is on rural social and economic development and topics include local elections, tax and fee reform, rural education and the cadre management system. He is currently examining local government evaluation of public services and specific health care issues. He has published articles in *The China Quarterly*, the *Journal of Contemporary China*, *Asian Survey*, the *Journal of Chinese Political Science*, *Asian Politics and Policy*, *Whitehead Journal of Diplomacy and International Relations* and *Political Studies*.

Measurement, promotions and patterns of behavior in Chinese local government

Graeme Smith

In this paper, the everyday politics of rural Chinese officials will be analyzed, largely drawing on my own fieldwork in central China, and studies by Western and Chinese scholars who have spent extensive periods of time working alongside rural officials. The dual identities of Chinese officials as market players and the implementers of government policy are outlined. The paper will examine the formal, informal and semi-formal dynamics of political power in rural China, and propose a matrix of incentives and constraints that can be used to analyze the behavior of rural officials across different regions of China.

Introduction

A striking aspect of the recent literature on rural governance in China is that it has begun to open the black box of how the local state actually works. This paper will attempt to distill this new research. However, engagement with the internal workings of the local Chinese state is still limited. In part, this arises from academic fashions (such as 'civil society') that dominate the discourse on China for periods of time. In part, the lacuna arises from the success of the Chinese regime in defining the sorts of 'problems' local scholars and media outlets are given license to pursue. In the largest part, however, it arises from the difficulty in accessing sites to undertake long-term fieldwork in rural areas.

Local governments, particularly cash-strapped ones in central and western China, are understandably reluctant to make themselves the subject of critical examinations that offer little in the way of material reward. Nonetheless, scholars have gained access by working for county (Feng 2010) or township (Gu 2006, Wu 2007) governments, by studying their hometown (Lam 2000), gaining access through a powerful CCP (Chinese Communist Party) affiliated organ, such as the Party School (Göbel 2011) or the State Council (Zhao 2007), or working with a non-governmental organization (NGO) with close local government connections (Hillman 2010, Smith 2010). For both Western and local scholars, the time required to develop a level of trust and cooperation through extended time and

The author would like to acknowledge the support and understanding of Kevin O'Brien, Ye Jingzhong, Juan Liu, Lu Pan and the other conference organisers. I am also indebted to Ryan Manuel, Hans Hendrischke and three anonymous reviewers for greatly improving the manuscript. Finally, I would like to thank the resilient citizens of Benghai County, who have weathered many storms. All errors are of my own making.

repeated interactions (Axelrod 1984), is nearly impossible to find within the pressures of an output-focused academic environment. Some quantitative social scientists dismiss such studies as lacking theoretical rigor, and being unrepresentative of China as a whole, but mixed methods and interdisciplinary emphasis have become accepted as the norm for studies of the local state. It is only over an extended period of time that one can gain useful access to the variety of actors needed to provide an accurate picture of how the local state works, be they leading officials or village cadres, local entrepreneurs or the managers of large agribusinesses, teachers or farmers.

Examining a broad set of qualitative and quantitative studies, this contribution will explore the behaviors[1] and informal norms of rural officials[2] in a variety of geographical and economic settings. I will attempt to eschew normative evaluations of cadre behavior, and rather focus on the more interesting question of how they use formal and informal institutions[3] to 'get things done' (Kennedy 2010) in an increasingly pluralistic society (Mertha 2008), and an economy that exhibits remarkable heterogeneity (Goodman 1997, World Bank 2002, Y.S. Huang 2008). When referring to 'local' government, for the purpose of this review it refers to county and township government. County and township governments are the two lowest rungs in the formal hierarchy of the Chinese state. A case could be made for including the prefecture (or prefecture-level city), a remarkably under-researched aspect of the Chinese state (Chien 2010), which is part of the official Chinese definition of 'local government'. However, aside from the dearth of literature, the prefecture is increasingly bypassed as provincial governments deal directly with counties, and it is likely that the prefecture will be abolished at some stage during the process of administrative reform. The divided loyalties and motivations of village officials are the subject of a rich literature, but are beyond the scope of this review (Oi 1989, Rozelle 1994, Oi and Rozelle 2000, A. Chen 2007).

John James Kennedy's contribution in this collection covers the debate on the evolving fiscal relations between township and county governments (Kennedy 2013) and, in keeping with his finding that townships largely act as administrative outposts or field offices of county governments, I refer to township and county collectively as 'local government'. How do local governments work in rural China? What motives, constraints and rationales shape the decisions made by officials? There are many different factors affecting rural officials' behavior across China. I propose that officials will be motivated to wholeheartedly support an initiative when it substantially meets the criteria below:

(1) the initiative is given weight and is measurable in the annual assessment system, enhancing the promotion prospects of leading officials;
(2) the initiative raises revenue for local government, either through levying fines, taxes or service fees, or by opening up revenue sources from higher levels;

---

[1]This analysis will steer clear of essentialist cultural explanations of cadre behavior, such as 'Confucianism' or the unique characteristics of *guanxi* networks (Goodman 2007).

[2]Following Zhou's (2010, 55) example, I will refer to 'local officials' and 'local governments' interchangeably, as the behaviors of officials are 'institutionalized practice on the basis of government organization'.

[3]In this contribution, I adopt Nee and Opper's (2012) definition of institutions as 'systems of interrelated informal and formal elements – customs, conventions, norms, beliefs, and rules – governing social relationships within which actors pursue and fix the limits of legitimate interests. They are self-reproducing social structures that provide a conduit for collective action by enabling, motivating, and guiding the interests of actors and enforcing principal-agent relationship ... fundamentally transformative does not simply involve politicians remaking the formal rules, but requires the realignment of interests, norms and power'.

(3) the initiative benefits individual cadres and their partners in the 'shadow state' financially;

(4) the initiative can be realized with the resources of existing power structures with institutionalized mobilization.

A brief survey of the literature on cadre motivations and norms suggests that much apparent disagreement comes down to what part of the proverbial Chinese elephant is being studied, and through which particular lens. Much of the debate as to whether the state is predatory or developmental, corporatist or mafia-like is, to a large degree, a function of whether the local state dominates the local economy or not. Or, viewed from another angle, the way cadres behave depends on whether they face an environment where local or foreign firms, either applying informal social pressures or lobbying through associations, have sufficient clout that officials must accommodate their needs. As Jonathan Unger (2002) has noted, much of the variance in rural governance is rooted in geography, which comes with an historical and economic legacy.

Unger (2002) divides rural China into four rough categories: areas where local government has taken the lead in establishing factories, counties where investment is dominated by foreign corporations (such as the Pearl River Delta), the hinterlands where very little industrialization has occurred, and areas such as Wenzhou where private industry dominates. These categorizations are still useful. Many disagreements about the motivations of local officials, led by Jean Oi and Victor Nee, came down to differences in sources (officials or entrepreneurs) and the location of their field sites. Within most rural counties in China, all four of the conditions described by Unger can be found and, with the passage of time, there are now very few rural counties untouched by industrialization. Hence, I argue that we are now able to make *some* generalizations about the motivations for local officials, while keeping the vast disparities in the fiscal means of the local state in mind. The following section gives a brief outline of the types of officials and the structure of government in rural China. Before discussing their behavior, the following section gives a brief outline of the types of officials and the structure of government in rural China.

Staff and the rural hierarchy

A full description of the formal structure of rural government and the promotion system for rural officials is beyond the scope of this contribution and has been well covered elsewhere (Brodsgaard 2002, Edin 2003, Zhong 2003, Landry 2008, Chung and Lam 2010). In short, the party trumps the government, and everyone wants to live in town. Table 1 maps the basic incentive structures for rural cadres.

Before outlining what I believe are the four salient influences governing the behavior of rural cadres, some important rules of the game will be outlined.

Table 1. Hierarchy of preferences for rural officials.

Preference	Least			Most
Government	Village	Township	District	County
Location	Remote village	Remote township	Proximate township	County seat
Employment	Casual/contract	Enterprise	Service unit	Gongwuyuan
Gongwuyuan status	Deputy section head	Section head	Deputy division head	Division head

Localism

The identity of leading cadres, and ordinary government staff, is tightly bound up with their county. Most of them will spend their entire career working inside its boundaries (Zhou 2010, 72). County-level cadres take up postings as township party secretaries and mayors, and tend to live in the county seat rather than the township during their tenure. Even the most powerful official in the county, the county party secretary, will typically serve a full term as county mayor, and will have spent time early in their career working inside the county at the township or village level. Compared to the townships, the county level of government has a relatively complete set of Party and government offices that mirror those of the central and provincial governments (Lam 2010), and the post-revolutionary trend has been one of continuous expansion in the complexity and power of county governments (Xu 2003, 177–94).

Hollow townships

While township governments are often characterized as increasingly 'hollowed out', acting as agencies of the county government, and often preoccupied with the basic task of paying their employees, it would be a mistake to characterize this level of government as ever enjoying a high degree of autonomy. Jean Oi (1999, 29), conducting fieldwork in 1996, when township governments were arguably at their peak of fiscal independence, found that some townships were still not able to share revenue, even though provision for township fiscal systems had been made by the State Council in 1985. As Lam (2000, 137–8, see also L.C. Li 2007, 103–5) noted of the situation in Hainan in the 1990s, the township

> is still a very weak level of government vis-à-vis the county level. The recent attempts to make the township level a separate level in fiscal management have not strengthened it much, in part because the fiscal arrangements were designed to the disadvantage of the township level, and in part because county level has retained other important resources.

The fiscal resources of rural townships vary greatly across China. A survey of 20 townships across China found that the 'miscellaneous' category of revenue in the richest township (in Zhejiang) surveyed was four times the total revenue of the poorest (in Ningxia) (Oi and Zhao 2007, 80).

Township government have also been weakened by a process that Mertha (2005, 793–4) described as 'soft centralization', whereby bureaus were shifted from 'area management' (*kuai-kuai guanli*) to 'line management' (*tiao-tiao guanli*). He found that the first bureaus to the provincial level were those organizations associated with 'administrative regulation, financial regulation and commodities management ... because of their vital role in China's economic development'. From the perspective of township leaders, the equation is simpler. Bureaus with 'money and power' (*you qian you quan*) are claimed by higher levels of government, while agencies with neither money nor power, known as 'clear water agencies' (*qing shui yamen*), are left under the area management of township leaders.

Leading cadres

The core group of 'leading cadres' in local government is the County Party Standing Committee, which performs a similar function to the Politburo at the national level. Although they control government functions, this small group can be considered as politicians, rather than government officials. Their motivations are different to government staff,

114

whose careers are generally confined to within the borders of the county. Often these leading cadres have been 'sent down' or parachuted in from higher levels – the prefecture, province or even the central government – in the hope that a period being 'tempered' in local government will lead to more rapid promotion. From the perspective of the central state, it is hoped that cadre exchange will lead to a less corrupt and more disciplined Party organization (Yang 2004).

The three-year terms of township and county Party secretaries were recently extended to five years, in the hope that it would lead to a less short-term approach to governance. However, many leading cadres hope to be promoted out before the end of their term, as age limits for promotion to different levels are strictly enforced. Once you exceed the age limit, you are stuck at that level. The Benghai County Party secretary was described by a number of interviewees as being 'depressed' during his final two years in office, because he hadn't managed to secure a post in the provincial government before the end of his tenure, and would be retiring instead. His spirits were even lower by mid-2013, when investigations into his investments in Zhejiang commenced. With his patron leaving the province, the retired cadre made an easy target for the anti-corruption drive being pursued by President Xi Jinping's new regime.

Gongwuyuan

Throughout China, attaining the position of an 'administrative cadre' or *gongwuyuan*, is highly prized, even in coastal regions where private industry flourishes. In this paper, I leave the term *gongwuyuan* untranslated. 'Administrative cadre' is misleading. Many administrative tasks are performed by 'service unit' (*shiye danwei*) staff. The commonly used 'civil servant' is inadequate, because it implies a Weberian concept that covers all staff working at all levels of the government bureaucracy. It also misses the dual allegiance of *gongwuyuan* to Party and government organs. In the central and western regions, where average wages are low, and opportunities in the private sector are limited, the prospect of a pension, comprehensive medical insurance and security of tenure are appealing. China is estimated to employ 8 million *gongwuyuan* and 30 million 'service cadres' (*shiye ganbu*) (Ren 2010). While it is difficult to bring *gongwuyuan* cadres directly onto the books, it is relatively easy to increase the number of service staff when there is an influx of funding to an area that is deemed to be a national priority, such as the grain for green program (Guo 2010). After several years working in local government, or as a 'sent-down' village cadre, they have the prospect of transfer to *gongwuyuan* status, often through the purchase of these posts (*mai guan* 买官) (Zhong 2003, 115–9, Sun 2012). These mutual debts form the basis of trust and loyalty that bind subordinates to their patrons. Those who become *gongwuyuan* via the examination system at the township or county level may find themselves initially stigmatized as 'outsiders', even if they are returning to their home county.[4]

The lost staff

The prospect of status change goes a long way to explaining how county and township governments continue to function. Given the compressed nature of pay grades within both the *gongwuyuan* system and the public service units (Burns and Wang 2010, 70–4), the failure of the evaluation system to function effectively in terms of serving the public (Edin 2003,

---

[4]For university graduates, working as a township official is generally a last resort (Bai 2006).

Zhao 2007, 64–73, Smith 2009, 49–52), leads cadres to rely heavily on the promise of a future promotion to motivate their staff. These workers aspiring to a change in status do not shirk, though the question of whether they are serving the public, serving their patron or serving their faction remains open. Those cadres with no prospect of promotion, such as deputy township leaders or mid-level county officials over the age of 40, are a different matter. In their own words, these cadres are 'Too old for promotion, too young to stop working, but just right for mahjong and drinking' (Zhao 2007, 71). As for ordinary township government staff, Zhao (2007, 72) sees them as 'loose and slack, dreaming neither of becoming officials nor of moving to the county seat', viewing their work as a 'casual job to be done in their spare time'.

The limits of generalizations

A 'monolithic' political system contains exceptions. Obtaining a remarkable breadth and depth of access, from the Party Secretary to ordinary villagers, Hillman's (2005, 2010) ethnographic study in southwest China presents many findings that challenge the prevailing orthodoxy about the nature of the local state and cadre motivations in rural China. The most startling counterfactual is the tendency of the county and township heads to trump county and township party secretaries (*cf.* Zhong 2003, 78–9, Landry 2008, 116–52, Thogersen 2008, 418). Since the tacit abandonment of the separation of Party and government functions in the 1990s, local party functionaries in most areas of China have increased their control over government agencies and personnel. However, in Hillman's study, party secretaries, being sent in from outside the area under the rule of avoidance,[5] lack the political connections to play the game of rural politics, leaving local county mayors and heads in charge of personnel appointments and the distribution of state largess. Barriers of ethnicity and language increase their isolation. Hillman (2005, 93) writes of one Han Chinese township Party secretary trying to make a point at a meeting, while being ignored by his subordinates who continue their discussions in the local Naxi language. If an incoming party secretary were stubborn, 'local power brokers' would adopt countermeasures, such as fomenting social instability, to bring them into line. The Party has a firmer hold on the county level, with county party 'deputies' able to use their veto power to influence Party and government work (Hillman 2010, 9–11), mirroring their influence in central China (Smith 2009, 34–7).

Informal networks

The practice of government jobs being allocated on the basis of politics and patronage, rather than merit, remains consistent in Hillman's study. However, in a region where potential appointees have limited financial resources, the main motivation for appointments is to strengthen one's faction, rather than immediate personal enrichment for the gatekeepers of these government appointments. Hillman (2005, 133) explains,

> Factions operate like executive placement agencies … Getting a member appointed as a leading cadre provides a faction with direct control over public resources and the power to make further

---

[5]The rule of avoidance (*huibi zhidu* 回避制度) is a long-standing practice in China of preventing leading officials, particularly the party secretary, from serving in their home townships, counties or provinces.

appointments. Appointing ordinary cadres is a way of building legitimacy in one's place of origin, and can be used as currency in the exchange of favors. Particularly in poor areas like Laxiang County any state job is like gold. It provides a secure income and access to resources that are beyond the means of ordinary rural citizens.

One factor that remains consistent in career success for officials throughout China, however, is the necessity of having a powerful patron (or patrons). Hillman (2005, 134) cites a county official:

> You can make a lot of mistakes in your career and still get promoted, and you can do everything really well and spend your career languishing at the bottom. It all depends on who your friends are.

As officials in Benghai County would often reflect, success is matter of who you have working for you 'backstage' (*hou tai* 后台). There is, however, also the formal system of assessment to deal with.

The cadre responsibility system and the grapevine

The most commonly cited influences on the behavior of rural officials are the performance targets (*kaohe zhibiao* 考核指标) in the assessments and inspections which form the basis of the cadre responsibility system (*ganbu zeren zhidu* 干部责任制度). The variation in fiscal capacity across China is not reflected in variation in the responsibility system. Surveys across a variety of townships find that very similar items are given priority, regardless of economic and social conditions (Zhao 2007, 64–73, Ong 2012, 76–86). Highest priority items are subject to the 'one-strike veto' (*yi piao foujue* 一票否决), whereby a failure to meet the target annuls all other achievements.[6] Maintaining social stability[7] and meeting family planning targets are nearly universally covered by this veto, which may also cover items such as fiscal income, workplace safety, attracting investment and, recently, environmental protection (A. Wang 2013). The range of items has expanded in recent years, perhaps diminishing the efficacy of the veto. Following this are 'hard targets' (*ying zhibiao* 硬指标), typically quantifiable achievements related to tax revenue and attracting investment, and 'soft targets' or 'ordinary targets' (*yiban zhibiao* 一般指标), which are typically more difficult to quantify, such as building up Party work style, or reducing peasant burdens. The nomenclature of the targets is a signal to rural cadres as to how to prioritize their work, and the relationship between the point-scoring system and the prospects for promotion of the leading cadres (most directly the township and county Party secretaries; to a lesser extent to county and township mayors) is made clear. Despite noises made in the direction of greater involvement by societal actors in assessing leading cadres, the decisive power of assessment lies in the hands of party officials one level above the cadre in question.

---

[6]This is not absolute. As Zhao (2007, 24–5) has pointed out, inspection teams that discover violations of a veto item, such as an out-of-quota birth, can be 'squared off' by local cadres. It does, however, influence the manner of grassroots governance. Wu (2007, 37) recalls a speech given to township government cadres by the head of the district work team for reducing peasant burdens: 'Comrades, over the past few days, the township government compound has been deserted, which tells me you've been diligent, stationed in the villages. The big news in the world is the Iraq War, but our greatest concern is the upcoming inspection. We're about to take Baghdad: the decisive battle is ahead! This inspection has four items that are subject to the one-strike veto. If one item doesn't meet the requirements, we've missed out completely, and all your hard work has been in vain'.
[7]For an excellent overview of social unrest in China, see Göbel and Ong (2012).

Control is less straightforward than lines of authority might indicate. Zhou's (2010, Zhou and Lian 2011) theory of local government behavior adapts Tirole's principal-supervisor-agent model to explain the institutional logic of governance in China, where in effect the agent, rather than the principal, can shape the outcome of higher-level policies. The tail often wags the dog. This effect arises from tension between insistence on the primacy of the centralized state, coupled with its inability to effectively govern a culturally and geographically diverse nation, and the imperative for agents at all level to 'get things done' with the limited resources that are available to them. Where the distance between the principal (such as the provincial government), the supervisor (county family planning centre) and the agent (township family planning office) is greatly increased, this tension is greater, and the scope for subverting central government policy increases. Collusion can occur at the same level, or between levels. Such behavior is not necessarily nefarious. While 'hidden' from official reviews and inspections, it is often practiced in the open and organized on the basis of formal bureaucratic authorities.

Pierre Landry's (2008) remarkable quantitative study of four counties in wealthy Jiangsu Province suggests that the appointment and promotion system is effective in instilling the Party's elite norms of how the institutions involved in the system should work, and that the local Party elite adapt rapidly to changes in the formal system, leading him to question whether 'informal politics reign supreme in authoritarian regimes'. Landry (2008) goes a considerable way to explaining a paradox explored by Whiting (2004), that a flawed appointment system has contributed to the resilience of the CCP regime. However, politics in rural China is not a zero-sum game, and much of it is played in the 'semi-formal' realm. Indeed, Landry's (2008, 145) ranking of county institutions in terms of their importance for county-level cadres' promotions[8] matches the level of culpability and size of payment received when cases of 'selling positions' make it to court (Smith 2009, 40–2). The man in charge of Jiangsu's appointment system, the head of the Provincial Party Organization Department, Xu Guojian, was arrested on corruption charges in 2004, and received a suspended death sentence in 2006 (Xinhua 2004, 2006). Landry (2008, 203–6) is on safe ground for much of China, however, when he asserts that access to political power

> Remains tightly correlated not only with CCP membership but, far more importantly, with specific professional experience in the Party institutions that effectively control the decision-making process throughout China's localities … In lieu of crude, costly, and unpredictable controls by diktat or reliance on 'interlocking directorates', the CCP has devised simple, easily monitored criteria that favor the promotion of cadres whose sequence of past professional experiences in CCP organs reinforce its control over local governments. Despite decentralization, the Party goes on.

The question of whether the local state in China is able to attract the 'best talents' is the subject of considerable debate within the Chinese literature. In central and western China, local government is seen as a desirable form of employment, and recruitment is largely based on nepotism, a widely acknowledged fact that, to my knowledge, only one scholar has been indiscreet enough to verify empirically by examining recruitment to different

---

[8]The rankings, in order of importance, are (1) County Party secretary, (2) County Organization Department, (3) County head, (4) County Party committee, (5) County Discipline Inspection Committee, (6) Municipal Organization Department, (7) Municipal Party committee, (8) County People's Congress, (9) Municipal Discipline Inspection Committee, (10) Provincial Organization Department and (11) Provincial Discipline Inspection Committee.

county government agencies and tracing the rise of different political clans within that county (Feng 2010). The deputy head of the city Party Organization Department explained the rise of certain families thus (Feng 2010, 159):

It's like a grape vine. As the trellis gets larger, the vine will expand with it. Branches proliferate, and there's also more sunlight and water available. But the children of ordinary citizens might not have a trellis, or they receive less sunlight and water.

A mid-ranking official in Zhong County explained the breakdown of the formal assessment system thus (Feng 2010, 160):

Why does the cadre assessment never find anything real? Because this county is too small and personal networks are too complicated. You can never be sure that someone in the assessment team isn't a relative of the official be evaluated. Who'd tell the truth?

Reforms to the recruitment system, including the revival of the examination system and wider public consultation, are in their early phase (Thogersen 2008), though even skeptics note that the written portion of the examination is relatively free from manipulation (Feng 2010, 78). The face-to-face interviews that follow, however, are less straightforward. As one mid-ranking county official in Benghai explained to me,

The exam is a matter of marks, but the interview, it's how you represent (*biaoxian* 表现) yourself. The crucial problem is the interview ... even if the three panelists find no fault, you can still be passed over. Who sees that you are passed over? People behind the panel. If you are admitted, the logic is the same. The exam is just a cover (*huangzi* 幌子).

While the appointment system can be manipulated, the targets set under the assessment system cannot be disregarded. Leading officials in the township or the county who hope to be promoted cannot afford to be found wanting by inspection teams from higher levels, particularly in key policy areas such as attracting investment, family planning or villagers petitioning higher levels of government. Resources and staff will be mobilized to achieve some of these targets, often very forcefully.[9] However, the question of which policies they implement remains. O'Brien and Li (1999, 168) wondered 'Why an implementer who is responsible for a range of policies executes some well and others poorly'. They argue that the 'one-level-down management' (*xiaguan yiji*) aspect of the cadre responsibility system,[10] which makes officials immediately responsible to the officials directly above them, insulates cadres from social pressure, and encourages them to implement unpopular rather than popular reforms. They identify an interesting correlation between the ability to quantify a target, and how 'hard' or 'soft' it is in the eyes of implementing officials.

This matches well with my experiences working with local agricultural technicians in Anhui and Sichuan, who were constantly frustrated that their ability to train farmers in labor-saving techniques such as the rice dispersal method was impossible to quantify, and thus invisible to higher officials with no knowledge of agriculture. This was seen as

---

[9]As Zhou (2010, 68) has noted, leading officials also cultivate informal ties with their superiors as a 'safety net', in the event that not all goes to plan.

[10]The one exception to this rule within the Party/state apparatus is the appointers themselves, the Party Organization Department. The Central Committee of the CCP (via the Central Organization Department) maintains nomenklatura authority over the directors and deputy directors of both provincial and municipal/prefectural Organization Departments (Landry 2008, 166).

a major reason why agricultural technicians were diverted into family planning work (Smith 2010). O'Brien and Li (1999, 171–6) argue that unpopular policies, such as birth control and revenue collection, tended to be quantifiable, while popular policies, such as fee limits and the implementation of fair and competitive village elections, are harder to quantify, and thus 'soft'. However, many popular policies, such as the abolition of the agricultural taxes (L.C. Li 2006, 2007), aspects of the 'building a new socialist countryside' program (Thogersen 2011, Schubert and Ahlers 2012, 2013) and the rural medical cooperatives (S.G. Wang 2009, Klotzbücher *et al.* 2010), have been effectively implemented in parts of rural China. So popularity does not necessarily mean the failure of an initiative.

Yet, as O'Brien and Li have observed, an obsession with quantification does have an impact on the way local governments operate, and how cadres allocate their time. Sociologist Wu Yi, in *Uproar in a small town (Xiao zhen xuanxiao)*, a detailed first-hand analysis of the workings of a township government in the early 2000s, describes his first encounter with the township family planning office (Wu 2007, 16):

> Everyone was hunched over their desks, and I asked what they were up to. They laughed, 'We're doing our homework'. They explained that although statistics were impossible to obtain, because many peasants couldn't be found, and those that could were reluctant to cooperate, family planning work nonetheless required a host of complex and precise indicators for 'high-quality service' to be met. The higher-ups were unconcerned with practicalities; they required systematic and standardized management.[11] They sent down a plethora of accounts, tables, cards and booklets, with an array of indicators to fill in, and a very short time to do so. Aside from crucial numbers, such as population or the number of pregnant women, which couldn't be dreamed up, all other numbers were estimates. This reminded me of the Shuang village party secretary, who said, 'You want a number? I'll give you that number. No matter how big you want it to be, I can make that happen'.

While the books may be cooked as local officials play these games on paper, government staff still need to be paid, cars still need to be fuelled, and investors and higher level officials still need to be entertained. A major influence on whether local officials give priority to an initiative is its ability to provide revenue for these needs.

The revenue imperative

From the perspective of many scholars, local officials in China have adopted pro-business policies largely to secure the fiscal health of their own constituencies. During the 1990s, the extraordinary growth rates experienced in parts of rural China led to a body of literature which saw the newly-decentralized local state as a positive force for economic development. The influential theory of 'local state corporatism', developed by Oi and others, persuasively explained the explosive economic growth in terms of the entrepreneurial behavior of local officials, particularly at the township level (Byrd and Lin 1990, Oi 1992, 1999, Oi and Walder 1999). This was supported by another influential study, which found that competition among sub-national jurisdictions had led to the development of 'market-preserving federalism' (Montinola *et al.* 1995). These studies have been critiqued for their optimistic view of the role of local officials (Zhao 2012) and their emphasis on crediting local

---

[11]This disconnect between edicts from 'higher ups' was common to many accounts. Gu Wenfeng (2006, 97) relates a popular ditty: 'Half of the spirit of higher level instructions is incomprehensible; half of what is comprehensible can't be understood; half of what is understood can't be done; half of what can be done is useless; half of what is useful isn't commended; and everything that is commended encounters the opposition of ordinary people'.

government with promoting economic development, rather than the entrepreneurs themselves, who form the informal networks that make private enterprise viable within an unsupportive, and often hostile, formal regulatory framework (Lam 2000, 47–8, Y.S. Huang 2008, Nee and Opper 2012).

The revenue imperative can manifest in many forms, particularly when there are many staff to feed. A recent study from a county in rural Gansu demonstrated that subsidies earmarked for education were diverted towards hiring 'administrators' for schools. By contrast, subsidies for agriculture and poverty alleviation were used to increase the operating budgets of county *gongwuyuan* cadres, largely funding for 'buying and using new cars, for food bills and paying for networking to obtain more earmarked subsidies' (Liu *et al.* 2009, 992, Zhou 2010, 49). While diversion of funds from higher levels tends to disproportionately benefit *gongwuyuan* cadres, the practice of 'knocking down the east wall to prop up the west wall' is accepted by service and enterprise staff at the county and township levels, because it is seen as vital for guaranteeing salaries and operating budgets (Gu 2006, 151, Smith 2010, 610, Zhou 2010, 58–9). Paradoxically, the stricter earmarking of funds, and greater centralisation of resources, leads to leading officials spending even more time cultivating the social networks that makes it possible to mobilize and protect resources (Zhou 2010, 70–1).

This interest on the part of local officials in attracting projects is not simply a matter of personal greed. At a macro level, H. Li and Kung (2012) demonstrate how local officials across China reacted to an exogenous change in the revenue-sharing arrangements to reduce their take of enterprise tax by shifting their efforts away from encouraging industrial manufacturing, and towards maximizing their return from land sales. This supplements the findings of a survey which found that inland provinces, such as Hunan and Sichuan, imposed heavy and rising informal and formal tax burdens on peasants during the 1990s. By contrast, in wealthier coastal provinces, such as Guangdong and Zhejiang, burdens actually decreased as a proportion of income during the 1990s (Tao *et al.* 2011). The authors explain this apparent anomaly in terms of incentives built into the taxation system, as 'collecting a unit of tax from an industrial firm was more cost effective than collecting a unit from individual farming households in China … [but] many officials simply had no viable alternatives' (Tao *et al.* 2011, 669–70). The interesting exception was wealthy Jiangsu, which registered one of the heaviest 'peasant burdens'. The authors surmise this is due to the continuing strength of rural collectives in this former revolutionary base area,[12] suggesting that the ability of local party organs to mobilize resources pairs well with the fiscal incentive.

Aside from the risky propositions of diverting earmarked funds, or selling off land to developers, since the abolition of agricultural taxes, diverting funds from construction projects is one of the few ways for township leaders to bring in revenue to keep their staff funded and their offices running (see Zhao 2007, 45–54). As one township mayor explained to Wu Yi (2007, 578):

'Since we stopped collecting taxes, the measure of whether or not an official has any tricks up his sleeve is whether or not he can bring in a project and some money to the village. If projects and money come in, the villages have hope and the township can develop. Compared with that, providing services is a low priority.' This was the mayor's firm position on the work of township and village officials. And the mayor stuck to his own principles. Aside from his basic

---

[12]Southern Jiangsu was one of the few areas in China that maintained a cooperative medical system during the 1980s (S.G. Wang 2009, 383).

work, all of his time was spent thinking of ways to bring in projects to prop up the township economy. He knew that agriculture could only fill the belly; it couldn't bring in the cash.

This emphasis on leading cadres achieving measurable outcomes within a short time frame has many potential impacts on rural governance (Zhou 2010, 62–6). As one township party secretary explained to Lynette Ong (2012, 84):

> Since cadres are assessed only by their achieved targets at day-end, rather than how much resources have been deployed to bring about the outcome, some of us borrow money from banks and other sources to build industrial parks, develop town squares in order to showcase our 'political achievements'. In the first two years as the township party secretary, I have been trying to resolve the debts my predecessors accumulated and left behind after being promoted to county-level positions.

Meeting fiscal targets largely forms the basis for punishments and rewards that are meted out during an official's tenure in the form of monetary bonuses, as well as non-monetary incentives, such as elevation to the leading decision-making body in the county, the county party standing committee, or publication of their ranking relative to other township party secretaries. In Benghai, this ranking now largely depends on the success of a township in attracting investment, a trend noted elsewhere (Zhao 2007, 66–73, Ong 2012, 82–4). Attracting investment, and bringing in projects, can benefit both leading cadres and their partners in the 'shadow state' (Harriss-White 2003). Hillman captures this connection when he observes a township Party secretary getting the first inkling of how to spur his staff to attract investment, an activity that is now the 'core task' of township and county officials throughout rural China. Hillman explains,

> The Party secretary came up with an idea to attract investment. He decreed that anyone who succeeded in attracting investment to the township would be given a 10 percent commission by the township government. When I interviewed him about this, he said 'it's just an idea. I figure you have to spend money to make money'.

Precisely the same logic underlies one of more persistent markets in rural China: the jobs of rural officials.

Self-interest: selling posts and the shadow state

One consequence of the concentration of the powers of appointment in the hands of the party secretary, and, to a lesser extent, the mayor and the Party Organization Department, is that it creates a market for party and government posts, which are now routinely bought and sold throughout China. In a court case from Jilin province, a former county Party secretary, Li Tiecheng, was charged with accepting 1.4 million RMB in bribes for following his version of the "normal procedure" (zhengchang chengxu 正常程序) for appointments (China Legal Daily 2003).

When reshuffling posts, the Organization Department would assess candidates, and then pass the list of names to the deputy party secretary for personnel, who would check it, show it to the county party secretary, and then to the county party standing committee for discussion … There was no chance Li Tiecheng would utter the names of gift-givers he wanted to promote. Before each assessment, his office would compose a 'tune' (*diaozi* 调子) which matched the description of his candidate, including age, years in work, qualifications, experience and post to pass on to the Organization Department, which searched with those criteria. If the desired name did not appear, the party secretary would decline, and the Organization Department would start over. The deputy party secretary, and the head of the Organization Department, would also devote time to fathoming the wishes of Li Tiecheng.

Securing a promotion, however, is not simply a market transaction. The dilemma of adverse selection remains: the principal cannot afford to choose an agent who is totally unqualified. The prospective candidate must be capable of doing the job, and numerous studies attest to the increasing importance of professionalism and education credentials (Zang 2006, Landry 2008, Thogersen 2008). The candidate may have spend time in the Party school (Pieke 2009), working in the Organization Department (Thogersen 2008, 418), and typically will supplement the payment with pressure from above, usually in the form of a phone call from someone in the level of government above the party secretary. At this point, the shadow state can come into play, either providing connections to apply pressure from above, by helping the candidate to raise the funds for his first payment, or by hosting banquets (Smith 2009, 40–2, Feng 2010, 166–7).

Efforts to widen the number of people involved in the decision-making process to encourage more 'democratic' decision-making have had an impact, but the effect has been to been to expand the number of players (and diners), rather than close down the market or change the rules of the game. Feng Junqi (2010) describes a 'vote-canvassing network' (la piao wang 拉票网) for leading positions, specifically township party secretaries, heads of county bureaus, and the county party and government leaders. A township party secretary with hopes of promotion to the county will assemble a campaign team based around the leading officials in the township and his own friends and relatives. In Zhong County, a section chief-level cadre can expect to receive a red envelope with 1000 RMB from a candidate, while a county government leader is worth at least 2000 RMB. The team will take potential supporters out for meals, often using the line ministry system to invite a number of cadres at once. A yearlong campaign is estimated to cost between 300,000 and 500,000 RMB. As one veteran of such a campaign explained,

> Canvassing in Zhong County is now common enough to be out in the open. You have to solicit votes; otherwise you might as well give up. And there are no guarantees: you've got to cover 100 percent of voters. My campaign was about seven years ago. We had a list of people at section chief rank, and we went from house to house. It was about 500 RMB each. (Feng 2010, 169)

One phenomenon that is consistent for rural cadres is the enthusiasm for construction projects, particularly roads. Such projects represent an opportunity for leading cadres to fulfill all three criteria mentioned to this point. They benefit politically under the assessment system, as their superiors see evidence of them getting things done – such projects are often referred to as 'political points projects' (zhengji xiangmu 政绩项目). Moreover, if funding is sourced from higher levels of government,[13] it brings in revenue, and there is a potential windfall for the 'shadow state', in the form of friends and relatives (or in western China, often the cadres themselves) who own local construction companies (Goodman 2007, 180–4). As Lam (2000, 179–80) observed,

---

[13]As Göbel (2011, 73) has noted, many rural counties sacrificed expenditure on basic infrastructure when faced with the fiscal pressure of 'tax-for-fee' reform during the early 2000s. Shifting transfers from the central government (even earmarked transfers which are supposed to be 'off limits') to cover the payment of salaries, and even to fund banqueting and the purchase of vehicles, is common practice (Liu et al. 2009).

Almost all new party secretaries proposed initiatives of one sort or other, all of them involving construction. I asked a number of interviewees why this was the case. Almost unanimously, they said that these were opportunities for the party secretaries to make money. And the projects were packaged as attempts to do something to benefit local people and presented by the party secretary to his superiors in the county government as evidence of his boldness and resolution. And because they did not involve direct levies from the peasants, they tended to be quite acceptable to the rural populace.

The formulation of the 'shadow state' suggests passivity, with contractors largely dependent on the political favors of their backers in the county government. It may be a tacit agreement that a certain bureau will dine at their restaurant, or a monopoly on providing certain building materials to bureaus that are involved in construction, be it windows to the county Education Bureau for the school consolidation project, or pavers for flood control works (in the case of Benghai, the first contract went to a brother of the owner of Sauna City, the county's favored hot pot restaurant/gambling house/brothel, while the latter went to the county Party secretary's favored driver). Often, however, the 'shadow state' will be proactive in seeking out such projects on behalf of their nominal political masters. Hillman describes a local road contractor being sufficiently confident in their connections within the provincial government to commence construction of a road, forcing the district[14] governor to accept an inflated figure for the cost of its construction (Hillman 2005, 139–40). Notions of 'predatory' and 'developmental' states are simplistic. In practice, the formal local state is both prey and predator of the informal, or shadow state. When a mid-level official complains that his brother, who owns a computer shop, has to hand out tens of thousands of dollars worth of supermarket cards every year to ensure that contracts from the official's department continue to flow, is there a victim? Moral categorizations make little sense.

While in many areas, the contractors' tail wags the dog of the local state, for construction companies with limited political connections, or when there is a great deal of competition for projects, companies will often assume the risk of 'making up the funds' (*dianzi* 垫资) to win a project bid, or to get the project off the ground. As Lam (2000,180) explained,

> The frequent rotations of party secretaries were also closely related to the practice of '*dianzi*' in government projects. When a government body did not have the revenue to start a project, but still wanted to do it, it could ask a construction team to pay for the cost of construction first, and would promise that the government body would pay it back when it had revenue … When a party secretary was transferred, the fiscal burden was passed to the successor.

Not surprisingly, a survey of township government debts found infrastructure projects to be significant, second only to funds borrowed to meet centrally mandated education quotas, and funds to pay the debts of the rural credit associations (Oi and Zhao 2007, 85–6). Despite the rhetoric demonizing township officials, most of these debts arise from actions at the county level or above.[15] Construction companies also face pressure from

---

[14]Due to the remoteness of Hillman's field site, the area had retained an additional level of government between the counties and the townships (Hillman 2005, 72–6), the district (*qu* 区). When streamlining required this level be abolished, they simply renamed the districts as townships, and renamed the townships as 'administrative villages'. Thus, Poshan is one of the few places in China where township elections take place.

[15]John James Kennedy's contribution in this collection discusses the effect of debt on township and county cadre behavior (Kennedy 2013). See also Zhao (2007), Wong (2009) and Ong (2012).

developers to provide an upfront discount of up to 20 percent, and in the real estate sector, 'contracting on loan' has become the norm (X. Li 2003).

In the context of a rural county or township, the balance of power in these transactions depends on local political connections, which are often heavily in favor of the developers. Being part of the shadow state is not without risk for local companies and their political patrons. Particularly at the county level, such links also create opportunities for entrepreneurialism (Walder 1995, Duckett 1998) or rent seeking (Smith 2009, 39–45, Hillman 2010, 7–11) by officials in the Construction, Transport, Land Management, Tax, and the Industry and Commerce Bureaus. When these large developers are from outside the county, or even the province, a trend encouraged by the recent enthusiasm for 'attracting investment', the possibility of institutional capture is real. In Benghai, county officials would joke that their government now answers to entrepreneurs from Zhejiang, not officials from Hefei.[16] Given that many of these developers are private, or effectively private companies, this undermines orthodox economic assumptions that 'polities significantly shape economic performance because they define and enforce the economic rules' (North 1994, 366). It also suggests that the traditional way of viewing private enterprise within China needs to be reassessed. It can no longer be assumed that governments have the upper hand in negotiating the rules in a game of constrained cooperation, where firms rely on the threat of exit and a shared interest in economic growth to keep the predatory tendencies of local officials at bay. A large real estate company from Zhejiang faces few constraints, and is unconcerned with the long-term economic prospects of Benghai County.

In the context of these informal rules and norms, it is worth emphasizing that although 'shadow state' many suggest negative or nefarious motives on the part of rural cadres, my main concern is to highlight that much of the real business of government goes on out of 'sight' of the formal structures of power. Economic rules and norms come in many guises: formal, informal and (more commonly in China) semi-formal, where the formal and informal identities of rural cadres coexist. Informal rules, as North notes, usually change only gradually. Indeed, the actors surrounding officials, such as 'secretaries, advisors and researchers' (Heilmann 2008, 9), can be vital in promoting local experiments and innovations that result in beneficial changes to the formal rules of the game. As Sebastian Heilmann has observed, in examining the development of policy in China, 'distinguishing between bottom-up ('spontaneous') and top-down ('mobilization-style') initiation of experimentation is nearly meaningless since there is a strong element of both local initiative and central sponsorship in the initial stages of major experimental efforts' (Heilmann 2008, 10). Regardless of whether officials act as pioneers, bandwagoners or resisters (Göbel 2011, 68–73), the interactions between the informal and formal sectors are crucial in initiating and shaping policy.

This is not to say that formal institutions have no influence – the fortunes of the shadow state constantly shift over time in response to changes in the makeup of the formal leadership and the formal regulations which apply to differ sectors of the economy (Krug and Hendrischke 2007, 91–3, Alpermann 2010, Smith 2011). Whether you are a local restaurant proprietor, a large property developer from another province or the manager of the local brewery that provides half of the county revenues, you have real 'skin in the game' when local tax regulations are being devised or scarce resources are being allocated. Likewise, the local officials and their agencies, who are the 'quasi-owners' of resources in any given

---

[16]Instances of local officials being bilked by entrepreneurs are numerous, but it is rarely officials who bear the costs (Smith 2009, 52–4). See also Sargeson (2013).

locale, can be viewed not merely as regulators or rent seekers, but more usefully as market players, whose influence is defined by their influence and success in the market. This influence, in large part, is determined by continuous negotiations with actors in the shadow state.

From this perspective, bureaucratic coordination, patronage, even the use of price control and social mechanisms, are part of local institution building among market players (Krug and Hendrischke 2012, 529–30). Local officials have some self-interest in achieving allocative efficiency, and maximizing their long-term fiscal returns from the business sector. They become partners with local firms in seeking new ways to organize production, and to coordinate business relations (Krug and Hendrischke 2012, 540). Other researchers have noted a trend for former officials to be employed as advisers by local enterprises (Ahlstrom *et al.* 2008). In the other direction, there is evidence that CCP membership on the part of an entrepreneur can enhance their firm's profitability (Li *et al.* 2008), but membership of higher-level bodies (such as the local People's Congress) and participation in informal networks across the firm seem to be more crucial (Z. Chen *et al.* 2012). Mutual dependence between the shadow and formal states is reinforced by peculiarities in central-local fiscal relations. Local governments get to keep all business taxes (*yingye shui* 营业税) and income taxes generated by locally-based service industries, but retain only one-quarter of the value added tax (VAT) generated by manufacturing industries, which have a higher factor mobility (Su *et al.* 2013).[17] Thus, relatively immobile local service businesses support the formal local state. While the new, informal side of the party is in business throughout rural China, its traditional side still influences the behavior of rural cadres.

The party in charge: an endless campaign

One consistent picture that emerges across many studies of rural China is the reassertion of CCP control at the local level during the post-Tian'anmen era. But what practical consequences does CCP dominance have for rural governance? This can manifest in a variety of ways. Zhao (2012) presents two examples from recent fieldwork. In one municipality, the Party leadership wants to take the lead in 'greening work', but the deputy mayor who should be in charge of this was deemed to have insufficient clout to push through the campaign. So the city Party secretary intervened and put the head of the Party Organization Department in charge, not only of the deputy mayor, but also directly instructing the relevant government departments and party organs in how to carry out their work. Similarly, another county decided to take the initiative in developing a dairy industry. Rather than entrusting the project to the deputy county head from the relevant *xitong* (portfolio), the head of the county Disciplinary Inspection Committee (a Party organ which is nominally in charge of investigating malfeasance by Party members) was given responsibility for allocating tasks.

Similarly, party branches have asserted their control over the local People's Congresses, which were once seen to hold the potential to fulfill a meaningful supervisory role in constraining the excesses of local courts and party officials (Almén 2013). It is now common for the local Party secretary to be the head of the People's Congress, and, since the passing of the 2006 Supervision Law, these Local People's Congresses no longer have the power to conduct appraisals of government department heads. This function, which was sometimes supplemented by telephone hotlines and public participation in appraisal meetings, had

---

[17]This contradicts a recent overview by Stephen Green (2013) of Standard Chartered Bank.

been one of the most effective checks on local abuses of power (Cho 2009). Oscar Almén speculates that the pluralization of Chinese society and politics may be the reason for this reactive step. With informal institutions growing in influence, the CCP is reluctant to relinquish its trump card: Party discipline and control over key political actors through the cadre appointment system (Almén 2013, 253–4).

A striking aspect of the expansion of Party authority is its impact on the manner in which governing happens. This in turn, I would argue, has a profound impact on the priorities of local government and the behavior of rural cadres. As Göbel (2011, 56) has argued, 'selective implementation is at least in part a function of the control mechanisms themselves'. In other words, *how* things can be done with the existing institutions to a large extent determines *what* things are done. During the late 1980s, when it seemed that government functions were becoming routinized, and separated from Party activities, White (1990) observed that, in implementing the national policies on family planning, 'the gap between official family planning targets and the capacity of the formal party/state apparatus to reach them undermined efforts to regularize grassroots implementation. Instead, the implementation process settled unstably into a pattern of institutionalized mobilization'.[18]

One of the consequences of the continued (and expanding) primacy of the communist party at the local level is the persistence of the practice of institutionalized mobilization at the local level, particularly in the townships. Weberian bureaucracies may like to govern, but political parties like to campaign. Wu (2007, 8–15) outlines the logic of campaigns and 'core tasks' (*zhongxin gongzuo*):

> The inspection [by the province to assess the reduction of the 'peasant burden'] acted as an accelerator for the campaign. The whole township and village apparatus revolved around it … After a while in the township, I got an idea of how much time, manpower and resources were consumed in coping with inspections and evaluations by higher levels. Party secretary Lin explained, 'The township is the lowest level of government. Everyone can inspect you, but you can't inspect anyone else. You just lump it … The number of inspections each year was at least ten, a major one each season, and a smaller one every month. No wonder cadres claimed that at least one third of their time was taken up receiving inspections of every flavor'.

Zhao (2012, 77) explains the current manifestation of this trend:

> Differences in interests and fragmentation of the formal system are characteristic of local governments. The response to fragmentation manifests in the way governments launch campaigns to complete tasks. Administrative work and the provision of public services have been replaced by mobilization. This campaign approach manifests itself by focusing all of the administration on a target. In as short a period of time as possible, all resources – personnel, material and finances – are employed like a torrent of wind and rain (*jifeng zhouyu* 急风骤雨). Campaigns don't harness civil society; the state employs administrative force to create a government movement. Leading cadres use inspections, assessments, and competition between officials; there's colour and movement as propaganda starts up; and

---

[18]White defines 'institutionalized mobilization' as involving efforts that '(1) temporarily intensify coercive and normative incentives; (2) vary from region to region in timing, intensity, and scope; (3) last for limited, predictable periods of time; (4) have as their primary goal behavioral control or 'practical results', not attitudinal or cultural change; (5) have a diminished scope of mass participation in favor of narrow mobilizations of the target populations; and (6) utilize extensive propaganda to shape public sentiment, but discourage disruptive mobilizational activities beyond the target population in order to insulate the project from economic production and other reform initiatives' (White 1990, 62–3).

the results are announced with fanfare. At the end, everything is wonderful on the surface, with lessons learned and breakthroughs achieved ... But in reality, it's either a formalistic breeze, with 'documents implementing documents' and 'meetings implementing meetings', or a forceful campaign that leaves scars. Some of the enormous debts of township and village governments are the 'substantial achievements' that such campaigns have delivered.

The scarring mentioned by Zhao not only impacts on local population, but also has effects on ordinary cadres, who are forced to collect taxes, enforce family planning laws and requisition the land of residents in the same township (depending on the resources of the township, these tasks may be 'contracted out'; see Sargeson 2013).[19] A frequently mentioned word in conversation with grassroots officials is 'numbness' (*mamu* 麻木), which could be added to the standard suite of responses of 'working, shirking and sabotage' (Brehm and Gates 1997), although it could be argued that numbness is a manifestation of 'dissent-shirking' (*ibid.*, 30), where the opposition to the policy is part of a broader frustration with the nature of grassroots government work.

While Zhao views the persistence of campaign-style governance as a negative development, reverting to mobilization tactics is a rational response on the part of local officials to the decline in political commitment among township government staff. It also draws upon traditions which villagers, activists, and government officials are familiar with (Thogersen 2009), it plays to the strengths of the CCP's organizational structure (Wu 2007, 40–3, Heilmann and Perry 2011), and can be tailored to fit with the agricultural cycle. From the perspective of building a modern government bureaucracy, particularly for weaker line ministries whose priorities rarely constitute the 'central work' (*zhongxin gongzuo* 中心工作) valued by the CCP, the cyclical campaigns weaken their authority relative to the Party. Money, personnel and equipment are diverted towards the priorities of the local Party branch – agricultural technicians find themselves working full time on family planning tasks; forestry officers with a talent for networking will spend large parts of the year outside the county, promoting its virtues as a value-for-money investment destination (Hillman 2010, Smith 2010, Zhao 2012).

Moreover, compared to the period White was examining, the late 1980s, there are more government staff at the township and county level who can be called upon by the campaign of the day, whether it be converting sloping land to forests (the grain for green project), family planning, or attracting investment. All three of these initiatives, which have been implemented with alacrity throughout rural China, meet White's criteria for institutional mobilization. Campaigns are not, however, one-off events. All three of these initiatives have been subject to nationwide campaigns of varying length, and are ongoing. Thus, while they rely on mobilization techniques to galvanize the limited resources of county, township and village governments, these campaigns also cause shifts in the traditional capital by which the political influence of government departments are measured: personnel, finance and materials (*ren, cai, wu* 人,财,物). In the context of a political system where your immediate superior measures success, this creates a positive feedback loop with both the formal assessment system and with access to fiscal resources. The campaign succeeds precisely because it brings bureaucratic resources to the party.

---

[19]For township and village cadres, these scars can be real. Wu (2007, 94) recalls a ditty summarizing tax collection. 'If they hit you, don't hit back. If they curse you, stay quiet. If they have money, take it. If they don't, off you go'.

Testing the matrix

This is not to claim that these four factors are the only influences on the behavior of rural cadres. It seems that not all criteria need to be met. What is currently lacking in the literature, however, is a way of determining how cadres might 'weigh up' these different factors in decision-making. Below, I propose a matrix to analyze rural officials' decisions, as a way of more systematically examining rural politics. Table 2 offers a conceptual summary of the likelihood of success of a given policy, given its interaction with the formal and informal incentives faced by rural cadres. To illustrate how this matrix might be applied, I describe the implementation of two priorities of the Ministry of Agriculture – the dissemination of hybrid rice varieties, and the promotion of labor-saving direct seeding techniques.[20] This also provides a brief outline of how local state agents 'get things done'.

The promotion of hybrid rice has been seen as one of the success stories of China's agricultural extension system, with very little conventional rice now sown in China. However, resistance to the technology was considerable. Although yield gains of 15 to 25 percent over conventional rice were promised (OECD 1993),[21] farmers would have to purchase seed every time a new rice crop was planted, undoing thousands of years of traditional cropping practice. Early varieties of hybrid rice also tended to be more susceptible to diseases and insect pests. In terms of the proposed matrix, the technology had certain advantages. It was given priority in the cadre assessment system, its uptake was easy to measure,[22] and the distinct, dark green dwarf varieties were easily visible to higher-level inspections. It also made it easier for township and village cadres during the 1990s to meet the 'hard constraint' of grain procurement targets, and potentially free up land to cultivate cash crops (such as cotton) for village industries, thus increasing local government revenues (Rozelle 1994). It increased the sales of seed vendors (hybrid varieties also tended to require more fertilizer and farm chemicals); such shops were initially intertwined with the formal extension system (many seed vendors are still former extension agents), and the owners of larger suppliers based in the county seat tend to be well connected. Thus there was considerable benefit to the shadow state. Perhaps most crucially, the promotion of these varieties was suited to a mass mobilization campaign, and limited to a specific time of year – the planting season. All township and village government officials were drawn into bringing seed to every household, combining traditional Maoist techniques of criticism and forced interventions with economic incentives and punishment – township extension stations in Hubei offered farmers fertilizer at below the market price in return for deciding to plant hybrid rice, and seed stations to this day simply refuse to purchase non-hybrid grain.

In contrast, a more recent initiative by the Ministry of Agriculture has struggled to gain widespread adoption, even though direct seeding (*paoyang* 抛秧) offers benefits to farm households in saving labor and reducing costs, with few downsides. Although a priority within the Ministry, it does not intersect with any hard constraints within the cadre assessment system, nor is it easily measurable, or impressive to higher-level inspection teams. If

---

[20]For an introduction to these technologies, see Pingali *et al.* (1997) and the website of the International Rice Research Institute, http://www.irri.org/.

[21]There are large geographical differences in yield gaps. In Sichuan, hybrids have been found to outperform conventional varieties by 54 percent, while in Zhejiang the gap was only three percent (J.K. Huang and Rozelle 1996, 342).

[22]A researcher of the system during the late 1980s noted, 'Goals were – without exception – stated in quantitative terms, mostly in per-unit yield increases' (Delman 1993, 167).

Table 2. Matrix of rural cadre decision-making.

Task	Quantifiable, high priority	Raises revenue	Benefits shadow state	Suits mass mobilization
Hybrid rice	+++	+	++	+++
Direct seeding	+	-	-	-
Attracting investors	++	+++	++	++

Note: The number of positive signs indicate the strength of an imperative for a given task, with - denoting negligible, + denoting a moderate effect, ++ denoting a strong effect, and +++ denoting a very strong effect.

anything, a directly-seeded field is less visually impressive than a traditional hand-planted (*chayang* 插秧) field, with its orderly rows. Yield increases are not spectacular, and since the abolition of agricultural taxes, grain is no longer a significant contributor to local government revenues. Little extra equipment is required, so there is no apparent benefit to the shadow state either. Moreover, since its dissemination requires technical knowledge of planting methods, dispersion technique and irrigation, it is unsuited to mass mobilization by local government staff. In the words of county and township extension agents, its formal dissemination largely depends on the attitude of individual township party secretaries. In practice, the farmers teach each other, or seek advice from their local seed vendor.

Even if all four conditions to stimulate the enthusiasm of rural cadres are in place, local governments are still prisoners of geography. The final panel of Table 2 suggests why attracting investment has been irresistible to local governments throughout China. Yet while all local governments might strive to attract investors, by offering cheap land, tax holidays and straight-out kickbacks, what hope do remote townships in rural Sichuan have of competing with towns in the Pearl River Delta? The party secretary may be king, but not all kingdoms are created equal. Without adequate fiscal resources, simply feeding government staff will remain the main concern.

Not all of these incentives are mutually exclusive, and often an incentive may encourage the implementation of a policy, but not in the manner intended by higher levels of government, as revealed in Zhou's (2010) study of the environmental protection system. To give a couple of commonly cited examples, local governments can raise revenue, and individual environmental protection officers are enriched, by collecting fees from polluters, to the detriment of the environment. The campaign-style approach is applied to environmental protection, but is notoriously ineffective, with shuttered factories or closed mines reopening soon after senior officials leave the area. Similar caveats apply to family planning and the 'grain for green' program. However, following Kevin O'Brien's observation, we should accept that policy misimplementation is baked-in and normal (O'Brien 2010, 81). Only then can we have a clear view of the pressures faced by local officials, and an understanding of their multiple identities as policy implementers, revenue collectors, market players and political campaigners.

References

Ahlstrom, D., G.D. Bruton and S.Y. Kuang. 2008. Private firms in China: building legitimacy in an emerging economy. *Journal of World Business*, 43, 385–99.

Almén, O. 2013. Only the party manages cadres: limits of local People's Congress supervision and reform in China. *Journal of Contemporary China*, 22(80), 237–54.

Alpermann, B. 2010. *China's cotton industry: economic transformation and state capacity*. London: Routledge.

Axelrod, R. 1984. *The evolution of cooperation*. New York: Basic Books.

Bai, L. 2006. Graduate unemployment: dilemmas and challenges in China's move to mass higher education. *China Quarterly*, 185, 128–44.

Brehm, J. and S. Gates. 1997. *Working, shirking, and sabotage: bureaucratic response to a democratic public*. Ann Arbor, MI: University of Michigan Press.

Brodsgaard, K.E. 2002. Institutional reform and the *bianzhi* system in China. *China Quarterly*, 170, 361–86.

Burns, J.P. and X.Q. Wang. 2010. Civil service reform in China: impacts on civil servants' behaviour. *China Quarterly*, 201, 58–78.

Byrd, W.A., and Q.S. Lin, eds. 1990. *China's rural industry: structure, development and reform*. New York: Oxford University Press.

Chen, A. 2007. The failure of organizational control: changing party power in the Chinese countryside. *Politics & Society*, 35, 145–79.

Chen, Z., Y.Z. Sun, A. Newman and W. Xu. 2012. Entrepreneurs, organizational members, political participation and preferential treatment: evidence from China. *International Small Business Journal*, 30(8), 873–99.

Chien, S.S. 2010. Prefectures and prefecture-level cities: the political economy of administrative restructuring. *In:* J.H. Chung and T.C. Lam, eds. *China's local administration: traditions and changes in the sub-national hierarchy*. London: Routledge pp. 127–48.

Cho, Y.N. 2009. *Local People's Congresses in China: development and transition*. New York: Cambridge University Press.

Chung, J.H. and T.C. Lam, eds. 2010. *China's local administration: traditions and changes in the sub-national hierarchy*. London: Routledge.

Delman, J. 1993. *Agricultural extension in Renshou County, China*. Hamburg: Institute für Asienkunde.

Duckett, J. 1998. *The entrepreneurial state in China: real estate and commerce departments in reform era Tianjin*. London: Routledge.

Edin, M. 2003. State capacity and local agent control in China: CCP cadre management from a township perspective. *China Quarterly*, 173, 35–52.

Feng, J.Q. 2010. Zhongxian ganbu [Zhongxian cadre]. PhD Dissertation, Peking University.

Göbel, C. 2011. Uneven policy implementation in rural China. *The China Journal*, 65, 53–76.

Göbel, C. and L. Ong. 2012. *Social unrest in China*. London: ECRAN.

Goodman, D.S.G., ed. 1997. *China's provinces in reform: class, community and political culture*. London: Routledge.

Goodman, D.S.G. 2007. Narratives of change: culture and local economic development. *In*: B. Krug and H. Hendrischke, eds. *The Chinese economy in the 21st century: enterprise and business behaviour*. Cheltenham: Edward Elgar, pp. 175–201.

Green, S. 2013. What makes 10 million local government officials tick? [online]. Hong Kong: Standard Chartered Bank. Available from: https://research.standardchartered.com/configuration/ROW%20Documents/China_%E2%80%93_Masterclass__What_makes_10_million_local_government_officials_tick__%5BCorrection%5D_06_05_13_01_38.pdf [Accessed 31 May 2013].

Gu, W.F. 2006. *Feichang zishu: yige xiangzhen shuji de meng yu teng* [*Extraordinary accounts: the hopes and troubles of a township Party secretary*]. Beijing. Xinhua chubanshe [Xinhua Publishing].

Guo, J. 2010. Policy learning and policy implementation in China: a case study of the grain for green project. PhD Dissertation, University of Hong Kong.

Harriss-White, B. 2003. *India working: essays on society and economy*. Cambridge: Cambridge University Press.

Heilmann, S. 2008. Policy experimentation in China's economic rise. *Studies in Comparative International Development*, 43, 1–26.

Heilmann, S. and E.J. Perry, eds. 2011. *Mao's invisible hand: the political foundations of adaptive governance in China*. Cambridge, MA: Harvard University Press.

Hillman, B. 2005. Politics and power in China: strategems and spoils in a rural county. PhD Dissertation, Australian National University.

Hillman, B. 2010. Factions and spoils: examining local state behaviour in China. *The China Journal*, 64, 1–18.

Huang, J.K. and S. Rozelle. 1996. Technological change: rediscovering the engine of productivity growth in China's rural economy. *Journal of Development Economics*, 49, 337–60.

Huang, Y.S. 2008. *Capitalism with Chinese characteristics: entrepreneurship and the state*, Cambridge: Cambridge University Press.

Kennedy, J. 2013. Local finance and rural governance: local characteristics, Challenges and changes. *Journal of Peasant Studies* 40(06). DOI: 10.1080/03066150.2013.866096

Kennedy, S. 2010. The mandarin learning curve: how China is reshaping global governance. Paper presented at the China Social Science Workshop, Stanford University, 13 May.

Klotzbücher, S., et al. 2010. What is new in the 'New Rural Cooperative Medical System'? An assessment of one Kazak county of the Xinjiang Uyghur Autonomous Region. *China Quarterly*, 201, 38–57.

Krug, B. and H. Hendrischke. 2007. Framing China: transformation and institutional change through co-evolution. *Management and Business Organization Review*, 4, 81–108.

Krug, B. and H. Hendrischke. 2012. Market design in Chinese market places. *Asia Pacific Journal of Management*, 29, 525–46.

Lam, T.C. 2000. The local state under reform: a study of a county in Hainan Province, China. PhD Dissertation, The Australian National University.

Lam, T.C. 2010. County system and county governance. *In*: J.H. Chung and T.C. Lam, eds. *China's local administration: traditions and changes in the sub-national hierarchy*. London: Routledge.

Landry, P. 2008. *Decentralized authoritarianism in China: the Communist Party's control of local elites in the post-Mao era*. Cambridge: Cambridge University Press, pp. 149–73.

Li, H., L. Meng, Q. Wang and L.A. Zhou. 2008. Political connections, financing and firm performance: Evidence from Chinese private firms. *Journal of Development Economics*, 87 (2), 283–99.

Li, H. and J.K.S. Kung. 2012. Fiscal incentives and policy choices of local governments: evidence from China [online]. Working paper, Hong Kong University of Science and Technology. Available from: http://ihome.ust.hk/~sojk/Kung_files/fiscal%20incentives_JK.pdf [Accessed 18 February 2013].

Li, L.C. 2006. Differentiated actors: central-local politics in China's rural tax reforms. *Modern Asian Studies*, 40(1), 151–74.

Li, L.C. 2007. Working for the peasants? Strategic interactions and unintended consequences in China's rural tax reform. *The China Journal*, 57, 89–106.

Li, X. 2003. Zhuizong jianshe gongcheng 'heise zhaiwu lian' [Tracking the chain of bad debts in construction projects] [online]. *Jingji Cankao Bao* [*Economic Information*]. Available from: http://finance.sina.com.cn/b/20030820/1308414250.shtml [Accessed 10 February 2013].

Liu, M.X., *et al.* 2009. The political economy of earmarked transfers in a state-designated poor county in western China: central policies and local responses. *China Quarterly*, 200, 973–94.

Mertha, A. 2005. China's 'soft' centralization: shifting tiao/kuai authority relations. *China Quarterly*, 184, 791–810.

Mertha, A. 2008. *China's water warriors, citizen action and policy change*. Ithaca, NY: Cornell University Press.

Montinola, G., Y.Y. Qian and B. Weingast. 1995. Federalism, Chinese style: the political basis for economic success. *World Politics*, 48, 50–81.

Nee, V. and S. Opper. 2012. *Capitalism from below: markets and institutional change in China*. Cambridge, MA: Harvard University Press.

North, D.C. 1994. Economic performance through time. *American Economic Review*, 84(3), 359–68.

O'Brien, K. 2010. How authoritarian rule works. *Modern China*, 36, 79–86.

O'Brien, K. and L.J. Li. 1999. Selective policy implementation in rural China. *Comparative Politics*, 31(2), 167–86.

OECD (Organisation for Economic Co-operation and Development). 1993. *Traditional crop breeding practices: an historical review to serve as a baseline for assessing the role of modern biotechnology*. Paris: OECD.

Oi, J. 1989. *State and peasant in contemporary China: the political economy of village government*. Berkeley: University of California Press.

Oi, J. 1992. Fiscal reform and the economic foundations of local state corporatism in China. *World Politics*, 45, 99–126.

Oi, J. 1999. *Rural China takes off: institutional foundations of economic reform*. Berkeley: University of California Press.

Oi, J. and S. Rozelle. 2000. Decision making in Chinese villages. *China Quarterly*, 162, 513–39.

Oi, J. and A. Walder. 1999. *Property rights and economic reform in China*. Stanford: Stanford University Press.

Oi, J. and S.K. Zhao. 2007. Fiscal Crisis in China's Townships: Causes and Consequences. *In*: E. J. Perry and M. Goldman, eds. *Grassroots political reform in contemporary China*. Cambridge, MA: Harvard University Press. pp. 75–96.

Ong, L. 2012. *Prosper or perish: credit and fiscal systems in rural China*. Ithaca: Cornell University Press.

Pieke, F.N. 2009. Marketisation, centralisation and globalization of cadre training in contemporary China. *China Quarterly*, 200, 953–71.

Pingali, P. L., M. Hossain and R. V. Gerpacio. 1997. *Asian rice bowls: the returning crisis?* New York: CAB International.

Prabhu, L., *et al.* 1997. *Asian rice bowls: the returning crisis?* New York: CAB International.

Ren, J.T. 2010. Gongwuyuan daiyu ying zouxiang 'pinminhua' [Cadres' benefits should become more 'civilian'] [online]. *Huanqiu Wang*. Available from: http://opinion.huanqiu.com/dialogue/2010-04/771539.html [Accessed 12 February 2013].

Rozelle, S. 1994. Decision making in China's rural economy: the linkages between village leaders and farm households. *China Quarterly*, 137, 99–124.

Sargeson, S. 2013. Violence as development: land expropriation and China's urbanization. *Journal of Peasant Studies*, 40(06).

Schubert, G. and A. Ahlers. 2012. County and township cadres as a strategic group: 'Building a New Socialist Countryside' in three provinces. *The China Journal*, 67, 67–86.

Schubert, G. and A. Ahlers. 2013. Strategic modeling: 'Building of a New Socialist Countryside' in three Chinese counties. *China Quarterly*, 216 (December).

Smith, G. 2009. Political machinations in a rural county. *The China Journal*, 62, 29–59.

Smith, G. 2010. The hollow state: rural governance in China. *China Quarterly*, 203, 601–18.

Smith, G. 2011. Franchising the state: farmers, agricultural technicians, and the marketization of agricultural services. *In*: B. Alpermann, ed. *Politics and markets in rural China*. London: Routledge. pp. 69–86.

Su, F.B., R. Tao and D.L. Yang. 2013. Rethinking the institutional foundations of China's growth. *In*: C. Lancaster and N. van de Walle, eds. *The Oxford handbook on the politics of development*. Oxford: Oxford University Press. Forthcoming.

Sun, Y. 2012. Cadre recruitment and corruption: what goes wrong? *In*: T. Gong and S.K. Ma, eds. *Preventing corruption in Asia: institutional design and policy capacity*. London: Routledge. pp. 48–63.

Tao, R., *et al.* 2011. Grain procurement, tax instrument and peasant burdens during China's rural transition. *Journal of Contemporary China*, 20(71), 659–77.

Thogersen, S. 2008. Frontline soldiers of the CCP: the selection of China's township leaders. *China Quarterly*, 194, 414–23.

Thogersen, S. 2009. Revisiting a dramatic triangle: state, villagers and social activists in Chinese rural reconstruction projects. Paper presented at European Conference on Agriculture and Rural Development in China, Leeds University, 3 April 2009.

Thogersen, S. 2011. Building a New Socialist Countryside: model villages in Hubei. *In*: B. Alpermann, ed. *Politics and markets in rural China*. London: Routledge. pp. 172–86.

Unger, J. 2002. *The transformation of rural China*. Armonk: M.E. Sharpe.

Walder, A.G. 1995. Local governments as industrial firms: an organizational analysis of China's transitional economy. *American Journal of Sociology*. 101, 263–301.

Wang, A. 2013. The search for sustainable legitimacy: environmental law and bureaucracy in China. *Harvard Environmental Law Review*, 37, 365–440.

Wang, S. G. 2009. Adapting by learning: the evolution of China's rural health care financing. *Modern China*, 35(4), 370–404.

White, T. 1990. Postrevolutionary mobilization in China: the one-child policy reconsidered. *World Politics*, 43(1), 53–76.

Whiting, S. 2004. The cadre evaluation system at the grassroots: the paradox of party rule. *In*: B. Naughton and D.L. Yang, eds. *Holding China together: diversity and national integration in the post-Deng era*. Cambridge: Cambridge University Press. pp. 101–19.

World Bank 2002. *China – national development and sub-national finance: a review of provincial expenditures*. Poverty Reduction and Economic Management Unit, East Asia and Pacific Region 22951-CHA. Washington, DC: The Bank.

Wong, C. 2009. Rebuilding government for the 21st century: can China incrementally reform the public sector? *The China Quarterly*, 200, 929–52.

Wu, Y. 2007. *Xiao zhen xuanxiao: yi ge xiangzhen zhengzhi yunzuo de yanyi yu chanshi* [*Uproar in a small town: interpretation of a township's political operation*]. Beijing: Sanlian Shudian.

Xinhua 2004. Xin jing bao: jiekai Jiangsu sheng yuan zuzhibuzhang Xu Guojian fubai guanxi wang' [Beijing News: uncovering the corrupt networks of Jiangsu's former Organization Department head, Xu Guojian] [online]. Available from: http://news.xinhuanet.com/legal/2004-07/12/content_1591623.htm [Accessed 11 February 2013].

Xinhua 2006. Jiangsu sheng weiyuan changwei, zuzhibu buzhang Xuo Guojian yi shenpan sihuan' [Jiangsu Province standing committee member and Organization Department head, Xu Guojian, receives suspended death sentence] [online]. Available from http://news.xinhuanet.com/legal/2006-01/24/content_4093987.htm [Accessed 11 February 2013].

Xu, Y. 2003. *Xiangcun zhili yu zhongguo zhengzhi* [*Rural governance and Chinese politics*]. Beijing: China Social Sciences Press.

Yan, Y. 2003. Dang maiguan yin zang zai 'zhengchang chengzu' zhi zhong. [When selling posts is hidden among 'normal procedures']. *Renmin fayuan bao* [People's Court Daily]. Available at http://www.people.com.cn/GB/guardian/26/20030411/969136.html [Accessed 13 December 2013].

Yang, D.L. 2004. *Remaking the Chinese leviathan: market transition and the politics of governance in China*. Stanford, CA: Stanford University Press.

Zang, X.W. 2006. Technical training, sponsored mobility, and functional differentiation: elite formation in China in the reform era. *Communist and Post-Communist Studies*, 39, 39–57.

Zhao, S.K. 2007. Rural governance in the midst of underfunding, deception, and mistrust. *Chinese Sociology and Anthropology*, 39, 8–93.

Zhao, S.K. 2012. Difang zhengfu gongsihua: tizhi youshi haishi lieshi? [The corporatization of local government: a strength or a weakness of the system?] [online]. *Wenhua zongheng* [*Beijing Cultural Review*], 73–80. Available from: http://www.21ccom.net/articles/zgyj/dfzl/2012/0427/58619.html [Accessed 31 October 2013].

Zhong, Y. 2003. *Local government and politics in China: challenges from below*. Armonk: M.E. Sharpe.

Zhou, X.G. 2010. The institutional logic of collusion among local governments in China. *Modern China*, 36(1), 47–78.

Zhou, X.G. and H. Lian. 2011. The institutional logic of collusion in the Chinese bureaucracy: further explorations. Paper presented at the Workshop on politics and autonomy in China's local state: county and township cadres as a strategic group. University of Tübingen, 1–3 July 2011.

Graeme Smith is a research fellow in the China Studies Centre, University of Sydney Business School and with the State, Society and Governance in Melanesia Program at the Australian National University. His research has explored the demand for organic produce in Chinese urban centres, the political economy of agricultural service delivery, the role of rural cadres in China's development, and the persistence of informal land markets in rural China. He also studies the political economy of Chinese outbound direct investment, migration and development assistance in the Asia-Pacific region, with ongoing projects in Yunnan, Fujian and Anhui, as well as Papua New Guinea, Samoa, Tonga, New Caledonia and Myanmar. He received the 2013 best article prize in the Journal of Pacific History, and was the 2011 winner of the Gordon White Prize for the best article published in China Quarterly, the leading journal in China Studies. He co-edited recent special issues of Pacific Affairs and Asian Studies Review. Dr Smith also holds a PhD in inorganic chemistry, and has written several guidebooks to China.

Rightful resistance revisited

Kevin J. O'Brien

James Scott (1985) placed 'everyday forms of resistance' between quiescence and rebellion. Others have noted that defiance in unpromising circumstances need not be quiet, disguised and anonymous if the aggrieved use the language of power to mitigate the risks of confrontation. How does 'rightful resistance' (O'Brien and Li 2006) relate to Scott's everyday resistance and other types of protest in contemporary China? Are rightful resisters sincere or strategic? Is their contention reactive or proactive? Does rightful resistance suggest growing rights consciousness or only a familiar rules consciousness? *Rightful resistance in rural China* has been criticized for (1) lacking 'peasantness', (2) shortchanging history and culture, (3) focusing on elite allies and one pattern of protest and (4) being overly rationalist, state-centric and caught in 'developmental thinking'. How do I respond?

In the 1980s, James Scott transformed peasant studies with his work on 'weapons of the weak'. Several years later, with Scott's (1985, 1990) books firmly in mind, Lianjiang Li and I started noticing something a bit different in China. Rather than cloaking their dissent in dissimulation, deniability and ambiguous gestures, people we called rightful resisters were challenging the powerful head-on. In particular, when faced with illegal extraction, rigged elections or corrupt cadres, villagers were deploying the policies, laws and commitments of the state to combat local officials who were ignoring those policies, laws and commitments. Whereas Scott's everyday forms of resistance were quiet, disguised and anonymous, rightful resistance was noisy, public and open. Whereas everyday resistance focused on relations between subordinates and superordinates, rightful resisters were engaged in a three-party game where divisions within the state and elite allies mattered greatly. Over the next decade, as we gained access to protest leaders in parts of Hunan where rightful resistance was especially vibrant, we concluded that this type of protest had four main attributes: it operated near the boundary of authorized channels, employed the rhetoric and

I am grateful to Jun Borras for suggesting that I 'revisit' rightful resistance and to Sara Newland for research assistance on other types of resistance. I would like to thank Tamara Jacka, John Kennedy, Sally Sargeson, James Scott, Rachel Stern and Emily Yeh for comments on an earlier draft. Generous financial support was provided by the Institute of East Asian Studies and the Center for Chinese Studies at the University of California, Berkeley. For a nearly career-long collaboration and for teaching me far more than I ever taught him, I am indebted to Lianjiang Li.

commitments of the powerful to curb the exercise of power, hinged on locating and exploiting divisions within the state, and relied on mobilizing support from the community (O'Brien and Li 2006, 2).

Our research built directly on Scott's and, like his, slotted an under-appreciated form of contention on the continuum between quiescence and rebellion. In particular, we sought to explain how skillful use of the language of power can at times allow the aggrieved to act up effectively without taking intolerable risks. As China scholars interested in both bottom-up and top-down sources of change, we also wanted to say as much as possible about the outcomes of protest, including its effects on rightful resisters, the community, and policy implementation (and sometimes policy itself).

In recent years, others have picked up the idea of rightful resistance and have explored aspects of its origins, dynamics and consequences. This research has taught us much but, alas, will receive little attention here. Instead, I will focus on several lines of criticism that have emerged, and their implications for taking the study of contention in China in new directions, both within the rightful resistance framework and outside or at the edges of it.

The 'peasantness' of rightful resisters

In a review in this journal, Susanne Brandtstädter (2006) recognized an important fact about *Rightful resistance in rural China*: the book was designed to speak to social movement studies and political science as much as to peasant studies, and we hoped that the concept would prove useful for understanding contention by various social groups in China and elsewhere. Still, the argument's empirical core is a treatment of protest in China's countryside, engaged in by rural folk. So Brandtstädter's critique, that the analysis suffers from a lack of 'peasantness', is overly rationalistic and shortchanges history and culture, deserves attention. Indeed, the book does have a certain 'what would I do if I was in this situation' deductive logic to it. We, by and large, reason from interests and assume a fairly straightforward weighing of costs and benefits by prospective rightful resisters. We do not problematize notions such as risk or payoff (or situate them deeply in the rural Chinese context) and instead imagine a rather simple mapping of opportunities onto actions. In our defense, we do bring in perceptions of opportunity, as a side-door way to consider culture and history, and we treat expectations of assistance from the Centre as rooted in long-standing understandings of an emperor's responsibilities. But to the extent that religion, morality or communal, 'little tradition' ways of thinking inform decisions by rightful resisters there is room for more discussion of what peasants bring to protest themselves. Despite our ground-level orientation, there is still a need for an ethnography of rural rightful resistance that focuses on 'peasants involved in these protests and their reality: how they live, who they are, and what they think', and 'the historical particularities of the Chinese situation' (Brandtstädter 2006, 712, 711).[1] In recent years, scholars have taken steps in this direction with research on 'contractual ways of thinking' (Pan 2008), the effects of neoliberal capitalism on agrarian contention (Walker 2008), 'righteous', moral economy-based resistance in the 1950s (Thaxton 2008,

---

[1] See also Ortner (1995, 190): 'Resistance studies are thin because they are ethnographically thin: thin on the internal politics of dominated groups, thin on the cultural richness of those groups, thin on the subjectivity – the intentions, desires, fear, projects – of the actors engaged in these dramas'.

H.Y. Li 2009), [2] the moral underpinnings of contemporary protest (Tong forthcoming), why residents of 'cancer villages' do not get angrier about the health consequences of pollution (Lora-Wainwright 2010), and our own work on the resilience of political trust (L.J. Li 2013) and the group dynamics of protest leadership (L.J. Li and O'Brien 2008; also Wang 2012). 'Isolated individuals' (Brandtstädter 2006, 712) are often not the best unit of analysis to understand collective action. Rightful resistance is more than a matter of tallying up choices made by individual *Homo economicus*, each of whom decides to take a chance on protest. It is also a product of collectivities living in a specific political economy and socio-cultural setting: of people with histories and moral understandings who are grappling (individually and together) with marketization, legal reforms, post-socialism, neoliberalism and many other challenges and opportunities.

State-centredness and the Chinese State

Embedding rightful resistance deeper in the habits and values of peasant life would also address a second issue Brandtstädter (2006) raises: the state-centredness of the analysis. In *Rightful resistance in rural China*, officials are essentially granted 'first mover' status. They emit signals and ordinary people respond. This assumption was carried over from the structural sociology that underlies social movement studies, and in particular its focus on the external aspects of opportunity. Awarding officialdom pride of place is also common in a country where the state figures so large and is always present when protest occurs. But signals are not just sent by officials and received and processed by rightful resisters. Protesters fill spaces that the state and its reforms create, but also push against boundaries in ways that cannot be read straight off an opportunity structure. Rightful resisters may, through persistence and probing the limits of the permissible, create their own opportunities. We need to learn more about how the aggrieved understand signals and also how they sometimes grab the initiative and wrong-foot their targets in unexpected ways. [3] Reducing state-centredness and moving toward a richer, more interactive account of state-society communication, [4] much like taking 'peasantness' more seriously, would accord greater agency to rightful resisters, embed their contention more securely in their world as well as the state's, and drain a dollop of initiative away from officialdom and the opportunities a divided state provides.

In a state-centric analysis, it is noteworthy that there is not a more explicit theory of the Chinese state standing behind *Rightful resistance in rural China* or, for that matter, much recent research on protest in China. [5] This is a lost opportunity, both for understanding how rightful resistance emerges and for making research on protest legible to China scholars who might not be overly concerned with contention but are certainly interested in the Chinese state (O'Brien 2011). As Rachel Stern and I have argued (Stern and O'Brien 2012), there is a partial, implicit theory of the state lurking in *Rightful resistance in rural China* and other work on activism in China that centres on mixed signals and words such as ambiguity, ambivalence and uncertainty. But this theory awaits fleshing out and research on the origins of mixed signals remains preliminary, speculative and deductive.

---

[2]For a historian's account of the path from righteous to rightful resistance in the 1950s, see H.Y. Li (2009).
[3]On elderly protesters in Zhejiang and tactics they deployed to keep the authorities off balance, see Deng and O'Brien (2013 forthcoming). On Chinese protest tactics more broadly, see X. Chen (2012).
[4]For 'experience-near', less state-centric accounts of the thinking that gives rise to protest, see G.D. Chen and Wu (2006), Chan and Pun (2009), Erie (2012), and Lora-Wainwright et al. (2012).
[5]For exceptions, see Perry (2002), Lee (2007), Stern and O'Brien (2012) and Stern (2013).

Nor does ruminating about the state typically take place at the level of abstraction of 'modes of accumulation and legitimation' that Lee (2007, 17) has observed is lacking in *Rightful resistance in rural China,* or commonly plunge into topics as grand as how post-socialism and state capitalism shape protest (Walker 2008).

Instead of starting from a theory of the Chinese state, *Rightful resistance in rural China* is better suited to help cobble together a theory of the state. By observing how state power is experienced by people testing the limits of the permissible, we can learn about the contours of an authoritarian regime and how the state appears from below. An indirect, bottom-up take on state power can be as revealing as examining political institutions, bureaucracies and 'modes of accumulation' directly. But to make a 'state reflected in society approach' (Stern and O'Brien 2012) more than a metaphor, we need to move beyond principle-agent logics about information-gathering and monitoring local officials characteristic of *Rightful resistance in rural China* and other research on signaling in China (Lorentzen 2013, Weiss 2008). When unpacking the state, it will be important, for instance, to disaggregate horizontally as well as vertically. This will confirm that territorial levels of government are not unified: they have as many divisions and conflicts within them as they have with superiors above and subordinates below. This is now taking place, as students of Chinese protest move away from notions like 'the Centre' and 'lower levels' to identify actors at each level who facilitate or inhibit contention (Mertha 2008, Sun and Zhao 2008, Cai 2010). Additional work of this sort will take us beyond informative but incomplete understandings of the Chinese state, including one embodied in a saying often heard in rural areas in the 1990s: 'the centre is our benefactor, the province is our relative, the county is a good person, the township is an evil person, and the village is our enemy' (O'Brien and Li 1995, 778). As Stern (2013) has noted, the standard story of parochial local officials who subvert the Centre's good intentions can paper over other divisions and leaves too little room for variation. We should avoid blaming grassroots cadres for more than they deserve, and be alert to moments when the conventional central-local narrative breaks down: moments in which simple dichotomies no longer hold and a deeper understanding of pressures and divisions within the state is called for (Stern 2013).

Variation-finding

Disaggregating the state along multiple axes reminds us that although conceptual work requires lumping rather than splitting, variation-finding becomes the next task once a concept is put forward. One of the referees of *Rightful resistance in rural China* offered dozens of suggestions that basically came down to: pay more attention to differences across time and space. This reader wanted a scant, two-page 'Note on variation' to become a new, second half of the book. Our universalizing comparison (Tilly 1984, 97, 108), however, sought to do something else: to show that phenomena seldom discussed together (e.g. the pay equity struggle in the United States, 'censoriousness' by Norwegian prison inmates, anti-Apartheid activism in South Africa, 'consentful contention' in East Germany, rural protest in China) (O'Brien and Li 2006, 15–24)[6] could be understood within one framework once we located acts of resistance – in China or elsewhere – that shared:

---

[6]In the most repressive regimes, resistance is largely limited to 'weapons of the weak' (Scott 1985). In slightly less controlled settings, one or more features of rightful resistance may appear. As sanctioned coercion diminishes further and partial inclusion is formally extended, cases of more complete rightful resistance become possible. In circumstances where numerous rights are guaranteed, the rule of law is

- reliance on established principles to anchor defiance
- use of legitimating myths and persuasive normative language to frame claims
- deployment of existing statutes and commitments when leveling charges
- recognition and exploitation of congenial aspects of a shifting opportunity structure
- importance of allies, however uncertain, within officialdom.

Finding unexpected similarities rather than variation was the goal. We opted not to distinguish among grievances or locations in China, and deployed evidence in a manner that Scott (1976, 1990, 1998) does so well: as examples rather than cases, as telling, ground-level illustrations that hammer an analytical point home.

But this of course leaves much undone. As of the early 2000s, direct rightful resistance appeared to be especially common in the provinces of Sichuan, Anhui, Hunan, Jiangxi, Henan, Shaanxi and Hebei. Why? Perhaps because the 'peasant burden' (*nongmin fudan*) problem was more acute in 'agricultural China' (Bernstein and Lü 2003) and people with deeper grievances were more likely to give up on elite allies and rely on themselves? But we did not explore this possibility or consider competing explanations. Focusing on similarities also downplayed different motivations for rightful resistance and responses to it. Examining election, tax, corruption and other protests together, and using evidence to clarify a concept rather than establish a causal pattern, can gloss over sources of variation. Which kinds of misconduct are most amenable to rightful resistance?[7] Are higher levels responsive to certain kinds of rightful claims but deaf to others? We are largely silent on these issues.

Our focus on one form of contention also obscures the obvious truth that rightful resistance is not the only type of protest in contemporary China. *Rightful resistance in rural China* is a conceptual book, where we select on the dependent variable unabashedly[8] and say little about how common rightful resistance is compared to other forms of popular action.[9] We stop following rightful resisters when they drop out or move on to other types of within-system resistance or violence. Many other patterns of protest in rural China deserve equal attention, and some are receiving it, including Yu Jianrong (2008) on violent, 'anger-venting incidents', Justin Hastings (2005) on Uighur protests in the shadow of a 'unified state', Paik and Lee (2012, 262, 270–71) on episodes where the aggrieved and local cadres join forces to challenge higher levels, and Ethan Michelson (2008) on 'justice from below' and local solutions.[10] As these research programs unfold and multiply, we will be able speak more knowledgeably about the full palette of popular

---

established and political participation is unquestionably legitimate, rightful resistance may still be viable and effective, when dissatisfied citizens try to make officials and business leaders prisoners of their own rhetoric and exploit the gap between rights promised and rights delivered.

[7]Following the abolition of the agricultural tax in 2006, extraction receded as a source of rural discontent and other issues came to the fore, including land grabs and environmental degradation. The time is ripe to explore how the nature and intensity of particular grievances influence the dynamics of contention, and to identify how different complaints can be linked, as happens when environmental claims are 'piggybacked' on land-related grievances (Deng and Yang 2013). On 'issue linkage' more generally, see Cai (2010, Ch. 4).

[8]For a defense of this practice, in certain circumstances, see Collier and Mahoney (1996).

[9]See Michelson (2008, 44, 45) on our focus on one strategy of contention and not seeking to evaluate the relative popularity of competing strategies. On selecting cases 'according the value of the dependent variables' and 'caution in generalizing from [our] analysis', see R.B. Wong (2007, 851).

[10]John Kennedy (conference remarks, 28 September 2012) suggested that our differences with Michelson (2008) may be temporal rather than conceptual: the aggrieved may try 'local solutions'

contention and the import of protest for China's future. For our part, we have been surprised there is as much rightful resistance as surveys show,[11] and we look forward to more research that addresses 'the frequency problem' (Michelson 2008, 51). This will involve taking aggrieved people as the starting point, observing how they move up (or exit) the dispute pyramid (Felstiner *et al.* 1980–81, Diamant *et al.* 2005, 7, Michelson 2007), and considering many types of contention, as well as giving up or never taking to the streets. This line of work promises insight into how often rural Chinese protest instead of giving up on a grievance, and how they act when they do.

The role of elite allies merits mention here. We of course are not the first to examine occasions on which ordinary people draw in the powerful to assist them in their struggles.[12] But the presence or absence of allies is central to *Rightful resistance in rural China* and underpins the analysis. We anticipate more surveys, like Michelson's (2008), on the extent to which the disaffected depend on 'justice from below' rather than (or in addition to) 'justice from above'. How crucial are elite allies for the aggrieved? What happens when elites at higher levels are targets rather than allies?[13] How available and effective are powerful allies if the targets are state-backed companies rather than governments? Are some third parties more useful for attacking certain kinds of misconduct? How are elite allies secured?[14] What types of assistance do they provide?[15] Perhaps owing to factional conflict, do officials sometimes seek out protesters rather than the other way around?[16] How often are run-of-the-mill disputes resolved locally, so that escalation of claims and venues does not occur (Michelson 2008, 51)?

We need more research on the importance and availability of allies, and also on the consequences of depending on them. Does turning to third parties channel protest into safe precincts, lessen the chance of disruption, moderate demands and tactics and lead to unsatisfactory half-solutions? Are rightful resisters realizing over time that cultivating allies diverts energy from more productive strategies, and that elites are only willing to help so long as they are ineffective (McAdam 1982, 25–29)? Woodman (2011), for instance, has observed that lawyers who aid protesters sometimes hijack the framing of a dispute and deprive activists of their subjectivity, as they suppress social justice and gender-based claims in favor of their own concerns with democratic and legal reform. A nuanced understanding of the role elite allies play is crucial because their participation is a key reason that rightful resistance can be more effective than everyday resistance but safer than protest that edges toward rebellion.

---

first and then move on to rightful resistance. Michelson (2008) also considers a wider range of disputes than we do.

[11]See Michelson (2008).

[12]See, among others, Santoro and McGuire (1997) on institutional activists who work as insiders on outsider issues, Stearns and Almeida (2004) on state actor-social movement coalitions and Rucht (2004) on movement allies.

[13]On villagers siding with different levels of government depending on the type of land expropriation they face, see Paik and Lee (2012, 273).

[14]Michelson (2006) emphasizes family ties to government officials. On support 'from individuals within the government, not "the government" more generally', see Spires (2011, 14).

[15]Shi and Cai (2006, 316) suggest that social networks with officials and media personnel provide information that helps protesters formulate and implement strategies, create a channel to influence decision-making and generate pressure, and offer access to the media.

[16]Lianjiang Li, personal communication, 30 September 2012. On village cadres leading protests, see Wang (2012).

Conceptual questions

Sincere or strategic?

After reading the first article on rightful resistance (O'Brien 1996), James Scott (personal communication, 14 December 1995) asked a vexing question: are rightful resisters sincere or cynical? Nearly two decades later, I still can't say. Consider this remark by a rightful resister: 'If I didn't believe in the Centre, what would I have left?' Does this man really have faith in the Centre or does he merely recognize it is prudent to frame his attack on local misconduct as an effort to ensure state commitments are carried out? My best guess is that the motivations of rightful resisters are a mixture of belief and calculation that differs from person to person, and that often a protester cannot readily say if resistance is sincere or strategic because it has elements of both impulses.[17] This is why outcomes and the passage of time are important: what a rightful resister believes today depends on the last round of contention (and events elsewhere) and few people will act on, or even form, sincere beliefs if they are strategically suicidal. In a changing China, popular consciousness is neither static nor one-dimensional. A rightful resister may hold beliefs that spring from multiple and mixed motives.[18]

Reactive or proactive?

Whether rightful resistance is reactive or proactive is another knotty question best not forced into an either/or answer. What is new and 'has no grounding in current rules' (L. J. Li 2010, 59) and what is old and 'based on existing political rules' (L.J. Li 2010, 58) is always hard to agree upon.[19] As Sally Sargeson has pointed out (conference comments, 28 September 2012), it is also somewhat artificial and ahistorical to speak of an initial action when a long series of events precedes any episode of claimsmaking. Instead of freeze-framing a given moment and asking if rightful claims are transgressive, it is better to live with the contradiction. That rightful resistance is both loyal and can ask for something new is at the oxymoronic heart of an oxymoronic concept. Despite Tilly's (1993) forsaking of his competitive, reactive, proactive framework, we agree with Sewell (1990) that the trichotomy remains useful to think with, as a means to examine claims and ironies that arise during the spread of citizenship norms (which need not be liberal democratic) in authoritarian circumstances.[20] For us, at least, the idea of cloaking proactive claims in reactive garb helped make sense of a string of characters scrawled on a Hebei storefront: 'We're citizens. Return us our citizenship rights. We're not rural labor power, even less are we slaves'. In one sense, the graffiti artist was using a familiar tactic (writing a wall poster) and seeking a 'return' of his rights, in another sense, he was demanding rights he had never enjoyed while making it appear he had just been deprived of them.

---

[17]Lee (2007, 17) sees rightful resistance as mainly a form of strategic framing. It was not our intent to leave readers with this impression.

[18]'Analytically, distinct types of political consciousness often co-exist in the mind of the same individual' (L.J. Li 2010, 65).

[19]On recognition that contention in China is evolving, but doubts that proactive protest is overtaking reactive protest, see Bianco (2001, 249–53).

[20]Perry (2002, 275–308) also found it helpful to categorize protest in China as competitive, reactive or proactive.

Developmental thinking?

One reason that Tilly (1993, 266) abandoned his competitive, reactive, proactive tri-chotomy and introduced 'repertoires of contention' is that he came to believe the framework was teleological and amounted to an inadvertent endorsement of modernization theory. Brandtstädter (2006) sees similar signs of 'developmental thinking' in *Rightful resistance in rural China*, while others have suggested that the book points to China's coming demo-cratization. In my view, the presence and spread of rightful resistance implies little about democracy or systemic change.[21] Although its outcomes can have implications for policy (e.g. ending the agricultural tax and providing resources for reformers), it does not foreshadow political convergence. In fact, if rightful resistance performs as intended, it could help legitimate the current regime by proving to those perched on the verge of rebel-lion that the system works well enough. That a group of complainants could be thrilled when an official wrote on their petition 'it's legal to disseminate central policies' reminds us that rightful resisters sometimes interpret tiny concessions as great victories.[22] Like everyday resistance, rightful resistance is a within-system form of contention that operates in the reform, not the revolution, paradigm. For a clumsy authoritarian regime that has long been struggling to grow fingers,[23] it offers a means for higher levels to learn about local misconduct and deal with social pressures. Should it continue to develop, it could dampen demands for far-reaching change and contribute to regime resi-lience rather than hastening the end of authoritarian rule.[24] Pulling back to consider 'larger implications' and imagining China's future are often encouraged by editors and can be tempting for authors, but searching for hints of democratization says more about China studies than it does about China.

---

[21]Lianjiang Li is more inclined to link rightful resistance and growing rights consciousness with regime change. 'Compared to those who only have rules consciousness, individuals who also have rights consciousness are more likely to press for institutional changes in the hope of converting revocable "state-endowed rights" into inalienable rights. If rights consciousness keeps a democracy healthy by turning citizens into active participants in governance, the mobilization of rights con-sciousness may help chart a course toward a more participatory political system in China' (L.J. Li 2010, 66).

[22]There are signs, however, that this is becoming less common, as small, often symbolic victories may lead to larger demands. The popular notion that 'no disturbance leads to no solution' while a 'large disturbance leads to a large solution' also points to the value of risk-taking, persistence, and an unwill-ingness to settle for partial solutions.

[23]On authority systems with 'strong thumbs, no fingers', see Lindblom (1977). This image refers to the ability of centralized, non-market systems to do homogeneous, repetitive activities well, but not to discriminate or adapt.

[24]Froissart (2007, 119) writes: 'By seeking intervention from higher levels of administration, this form of resistance helps the Centre respond to conflicts through ad hoc solutions that expand its capacity to manage contradictions while withholding political reform... It would thus appear that this form of resistance is an integral part of the regime's dynamic stability, though that does not rule out the possibility of the balance being upset some day'. Perry (2008, 45) also suggests that today's pattern of protest 'may prove more system-supportive than system-subversive. In an authoritarian polity, where elections do not provide an effective check on the misbehavior of state authorities, protests can help serve that function – thereby undergirding rather than under-mining the political system'. In a highly hedged passage (O'Brien and Li 2006, 123–29), we mention that rightful resistance 'could evolve into a more far-reaching counter-hegemonic project', but I do not believe it is 'poised to mount a counter-hegemonic project' (Perry 2010, 24).

Rights consciousness or rules consciousness?

Elizabeth Perry (2007, 2008, 2009, 2010) has been at the forefront of a debate over whether rightful resistance and similar activities reflect growing rights consciousness or a rules consciousness that has existed for centuries. Lianjiang Li (2010) has responded to Perry's discussion of protesters' sensitivity to government rhetoric with a creative survey that finds both types of consciousness and suggests they can be distinguished by assessing whether claims are focused on rule-making or rule enforcement. Peter Lorentzen and Suzanne Scoggins (2012) have deployed formal modeling to show that findings about rights consciousness may refer to changes in values, rhetoric or beliefs about government policy. Both qualitative (Goldman 2007) and survey researchers (L. Wong 2011) continue to produce studies that document growing rights consciousness.

I find myself mostly unmoved by a 'debate' in which the parties agree on the empirics and disagree mainly about what to label their findings (O'Brien 2011, 536). Once we recall that there are many forms of citizenship (Mann 1996),[25] based on both individual and collective notions of rights,[26] and that rights consciousness need not evoke Locke and Jefferson (Perry 2008, 43) or imply convergence or 'something resembling an Anglo-American "rights revolution"' (Perry 2008, 47), the heat in this debate dissipates. There is no need to make a choice between rights or rules consciousness (O'Brien 2001, 408, 426–29) or to argue that one precludes the other, either logically or historically. As L.J. Li (2010) and Lorentzen and Scoggins (2012) demonstrate, both may be present at the same time, and my hunch is that coexistence is the path along which citizenship rights appeared in many places. In fact, it may well be ahistorical to suggest otherwise, at least for China. As a legal status that draws boundaries and ranks a population, and a set of practices that implies a willingness to confront the powers-that-be, citizenship and the expectations that surround it are a product of historical processes that bring forth changes in popular thinking and claimsmaking. In this regard, rightful resistance is a chapter in the history of Chinese rules and rights consciousness that demonstrates changes in modes of thought do not take place by leaps but through intermediate steps. Rules-based, righteous and moral economy claims coexist with (and inform) rightful ones, and the crucial issue is the mixture of claims in a game that is still in early innings.

References

Bernstein, T.B. and X.B Lü. 2003. *Taxation without representation in contemporary rural China.* New York: Cambridge University Press.

Bianco, L. 2001. *Peasants without the party: grassroots movements in twentieth century China.* Armonk, NY: M.E. Sharpe.

---

[25]Mann (1996) identifies five varieties of citizenship, only one of which is associated with free association, strong legislatures and liberal democracy.

[26]'Although rights talk always implies a certain assertiveness, in contemporary American discourse, for example, it tends toward the absolute, individualistic, and ontological in that individuals possess rights by virtue of being human beings. In contrast, Chinese rights talk tends to be relative, social and phenomenological in that subject-citizens of a given state only have rights defined in relation to obligations to each other and their relationship to rulers' (O'Brien and Li 2006, 117). 'Unlike the rights discourse employed by some liberal intellectuals, there is little evidence that most rural rightful resisters consider rights to be inherent, natural, or inalienable; nor do most of them break with the common Chinese practice of viewing rights as granted by the state mainly for societal purposes rather than to protect an individual's autonomous being' (O'Brien and Li 2006, 122).

Brandtstädter, S. 2006. Book review of *Rightful resistance in rural China*. *Journal of Peasant Studies*, 33(4), 710–2.

Cai, Y.S. 2010. *Collective resistance in China: why popular protests succeed or fail*. Stanford, CA: Stanford University Press.

Chan, K.C.C. and N. Pun. 2009. The making of a new working class? A study of collective actions of migrant workers in south China. *China Quarterly*, 198, 287–303.

Chen, G.D. and C.T Wu. 2006. *Will the boat sink the water?: The life of China's peasants*. New York: Public Affairs.

Chen, X. 2012. *Social protest and contentious authoritarianism in China*. New York: Cambridge University Press.

Collier, D. and J. Mahoney. 1996. Insights and pitfalls: selection bias in qualitative research. *World Politics*, 49(1), 56–91.

Deng, Y.H. and K.J. O'Brien. Forthcoming. Societies of senior citizens and popular protest in rural Zhejiang. China Journal.

Deng, Y.H. and K.J. O'Brien. 2013. Relational repression in China: using social ties to demobilize protesters. *China Quarterly*, 215.

Deng, Y.H. and G.B. Yang. 2013. Pollution and protest: environmental mobilization in context. *China Quarterly*, 321–36.

Diamant, N.J., S.B. Lubman and K.J. O'Brien. 2005. Law and society in the People's Republic of China. *In*: *Engaging the law in China: state, society and possibilities for justice*. Stanford, CA: Stanford University Press, pp. 3–27.

Erie, M.S. 2012. Property rights, legal consciousness and the new media in China: the hard case of the 'toughest nail house in history'. *China Information*, 26(1), 35–60.

Felstiner W., R. Abel and A. Sarat. 1980–81. The emergence and transformation of disputes: naming, blaming, claiming. *Law & Society Review*, 15(3–4), 631–54.

Froissart, C. 2007. Book review of *Rightful resistance in rural China*. *China Perspectives*, 4, 117–20.

Goldman, M. 2007. *Political rights in post-Mao China*. Ann Arbor, MI: Association for Asian Studies.

Hastings, J.V. 2005. Perceiving a single Chinese state: escalation and violence in Uighur protests. *Problems of Post-Communism*, 52(1), 28–38.

Lee, C.K. 2007. *Against the law: labor protests in China's rustbelt and sunbelt*. Berkeley, CA: University of California Press.

Li, H.Y. 2009. *Village China under socialism and reform: a micro-history, 1948–2008*. Stanford, CA: Stanford University Press.

Li, L.J. 2010. Rights consciousness and rules consciousness in contemporary China. *China Journal*, 64, 47–68.

Li, L.J. 2013. The magnitude and resilience of trust in the center: evidence from interviews with petitioners in Beijing and a local survey in rural China. *Modern China*, 39(1), 3–36.

Li, L.J. and K.J. O'Brien. 2008. Protest leadership in rural China. *China Quarterly*, 193, 1–23.

Lindblom, C.E. 1977. *Politics and markets: the world's political-economic systems*. New York: Basic Books.

Lora-Wainwright, A. 2010. An anthropology of 'cancer villages': villagers' perspectives and the politics of responsibility. *Journal of Contemporary China*, 19(63), 79–99.

Lora-Wainwright, A. *et al.* 2012. Learning to live with pollution: the making of environmental subjects in a Chinese industrialized village. *China Journal*, 68, 106–24.

Lorentzen, P.L. 2013. Regularized rioting: permitting public protest in an authoritarian regime. *Quarterly Journal of Political Science*, 8, 127–58.

Lorentzen, P.L. and S.E. Scoggins. 2012. Rising rights consciousness: undermining or undergirding China's stability [online]. Working paper. Available from http://papers.ssrn.com/sol3/papers.cfm?abstract_id=1722352 [Accessed 12 June 2012].

Mann, M. 1996. Ruling class strategies and citizenship. *In*: M. Bulmer and A.M. Rees, eds., *Citizenship today: the contemporary relevance of T.H. Marshall*. London: UCL Press, pp. 125–44.

McAdam, D. 1982. *Political process and the development of black insurgency, 1930–1970*. Chicago, IL: University of Chicago Press.

Mertha, A.C. 2008. *China's water warriors: citizen action and policy change*. Ithaca, NY: Cornell University Press.

Michelson, E. 2006. Connected contention: social resources and petitioning the state in rural China. Unpublished paper.

Michelson, E. 2007. Climbing the dispute pagoda: grievances and appeals to the official justice system in rural China. *American Sociological Review*, 72, 459–85.

Michelson, E. 2008. Justice from above or below? Popular strategies for resolving grievances in rural China. *China Quarterly*, 193, 43–64.

O'Brien, K.J. 1996. Rightful resistance. *World Politics*, 49(1), 31–55.

O'Brien, K.J. 2001. Villagers, elections and citizenship in contemporary China. *Modern China*, 27(4), 407–35.

O'Brien, K.J. 2011. Studying Chinese politics in an age of specialization. *Journal of Contemporary China*, 20(71), 535–41.

O'Brien, K.J. and L.J. Li. 1995. The politics of lodging complaints in rural China. *China Quarterly*, 143, 756–83.

O'Brien, K.J. and L.J. Li. 2006. *Rightful resistance in rural China*. New York: Cambridge University Press.

Ortner, S.B. 1995. Resistance and the problem of ethnographic refusal. *Comparative Studies in Society and History*, 37(1), 173–93.

Paik, W. and K. Lee. 2012. 'I want to be expropriated'. The politics of xiaochanquanfang land development in suburban China. *Journal of Contemporary China*, 21(74), 281–98.

Pan, C.X. 2008. Contractual thinking and responsible government in China: a constructivist framework for analysis. *China Review*, 8(2), 49–75.

Perry, E.J. 2002. *Challenging the mandate of heaven: social protest and state power in China*. New York: M.E. Sharpe.

Perry, E.J. 2007. Studying Chinese politics: farewell to revolution? *China Journal*, 57, 1–24.

Perry, E.J. 2008. Chinese conceptions of 'rights': from Mencius to Mao – and now. *Perspectives on Politics*, 6(1), 37–50.

Perry, E.J. 2009. A new rights consciousness? *Journal of Democracy*, 20(3), 17–20.

Perry, E.J. 2010. Popular protest in China: playing by the rules. *In*: J. Fewsmith, ed., *China today, China tomorrow: domestic politics, economics and society*. Lanham, MD: Rowman and Littlefield, pp. 11–28.

Rucht, Dieter. 2004. Movement allies, adversaries, and third parties. *In*: D.A. Snow, S.A. Soule and H. Kriesi, eds. *The Blackwell companion to social movements*. Malden, MA: Blackwell, pp. 197–216.

Santoro, W.A. and G.M. McGuire. 1997. Social movement insiders: the impact of institutional activists on affirmative action and comparable worth policies. *Social Problems*, 44(4), 503–19.

Scott, J.C. 1976. *The moral economy of the peasant: subsistence and rebellion in Southeast Asia*. New Haven, CT: Yale University Press.

Scott, J.C. 1985. *Weapons of the weak: everyday forms of resistance*. New Haven, CT: Yale University Press.

Scott, J.C. 1990. *Domination and the arts of resistance: hidden transcripts*. New Haven, CT: Yale University Press.

Scott, J.C. 1998. *Seeing like a state: how certain schemes to improve the human condition have failed*. New Haven, CT: Yale University Press.

Sewell, W. H., Jr. 1990. Collective violence and collective loyalties in France. Why the French revolution made a difference. *Politics & Society*, 18(4), 527–52.

Shi, F.Y. and Y.S. Cai. 2006. Disaggregating the state: networks and collective resistance in Shanghai. *China Quarterly*, 186, 314–32.

Spires, A.J. 2011. Contingent symbiosis and civil society in an authoritarian state: understanding the survival of China's grassroots NGOs. *American Sociological Review*, 117(1), 1–45.

Stearns, L.B. and P.D. Almeida 2004. The formation of state actor-social movement coalitions and favorable policy outcomes. *Social Problems*, 51(4), 478–504.

Stern, R.E. 2013. *Environmental litigation in China: a study in political ambivalence*. New York: Cambridge University Press.

Stern, R.E. and K.J. O'Brien. 2012. Politics at the boundary: mixed signals and the Chinese state. *Modern China*, 38(2), 174–98.

Sun, Y.F. and D.X. Zhao. 2008. Environmental campaigns. *In*: K.J. O'Brien, ed., *Popular protest in China*. Cambridge, MA: Harvard University Press, pp. 144–62.

Thaxton, R.A., Jr. 2008. *Catastrophe and contention in rural China: Mao's great leap forward and the origins of righteous resistance in Da Fo Village*. Cambridge, NY: Cambridge University Press.

Tilly, C. 1984. *Big structures, large processes, huge comparisons*. Cambridge, MA: Harvard University Press.

Tilly, C. 1993. Contentious repertoires in Great Britain, 1758–1834. *Social Science History*, 17(2), 253–80.

Tong, Z.H. Forthcoming. Zhongguo nongmin kangzheng de daoyi yishi (Moral consciousness of Chinese peasant protest). *Rural China: An International Journal of History and Social Science*.

Walker, K.L.M. 2008. From covert to overt: everyday peasant politics in China and the implications for transnational agrarian movements. *Journal of Agrarian Change*, 8(2–3), 462–88.

Wang. J. 2012. Shifting boundaries of state and society: village cadres as new activists in collective petitioning. *China Quarterly*, 211, 697–717.

Weiss, J.C. 2008. Powerful patriots: nationalism, diplomacy and strategic logic of anti-foreign protest in China, 1978–2005. PhD Dissertation, University of California, San Diego.

Wong, L. 2011. Chinese migrant workers: rights attainment deficits, rights consciousness and personal strategies. *China Quarterly*, 208, 870–92.

Wong, R.B. 2007. Book review of *Rightful resistance in rural China*. *Perspectives on Politics*, 5(4), 850–51.

Woodman, S. 2011. Law, translation, and voice. Transformation of a struggle for social justice in a Chinese village. *Critical Asian Studies*, 43(2), 185–210.

Yu, J.R. 2008. Emerging trends in violent riots. *China Security*, 4(3), 75–81.

Kevin J. O'Brien is the Alann P. Bedford Professor of Asian Studies, Director of the Institute of East Asian Studies, and Professor of Political Science at the University of California, Berkeley. His books include *Grassroots elections in China* (2011) (with Suisheng Zhao), *Popular protest in China* (2008), *Rightful resistance in rural China* (2006) (with Lianjiang Li), and *Engaging the law in China: state, society and possibilities for justice* (2005) (with Neil J. Diamant and Stanley B. Lubman).

Violence as development: land expropriation and China's urbanization

A review of the literature on expropriation violence in China shows that most analysts explain violence instrumentally, as a means by which competing actors attempt to capture, redistribute or defend income from land development, an indicator of different spatial political ecologies, or a catalyst of villagers' politicization. But these explanations of violence assume (1) antagonism between rational, unitary collective actors and (2) that violence is of limited temporal duration, spatial and social reach. This paper builds on Escobar's proposition that violence is constitutive of development, to argue for an alternative view: violence authorizes and constitutes an inclusive, ongoing project of urbanization in China. Violence authorizes development, because the rural spaces surrounding cities and towns are characterized as institutionally insecure, disorderly, economically under-productive and incompatible with modernity. It comprises development, because it involves the forced urban improvement of the nation, rural property, governance, people and livelihoods. The concluding section of the paper briefly demonstrates the generalizability and analytical and methodological utility of the concept of violence as development by applying it to three 'most different' cases of land expropriation in China.

Introduction

One morning in early March 1999, hundreds of residents in Baileqiao village, on the outskirts of Hangzhou, the capital of Zhejiang province, received notices informing them that their land would be expropriated and their houses demolished to make way for the integration of the area into the West Lake tourist precinct. Within a fortnight, they should present their residential registration documents and property deeds to the head of the demolition company, who would determine their eligibility for compensation. Everyone had to leave the village by 30 March. That evening, neighbours gathered together to discuss the notice:

> It was all announced so suddenly we haven't had time to prepare ourselves. Granny and grandfather next door have lived here their entire lives, still grow tea, but they have to leave too'.
>   Granny's quavering voice comes from the doorway: 'I won't sleep for worrying. I can't make sense of all the calculations. We went to him straight away and said it isn't fair, it doesn't take us old folk into account. We were members of the commune, got our house and land when it broke up, but we don't have documents for them. "Not fair? Not fair?" he says to me. "You will get compensated for 48 square metres of floor space but some will get nothing at all! How can you complain about fairness?" I'm so worried'.

This article was originally published with errors. This version has been corrected. Please see erratum (http://dx.doi.org/10.1080/03066150.2014.882610)

Wang interrupts: 'For our houses, it'll be calculated according to the floor area and quality of the building. You watch, everyone will hurry to enclose their storerooms, tile floors, fix windows that have been broken for years, to get more compensation for their houses'.

'I don't trust them. Some say this demolition company has no money, it's an empty shell. Us ordinary folk will get screwed. After our houses are torn down, maybe we won't get any compensation. Who knows what they'll do once we all move out'.

'Fifteen years ago we came from Dongyang and bought our house, but we didn't get any deeds and our registration is still back there. We won't be eligible for any compensation at all! Where are we four to go? We should all refuse to move!'

Old Zhang shakes his head. 'What can we do? Who has all the power, the guns? Earlier I was speaking with a fellow just come back from Taizhou. He'd been there farming with his brother, but last year the district government announced that his brother's village was going to be demolished so they all had to go. The villagers protested, said they wouldn't move, and when the bulldozers arrived they formed a line across the road, started throwing rocks at the construction workers. The local cops came, and still they held out. But then the public security police arrived with guns, shot at people. A few were killed, lots were badly injured, more than a hundred arrested and taken away. All so the government could make money from their land …
It wasn't on the news. They don't tell us what happens. They're like shepherds with a big stick, and we're the sheep who have to go wherever they want.[1]

In the decade after I recorded this conversation, the expansion of China's urban built-up areas dispossessed tens of millions more villagers, many of whom suffered similar anxieties to the people of Baileqiao or the physical harms described in old Zhang's cautionary tale from Taizhou. Over that period, Yu Jianrong estimates that 65 percent of rural 'mass incidents' were triggered by 'governments' forcible, violent expropriation of villagers' land' (Yu 2010b, 45, see also Fan 2011, Jiang 2012). Petitioners and protesters experienced additional, retaliatory, harms including threats, imprisonment and destruction of their homes (Yu 2010a, 220–1). Nor was there any sign that the violence was waning at the end of that decade. Despite crackdowns on journalists, as the non-exhaustive list in Table 1 shows, between 2010 and 2012 open source media reported that in at least 17 provinces conflicts over land expropriation resulted in people being shot, beaten and crushed to death, five cases of self-immolation (two involving three people), three bombings, thousands requiring medical treatment and unknown numbers being detained.[2] Construction sites were besieged by armed people, and buildings, infrastructure and vehicles were destroyed.

This paper examines the violence at work developing China's urban spaces. My use of the words 'at work' here is deliberate, and ironic. Deliberate, in referencing both Arendt's (1970) conclusion that violence is always instrumental, and the rational instrumentalism that underpins explanations of land expropriation conflicts in China. Ironic, because I wonder whether this analytic advances our understanding much more than the comments recorded above in Baileqiao: recall Wang's prediction that people opportunistically would try to maximize compensation for demolition of their houses, and Zhang's conclusion that Taizhou villagers were killed 'so the government could make money from their land'? Alike, these explanations direct attention to a crude distributive calculus

---

[1]Author's notes HZ 7 March 1999. Field research informing this paper was conducted on a dozen visits between 1999 and 2013, and primarily funded by the British Academy SRG 40650, Ford Foundation grant 1075 0591, Leverhulme Foundation Research Fellowship 2006/ 0381, and ARC DP 120104198.

[2]Amnesty International (2012, 56–63) documented nine deaths and 41 individuals self-immolating as a result of forced evictions between 2009 and 2011. However, the majority of these cases involved home evictions, not land expropriation.

Table 1. Media reports of land expropriation violence, 2010–2012.

Year	Location	Harms	Source
2010	Jiangxi, Yihuang	3 self-immolations	*China Daily* 6/11/2010
	Hubei, Wuhan	1 death	Amnesty International 2012
	Yunnan, Zhaotong	23 injuries and vehicles destroyed	*Yunnan zhengxie bao* 15/11/2010
	Yunnan, Kunming	1 bombing	*Yunnan zhengxie bao* 15/11/2010
	Zhejiang, Zaiqiao	1 death and multiple injuries	Watts, *The Guardian*, 2010
	Inner Mongolia, Huhehaote	1 death, 1 injury, detentions	Amnesty International 2012
	Guangxi, Wuzhou	multiple injuries, detentions	*Global Times* 14/10/2010
	Jiangsu, Suzhou	multiple injuries, detentions and vehicles destroyed	Yan, *Caijing* 22/7/2010
	Jiangsu, Pizhou	1 death, injuries	*Guotu ziyuan* 2, 2010
	Jiangsu, Lianyungang	2 self-immolations	*China Daily* 6/11/2010
	Jiangsu, Guantangqiao	3 injuries	Amnesty International 2012
	Shanxi, Guzhai	1 death	Amnesty International 2012
2011	Yunnan, Zhiliang and Shizong counties	multiple injuries, detentions	*Yunnan zhengxie bao* 25/7/2011
	Jiangxi, Ganxian	1 death, injuries and detentions	*Radio Free Asia* 12/7/2011
	Hubei, Yongnian	3 self-immolations	*Asia News* 8/11/2011
	Guangxi, Guigang	injuries, detentions and blocked roads	*Asia News* 13/4/2011
	Jiangxi, Fuzhou	3 deaths in bombing and 7 injuries	Areddy, *The Wall Street Journal*, 27/05/2011
	Jiangsu, Pizhou	injuries and detentions	Amnesty International 2012
	Gansu	21 injuries	*Fazhi ribao* 15/11/2011
	Hunan, Zhuzhou	1 self-immolation	Lasseter, *McClatchy* 27/5/2011
	Guandong, Wukan	1 death, injuries and detentions	Amnesty International 2012
	Inner Mongolia	dozens of injuries	*Asia News* 26/7/2011
	Henan	1 death and injuries	*Australian Financial Review* 16/12/2011
2012	Yunnan, Xishuangbanna	1 death, detentions and injuries	Branigan, *The Guardian* 3/4/2012
	Yunnan, Zhaotong	4 deaths in bombing and injuries	*Los Angeles Times* 10/5/2012
	Heilongjiang, Mudan	detentions and injuries	*Asia News* 13/4/2012
	Guangxi, Lingui	detentions and injuries	Amnesty International 18-23/4/2012
	Guangdong, Shunde, Zuotan	injuries, detentions, vehicles destroyed	*Caijing* 26/6/2012
	Shandong, Yuncheng	injuries, detentions	Amnesty International 18-23/4/2012
	Henan, Xuchang	injuries	Amnesty International 18-23/4/2012
	Chongqing	1 death	Amnesty International 6/12/2012
	Liaoning	self-immolation, shooting resulted in 1 death and injuries	*China Daily* 24/9/2012
	Hunan, Changsha	1 death	Parsons, *Mail Online* 25/9/2012
	Hubei, Luoshan	attempted suicide, injuries	*Radio Free Asia* 16/10/2012
	Hubei, Suixian	1 death, 2 injured	*Refworld*, Radio Free Asia 26/9/2012

motivating political action, which in Lasswell's (1950) memorable phrase centres on 'who gets what, when and how'.[3]

To be sure, violence in land expropriation in China often *is* motivated by desires to gain more of the income from urban growth. Nevertheless, I find these explanations of violence of limited explanatory value. Limited ontologically, because they are underpinned by questionable assumptions about the existence and motives of unitary collective actors, be they levels of government, village cadres or villagers. Limited conceptually, because violence is viewed as a behaviour whose short temporal duration and spatial and social reach can be defined objectively – a view that surely is inadequate for theorizing events that emerge from, and trail behind them, many intersecting histories of harm. And limited morally, by reasoning in which violence is understood as simply a moment, or a means to an end in conflicts that are of greater political consequence.

In trying to conceptualize violence in a way that is not solely instrumental and/or epiphenomenal, I build on Escobar's argument that as development essentially involves the displacement of people from places or prior ways of living in order to 'improve' them, 'the level of violence entailed by development [is] not secondary and temporary but actually long lasting and structural .... Violence is not only endemic but constitutive of development' (Escobar 2004, 16, see also Nixon 2011). Tania Murray Li (Li 2007), among others, similarly suggests that 'de-agrarianization' involves 'cumulative and permanent' displacements. My aim in this paper is to highlight how the violence of land expropriation in China constitutes urban development.

The argument is set out as follows. The next section explains why, and with what consequences, land became central in state strategies of accumulation in China. Section three reviews explanations of land expropriation violence in both the Chinese and English language literature. As it is beyond the scope of the paper to address all the arguments raised in such a voluminous body of work, for expository purposes I focus on debates that, from a Chinese perspective, speak to issues long discussed in *The Journal of Peasant Studies*. Specifically, I examine three influential arguments: that violence is instrumental in capturing income from urbanized land, reinforcing urban and rural political ecologies, and politicizing peasants. Having been thus 'normalized' as an instrument in political struggle, violence then disappears from the analytic.

The fourth section of the paper elaborates the proposition that violence is constitutive of an ongoing project of urban development. Violence authorizes development, because the rural spaces surrounding cities and towns are characterized as institutionally insecure, disorderly, economically under-productive and incompatible with modernity. It comprises development, because it involves the forced urban improvement of the nation, rural property, governance and people's subjectivity. The concluding section of the paper briefly demonstrates the generalizability and analytical and methodological utility of the concept of violence as development by applying it to three 'most different' cases of land expropriation in China.

State land-based accumulation in China: strategies and impacts

Inquiry into the violence of peri-urban land expropriation in China needs to be informed by an understanding of how state-directed accumulation strategies have changed since the mid twentieth century. In the 1950s, China's Communist Party (henceforth, CCP) established

---

[3]This criticism can be made of my own work. See, e.g., Sargeson (2008).

the institutional framework to support a centrally planned system of socialist primitive accumulation based on nationalization of the key factors of production – the foremost of which were land and labor. The extraction of agricultural produce to fuel state-building and low-cost industrialization was facilitated by four mechanisms: (1) the creation of a dual property regime under which the state owned all land occupied by urban settlements, mines and major infrastructure, and rural collectives owned all rural land. In 'the public interest', governments could expropriate land from rural collectives, and then allocate use of the state-owned land to state agencies and enterprises; (2) the collective organization of agricultural production under communes, brigades and production teams; (3) the creation of an artificial 'price scissors' between the rural and urban economies. This allowed the state to procure agricultural output at below-market prices and sell industrial goods to farmers at above-market prices; and (4) registration of the population according to their rural or urban residence, constraints on rural-to-urban migration, and the territorial compartmentalization of all expenditure on public goods. Rural collectives had to fund their own infrastructure, industrialization and welfare from whatever surplus they managed to retain, while the state financed these items in urban areas.

Since land ownership became the basis of organizing rural collectives and extracting agricultural surplus, it seems paradoxical that it was by no means clear who actually owned farmland. As Peter Ho (2001) has pointed out, the owner might be either the production teams which, after the de-collectivization of agriculture between 1977 and 1982, were referred to as village 'small groups'; the brigade, or their post-decollectivization successors, administrative villages; or, in some cases, the communes, whose governing functions and assets were assumed by townships when the communes were dismantled. Nor was the meaning of 'collective ownership' well defined. Collectives held an inalienable bundle of property rights on behalf of all members of the collective economic organization. The collectives were not permitted to sell the land or alter the land use without state planning permission. Hence, rural, collectively-owned land had use value but no exchange value.

Property relations in land became more complex after de-collectivization. Rural land ownership was separated from use rights. Collective owners contracted the land use rights to village households and, eventually, to individuals from outside the village. Between the 1980s and passage of the Rural Land Contract Law in 2002, land contractors gained more secure, longer-term tenure. Their bundle of rights also expanded: they were permitted to sub-lease, sell and bequeath their contracted use rights under the proviso that the land use remained unchanged. On rural land approved for construction, collectives could allocate sites for privately owned housing and businesses. Prohibitions against collectives' leasing construction land for development purposes were relaxed in some experimental sites. For example, in Guangdong's special economic zone, collectives were encouraged to attract investment by leasing construction land directly to foreign businesses.

Under the centrally planned economy, villagers rarely resisted land expropriation. On the contrary, because the collective owner was compensated with urban residential registration and jobs in public enterprises for a commensurate number of villagers, and those villagers then became eligible for state-subsidized food, housing, pensions and medical treatment, it was like winning a set of 'iron rice bowls'. Natural disasters, rather than expropriation, were the main cause of landlessness in this system.

This land compensation strategy began to unravel with the introduction of markets. To improve the public sector's competitiveness relative to private and foreign businesses, in the mid 1980s the state gave enterprise managers more control over hiring, firing and employee's remuneration. In 1988, it phased out the bureaucratic assignment of urban people to jobs. Employers' welfare obligations were reduced, co-contribution and user-pays principles were incorporated into urban medical and social security schemes, and urban public housing was privatized. Within a very short period of time, governments lost their capacity to provide jobs to land-losing villagers, and urban residence and employment became less attractive forms of compensation for expropriation. At the same time, a market for urban land leases opened up, and demand for land skyrocketed.

Impelled by revenue shortfalls and targets for economic growth, between 1990 and 2008, governments expropriated more than 4.2 million hectares of rural land for urban growth (Lin 2009, 126–7, 178, Rong 2010, 119–21, H.Z. Li and Zhang 2011, 56, World Bank and China Development Research Centre 2012, 142).[4] Much of the profit from the conversion and leasing of this land, as well as the deed taxes, urban construction and maintenance taxes, property taxes, land value-added taxes, urban land-use taxes and farmland occupation taxes, was captured by sub-provincial city, county and district-level governments (Wong 2012, 34). As levels of urbanization and industrialization grew, so did the percentage of local government revenue coming from land: in 2000, it was 10.5 percent; in 2009, 40.3 percent; in 2012, it reached 60 percent (Tang *et al.* 2012, 391, C. Wang and Xing 2013). In 2009, the office that expropriated Baileqiao, Hangzhou city, collected more than 105 billion *yuan* from land taxes alone ('2009 nian Zhongguo tudi churang jin da 15000 yi yuan', 2010).[5] According to one highly-placed informant, by 2012 land profits accounted for 80 percent of the off-budget income of some Zhejiang governments.[6] Land expropriation also propelled urban capital accumulation. Local government-established investment corporations used newly expropriated land as collateral to finance urban infrastructure and renewal projects. Private businesses benefited from discount pricing of urban land leases. Far more money was channeled from the rural to the urban economy by land-based government financing and accumulation than by the socialist central planning system (T. Zhou 2004, Rong 2010, 120).

But, unlike the centrally planned system of socialist primitive accumulation, this new strategy created mass landlessness. There is no reliable data on the number of villagers made landless by expropriation. According to the rough formula used by the Ministry of Land and Resources and Chinese academics, each hectare of rural land taken dispossesses around 21 villagers. On the basis of the above-mentioned figure of 4.2 million hectares of expropriated land, I estimate some 88 million became landless between 1990 and 2008. Between 2009 and 2030, the Ministry predicts that another 2.4 million hectares of land will be used for urban development, potentially dispossessing 50.4 million more (Rong 2010, 119–20, Fan 2011).

How are land-losers now compensated? Primarily with money and social insurance contributions. But the amount of compensation is limited by law, and by the revenue interests of local governments and businesses. The Land Administration Law (1986, revised 1998, 2004) stipulates that part of the income from developed land must be given to the original collective owner so it can invest in new income-generating ventures, and fund villagers' training and social insurance. Reflecting the socialist principle that land only has use value, however, Article 47 of the Law sets maximum compensation for these two items at 30 times the average value of agricultural output from the land over the previous three

---

[4]Yu Jianrong estimates that since 1990, more than 14 million hectares of land have been expropriated. Much of that land has been used for infrastructure and mining (China Daily, 2010).

[5]In December 2009, US$1 = 6.81 yuan. In 2010, total national revenue from land exceeded 7 trillion yuan (China Daily 2011).

[6]Author's interview HZ 22 September 2012.

years. In addition, collective members are compensated for their loss of crops, equipment, buildings and businesses on a 'replacement cost' basis. To reduce resistance to expropriation, State Council Documents 28 (2004) and 32 (2006) instructed that irrespective of Article 47, land-losing villagers must be left 'no worse off', and legislators began revising the Law in 2008 with the aim of deleting Article 47 and linking expropriation compensation to the market value of the land (X.Q. Wang 2013). But these efforts have been blocked by concerted opposition from local governments and industry. Consequently, collective owners have received less than 20 percent of the price paid for the land at auction (Cheng *et al.* 2009, 284, Rong 2010, 119–22). Many individual villagers received payments worth less than six years' average per capita income (Rong 2010, 121). According to one survey, 22.5 percent of people expropriated since 2001 had received no monetary compensation whatsoever (Landesa 2012, 2–4). These aggregate figures mask locational, temporal and demographic variations. Government data has confirmed that wealthy jurisdictions pay much more compensation than poorer ones, and in most jurisdictions compensation payments vary with the anticipated market demand for the site (Zhengdi zhidu gaige yanjiu ketizu, 2002). Payments have increased over time. And within villages, individuals' compensation might differ according to their gender, age, kinship and length of residence (Sargeson and Song 2010).

There is little doubt, though, that after expropriation many land-losers are 'worse off'. Looking at the impact of expropriations on villagers' subsequent incomes, H.Z. Li and Zhang (2011, 89) found that although 10 percent earned more, around half earned less. Low compensation payouts and reduced earnings combine with rising food prices to compromise the food security of some expropriated villagers. Consider the plight of a woman I interviewed on the outskirts of Changsha, Hunan, in 2008: pointing to the expressways and apartment blocks that covered her former fields, she said:

> We used to grow vegetables, raise livestock here. We could eat fresh food every day, we had enough of everything – chickens, ducks, fish, vegetables, all of that. Now we've no secure source of income and food is so expensive. There are lots of things we can't afford to eat.[7]

Besides, much of what expropriated villagers lose cannot be compensated for monetarily. In 2013, for example, when I visited Xu in the modest fifth-floor apartment he purchased after the demolition of his home at Baileqiao in 1999, I asked whether he felt 'better or worse off' after the expropriation. Xu paused, staring into the middle distance, then said:

> I thought it was good here to begin with, clean and convenient. But you know, what I really miss is wandering round the village in the evening smelling what people were cooking for dinner, chatting with them. If they were cooking something nice, they'd give you a taste, tell you how to cook it. It was so friendly. And I learned everything I know about cooking there. I really miss that. It was my favourite time of day .... Here it never happens. Everyone is busy, rushes home from work and shuts the door. Now if I want to cook something different I have to search on Baidu.[8]

Xu's words should not be dismissed as nostalgia. Contemporary land-based accumulation strategies are wiping out ways of producing, learning about and socializing over food. They are wiping out everyday ways of being a community.

---

[7] Author's interview CS 10 May 2008.
[8] Author's interview HZ 12 April 2013.

State accumulation strategies have figured strongly in interpretations of rural violence in China. Between the mid 1950s and end of the 1990s, violence was predominantly understood to be reactive, a last resort by peasants trying to prevent the state's extraction of agricultural surplus and taxes (Lu 1997, Bernstein 1998). Since the late 1990s, however, attention has focused on land as the key theatre of conflict over accumulation. Some have argued that violence is a strategy used by both governments and villagers to capture the value that urban markets add to the land. Others view it as a catalyst of urban-rural differentiation in China's political ecology or a spur to peasants' politicization. These arguments merit careful attention, for they are informing changes in institutions and practices that, I will later demonstrate, are implicated in violence *as* development.

Three perspectives on violence in land expropriation

Violence as gaming

A growing body of research uses a game theoretic approach to explain violence in land expropriation. Authors presume that there is a zero-sum competition between collective actors whose identities and interests aggregate around distribution of the profits from expropriating and leasing land for urban development. The primary actors are governments, village leaders and villagers. In China's weakly institutionalized market, violence is a calculated bargaining strategy that can be adopted by any of these actors.

The simplest two-actor variant of this model – in which an homogenous local government is pitted against a similarly undifferentiated collective, 'all villagers' – is set out in H. J. Wang *et al*. (2011, 77–86) and Yep (2013). Local government interests are twofold: As an agent of the central government charged with policy implementation, it must promote economic growth, urban expansion, employment and social stability in its jurisdiction (D.J. Yang 2003). Achievement in these areas is the basis on which officials' performance is evaluated (Yew 2012, 287). As a principal in the local land market, however, local government also is driven by fiscal incentives to expropriate as much land as possible, at the lowest cost. Confronted with villagers' demands for higher compensation, it can either concede and fall short of its economic development and revenue objectives, or engage in coercion and risk sanctions from the central government (Jiang 2012, 62). 'All villagers' face a similar choice: cooperate and lose out monetarily, or resist and risk repression. The latter risk is usually starkly underlined by expropriating agencies. For example, interviewees in Yuxi, Yunnan, told me that no sooner had they been informed that their village's remaining farmland would be expropriated, than officials took away a member of each family and detained them until the household head had signed an agreement 'voluntarily' accepting the compensation offered; 'Then, the day the bulldozers arrived, we were outnumbered by hundreds of men who arrived in trucks, police vehicles, they even brought ambulances! We were surrounded'.[9] Nevertheless, one study of 789 expropriated villagers in Guangzhou found that 61 percent had 'resisted', and 97 percent of the resistance was motivated by a desire for higher compensation (Wang Yang, Zhou and Huang, 2011, 79).

Other Chinese scholars have critiqued the two-actor model for failing to disaggregate collective actors. Ma Jianbo and He Xuefeng, for example, agree that land expropriation violence is driven by distributive competition. But Ma (2013) points to conflicts between the central government, and coalitions comprising local authorities, developers and villages

---

[9]Author's interview YX 2 May 2008.

who collude to conceal large land developments, while He (2010) identifies three actors, rather than two. He's third actor, village leaders, are ambivalent. They have an incentive to mobilize the villagers to agitate for higher compensation for all. But because local governments rely upon the leaders to mediate with the villagers, they also have an opportunity to sell their services to governments for a greater sum than they might get by bargaining on behalf of all villagers. As government agents, village leaders withhold information, urge villagers to accept low compensation, prevent them from mobilizing collectively by buying off weak households and threatening others that their family members will be sacked or excluded from tendering for lucrative public sector business, isolate potential 'troublemakers', and embezzle compensation funds. In a survey in Shandong, H.Z. Li and Zhang found that 94 percent of expropriated villagers believed their leaders had harmed collective interests for the sake of personal gain (H.Z. Li and Zhang 2011, 83–5, see also Sun 2010, 118).

The breakdown of trust between central and local governments, local authorities and villages, and village leaders and villagers complicates game theoretic explanations of violence. Actors' resentment at being betrayed, threatened and deprived unjustly might be just as important contributors to violent conflict as their desire for material gain (F. Li 2011). Such convictions certainly contributed to brutality and destruction in the Guangdong villages of Dongzhou and Taishi in 2005, and Wukan in 2011.

Moreover, although most game theorists interpret violence as the last resort in a brief one-off game, land expropriations usually are piecemeal and accretive. Even in China's breakneck rush to urbanise, cities and towns grow over years, or even decades. Besides, as the State Council must approve large-scale developments, local governments try to avoid oversight by expropriating many small parcels of land. Actors' strategies alter in iterated, drawn-out games (D.Y. Zhao 2009, 114–20). Confronted with the nth instance of betrayal or threat over land taking, villagers are more likely to be motivated by pent-up rage and a desire to publicly shame those they blame. A case in point occurred in Longquan, southern Zhejiang, after officials warned villagers that, 'The bird in the lead will get shot first'. Villagers responded by placing their most vulnerable, venerable members at their front line of the construction site, then disseminating film of 'women and old folks hanging on to raised excavator shovels' (Fujian [*sic*] Longquan tudi douzheng quan jilu 2011, 1).

Individuals' divergent temporal and issue framing of expropriation presents another difficulty for game theoretic explanations of violence. To adequately account for participants' motives, calculations and 'plays', game theorists need to ensure that their chronology of the game and what it is that is being taken coincide with participants' perceptions. However, participants frame the game according to their own life histories. In Baileqiao, for example, one interviewee worried that the compensation would ignore her family's informal migration and private purchase of a house 15 years previously, while Granny felt she was being dispossessed of recognition for a lifetime of toil on the land, her contributions to the commune, and her share from decollectivization.

In short, although the game theoretic literature illuminates what is at stake materially for abstract collective actors, it has little to tell us about dynamic, protracted conflicts that arise from ongoing economic transformations, iterated breaches of trust, threats and ancillary harms, and individual life histories. Paul Pierson suggests that such blind spots are common to rational choice approaches:

> Among the things that drop out of such exercises are issues of macro-structure, the role of temporal ordering or sequence, and a whole host of social processes that play out only over extended periods of time and cannot be reduced to the strategic 'moves' of actors. (Pierson 2004, 9)

Violence in the spatial ecology of urbanization

Whereas time presents difficulties for game theorists, space is the central problem in You-Tien Hsing's explanation of land expropriation violence. Hsing sets out an innovative framework in which the outcomes of expropriation are determined partly by communities' proximity to urban centres and their consequent capacity to engage in what she dubs counter-strategies of 'civic territoriality'. In Hsing's words:

> local governments use urban redevelopment powers to destroy, displace and rebuild, while inner city protesters make legal, historical, and moral claims over their rights to property, housing and livelihood in the city. Similarly, as an urban government initiates expansion in neighboring villages, villagers at the rapidly growing urban fringe strategize to avoid displacement, take advantage of urban real-estate markets, and even manage to secure a relative territorial autonomy. Meanwhile, in the remote rural fringe areas, large numbers of displaced villagers lose economic, social and cultural resources and become deterritorialized. (Hsing 2010, 14)

Hsing's argument is illustrated by a series of case studies. Pointing to collective struggles in the inner city in Beijing, she highlights activists' skill in using historical memory and legal argument to mobilize cross-neighbourhood networks, initiate petitions and lawsuits and delegitimize demolition and relocation programs. Exemplifying corporatist politics in the 'rapidly growing urban fringe', lineages in her Guangzhou case study site transferred their collective land to a shareholding company that then negotiated with government to retain rents from part of the land development. All this implies that it is only in the rural hinterland, where villagers lack legal, financial, organizational and discursive resources, that violence erupts.

Hsing's study makes explicit three assumptions that are implicit in many analyses of expropriation violence. First, the frontier of violence is assumed to be between a predatory, urban expansionist state and communities defending (or being dispossessed of) their property in land. Yet as Kalyvas (2003) points out, in granting analytical primacy to this cleavage, we risk overlooking something to which Zhao and Webster (2011) allude: cooperative interactions between governments and villagers. Around Beijing and Xiamen, they found that such developmental alliances produced village 'share-holding micro-governments that look like an odd hybrid between nineteenth-century company towns and modern gated communities' (Zhao and Webster 2011, 550). In fact, as George Lin (2009) shows, villages and townships have used even more farmland for industry, infrastructure and housing than cities, and much of this development has been supported by government development plans and loans. On the other hand, such land developments can also create significant inequalities in wealth and residential status, and pit kin and community members against one another (Xu and Yeh 2009, Ma 2013). Cross-cutting alliances and conflicts dissolve the assumed frontier of violence between binary actors.

Second, assumptions about the spatial distribution of expropriation responses and outcomes mimic a modernization trajectory, with a rational citizenry securing the institutional foundations for secure property and political participation in the urban centre, and resistance by an inchoate 'peasantry' being routed by the state in the hinterlands. Violence thus appears to be an historical-spatial residual. Yet there is no clear evidence that communities that vary in distance from urban centres systematically differ in the strategies used to resist expropriation, the frequency and severity of violence, or in their sharing of land profits.

A third, related, assumption is that in the course of urbanization, spatially discrete processes of 'civic territoriality' and 'de-territorialization' become self-reinforcing determinants of actors' resources and political capacities. However, as Hsing notes, the frontier of expropriation is continually shifting. The site of yesterday's de-territorialization becomes the venue for urban accumulation and rentier capitalism tomorrow. But this assumption leaves little room for the possibility that yesterday's de-territorialized peasants will collectively strategize to retain land and autonomy, become landlords and emerge as legally savvy urban activists in the future. It is precisely this trajectory of peasant politicization that animates the third explanation of expropriation violence.

Violence as a catalyst of political maturation

The proposition that conflicts over property propel much larger political revolutions is not novel: it informs some of the most influential scholarship on European history, including that of Karl Marx and Barrington Moore. Li Chenggui, among others, has argued that in China, expropriation violence is an expression of popular demand for real property rights in land (C. Li 2007a, 2007b). For Yu Jianrong, Li's colleague at the Rural Development Institute in the Academy of Social Sciences, the main significance of expropriation violence lies in villagers' growing political and legal consciousness, organizational capacity and, ultimately, the strengthening of China's civil and political institutions (see also F. Li 2011). As Yu's explanation also addresses the arguments of game theorists and those who analytically privilege urban activism, it is worth quoting at length:

> As land conflicts involve the interests of most villagers, they are more willing to mobilize to achieve their common objectives. When they confront governments' forceful land expropriation, they establish 'Land Defence Teams', 'Home Protection Associations' and other informal, temporary organizations. Most are highly organized, with a clear division of labour, appropriate tactics and just objectives. Members of the rural elite emerge to lead the rights defence. They typically are local entrepreneurs, intellectuals or retired military who have some economic capacity, worldly experience, legal knowledge, a good reputation or external resources, and who take responsibility for strategizing, commanding, and representing villagers in negotiations. Often, to increase their chance of success villagers seek external assistance (lawyers, academics, journalists, non-governmental organizations or other popular organs), or they study law and policy themselves. These become their weapons, giving effect to the saying, 'use law to defend rights'.
>
> Because land ownership is incomplete and expropriation is state action, villagers' resistance ultimately fails – even if they win, they will only revert to the starting point of the expropriation 'game'. But having been baptised by land conflict, their capacity for organization is enhanced and the protest leaders gain in authority. These new knowledges, capacities and leadership will not dissipate just because this or that land dispute is resolved. They sink into villagers' blood and lives, become living things, and having taken root they sprout, so even if they become dormant for a season, inevitably they will influence all aspects of rural society and government in the future. (Yu 2010b, 47)

Here, villagers' mobilization to resist expropriation becomes instrumental in their transformation into politically astute, collectively organised actors. Certainly, many villagers who have experienced expropriation, such as Chen from Baileqiao, have told me that it did indeed transform their attitudes and actions. Yet Chen's explanation of what these attitudinal and behavioural transformations entailed directly contradicts Yu's logic:

> You could say it was my first real lesson in how things work, how unfair and murky most processes are. Since then, I've always been suspicious, thought that government, people, they all

just look out for their own interests, so I have to as well. It gave me new insights into human nature, I guess.[10]

Individualism, rather than collective organizational capacity, might be strengthened during land conflicts.

Nor should we too readily accept Yu's assertion that land conflicts are cohesive movements whose participants share a common identity, grievance, goals and strategies. Empirical research by Bai (2012), D.G. Zhou (2012) and others has shown that land expropriation conflicts tend to gain in complexity over time, as people enraged by different issues join in. Participants protest because the compensation they received in the past was calculated according to different criteria to the methods used in the present, or because their neighbours received more, and they demand procedural equality. They protest because homes they were resettled in are of sub-standard quality, they were laid off from jobs to which they were assigned or, as in a case Song and I witnessed, their 'dole' payments were terminated (Sargeson and Song 2010). Complaints spill beyond dissatisfaction with current procedural and distributive issues to condemn violences that have been caused by other perpetrators, at other times, in other places. Then there are those who attach to expropriation conflicts to protest about other negative externalities of land development, such as pollution and losses of 'amenity'. Protest strategies that backfire can inspire further protests. For example, villagers complaining against inadequate compensation have been incensed, and moved to demonstrate by press coverage depicting them as selfishly raising the costs of urban development and housing, and using violence to obstuct officials performing lawful expropriation duties (*China Daily* 24 September 2012). Appeals to laws and the courts that have served to dispossess, rather than protect, villagers' property have embittered, and made people distrust state institutions. D.Y. Zhao (2009, 118) describes one court action brought by government against village protesters that lasted five years, exhausting the defendants' resources and patience.

In light of these criticisms, how can we avoid the pitfalls that, to my mind, are inherent in instrumentalist explanations of violence? In what follows, I attempt to show that expropriation violence possesses a temporal duration, scale and range of meanings that are ignored in the explanations reviewed above, and that it actually constitutes a vast new project of urban development.

Violence as development

Violence as development causes many forms of harm, including material, somatic, emotional, psychological and social harms. In China now, violence as development entails the forced urban improvement of the nation, rural property, governments, people and livelihoods. Like Rob Nixon's concept of 'slow violence' (2011), violence as development occurs over long arcs of time. Unlike 'slow violence', however, violence as development does not occur gradually, out of sight. Its tempo is uneven, and often it is all too visible. Violence as development occurs on multiple spatial and organizational scales. Because huge numbers of people are implicated in this iterated, cumulative process of forced urban improvement, and because it is built into the rationale, languages, rules, bureaucratic structures and historical narratives of displacement into urban ways of being

---

[10]Author's interview HZ 12 April 2013.

and doing, it resembles what Johan Galtung (1969) called 'structural violence'. But in contrast to 'structural violence' where the agents of harm are not readily identifiable and the outcome is not always intended, violence as development involves many different actors, purposefully engaged in a wide array of brutal, administrative, pedagogic and practical urbanizing tasks. Time, scale and complicated, collectively forced urban improvements all characterize violence as (urban) development.

We could explore violence as urban development from many perspectives. Here, I focus on four dimensions of violence that are constitutive of urban development in China: ideology, property, government and residents' economic subjectivity. I then demonstrate the generalizability and analytical and methodological utility of the concept by applying it to three very different cases of land expropriation.

Ideology

'By the mis-2000s', writes Hsing (2010, 18), 'urbanism largely took over industrialism as the basis for political legitimacy and policy discourse' in China. Leaders' pronouncements and policies stress the state's nationalist, developmental obligations to make land available for urbanization. These are long-term, nationally inclusive goals. Urbanization is said to be critical to national economic restructuring, the elimination of the rural/urban divide, creation of a more equitable, harmonious society, and the projection of an appropriate global image. In the wake of declining export growth following the 2008 global financial crisis, urbanization was presented as a means of boosting domestic consumption and employment. Even critical studies of land taking in Chinese typically open with a statement that it is inevitable, and necessary for China's modernization, to convert agricultural land to urban uses, a sentiment that is echoed by land-losing villagers protesting against expropriation terms, practices and outcomes. As an official from Yihuang county in Jiangxi wrote after an expropriation incident lead to three self-immolations, '[Our] county is a microcosm of China's rapid industrialization and urbanization, and a successful example of underdeveloped regions in central and western China catching up and surpassing coastal areas... While rebuking policies of forced demolition, people seem to ignore the basic fact that everyone is actually a beneficiary of such policies. Without forced demolition, there is no urbanization in China; and without urbanization, there is no brand-new Chinese society. As a result, we can say that without demolition, there would be no new China'. (Hui 2010). Violence as urban development therefore is ideological, because it is rhetorically presented and popularly understood to be progress for 'the common good'.

A second ideological dimension of violence as development is, as the foregoing review of the literature on gaming shows, that the harms associated with expropriations become justifications in arguments for institutional reforms that would make the displacement of agrarian places and ways of living more economically efficient and less politically costly. (Sargeson 2012) For example, policy advisors argue that the costs of urbanization would be reduced by property and land market reforms that could more accurately price land, grant villagers a larger share of land income and provide them with capital to invest in acquiring the property, skills and insurances needed in urban economies. More inclusive deliberative procedures would facilitate bargaining and reduce the incidence of and severity of conflict. These arguments simultaneously popularize an instrumental understanding of violence, and set out alternative instruments to achieve precisely the same, urbanizing, end.

Property

When speaking of land expropriation, we draw distinctions between 'mine and thine', between state, collective and individual property. As Blomley (2008, 133) argues, property legislation entails both the violence of definitional exclusion, and the violence of sanctioned enforcement. Indeed, the sovereign's legitimate use of force is immanent in the principle of 'eminent domain', which underpins expropriation legislation. Land laws identify which sovereign authorities can justly assert state ownership and enforce exclusive property relations. In mediating conflicts over expropriation and debates over proposed land reforms, China's central leaders consolidate and popularize their rational-legal credentials relative to lower levels of government and other land claimants. In physically removing expropriated villagers and defending the exclusive property rights of urban lessees, lower-level governments implement the sovereign's law of the land.

While the ideology of urbanism requires that land be expropriated and transferred to those whom Locke (1960, 291, cited in Blomley 2008) describes as 'the industrious and rational', the state's zoning of land for urban use, construction of publicly funded infrastructure and enforcement of urban leases dramatically increases land values and guarantees investment returns. Ironically, the greater security and profitability of this state-owned urban land then serve as measures against which collective land ownership is judged to be sub-optimal. Re-zoning land for urban construction and expropriating it thus become means of resolving the purported problems of collective ownership, of transforming rural land and housing from dead capital into fungible assets that can be sold, leased and mortgaged, and spurring cycles of building, refurbishment, demolition and rebuilding. The violence of property definition, exclusion, land use regulation, zoning and expropriation constitutes urban development.

Government

The restructuring of government is another intrinsic aspect of urban development. Lin (2009) traces Beijing's formation of the Ministry of Land and Resources (MLR), merger of township governments in the late 1990s, reorganization of the MLR and line management relations in 2003, satellite monitoring of land use and imposition of a target for the conservation of 120 million hectares of farmland, to inter-agency and central-local conflicts over land use. These centralizing moves were countered by provincial and city governments' movement of their subordinate jurisdictions' boundaries to enable hostile takeovers of potential development sites (Yew 2012). Cities integrated land expropriation compensation packages into their vocational training, pension and social insurance budgets, thereby shifting part of the cost of compensation from investors to the public, and allowing them to advertise cheap land and workforces as competitive advantages. Government at the grassroots level, too, is being reorganized. In many locations, when all the land of a village is expropriated, the communities and residents are redefined as urban. Villagers' elected committees are replaced by urban community organizations comprising staff whose appointments are approved, whose wages are paid by and who are answerable to the government for their provision of a generic suite of urban utilities and services to local households. Even where the elected village committees remain to manage residual collective assets, they become marginal in government of the community. Violence as urban development involves the dual centralization of state power, and extension of governments' authoritarian reach into neighbourhoods. It causes anxiety among officials, as well as among those they govern.

Villagers' economic subjectivity

It is not just the utility of land, but also the utility of the people resident on the land that is reassessed in the course of urban development. Few rural knowledges and skills are culturally or economically valued. And, although many rural households are involved in large-scale or intensive, highly specialised commercial production and some of their members work in industry and services, in discussions of urbanization they tend to be represented as reliant upon small-peasant agriculture. That representation supports the argument that urbanization will be their salvation, by providing opportunities to transform the younger generation from the spectral statistical designations 'off-farm' and 'migrant' peasant-worker into skilled workers and self-employed entrepreneurs. But village women aged over 40, and men older than 50, are generally considered to be unemployable in the urban economy. (Sargeson and Song 2010; Sargeson 2008).

Officials admit that this new demographic division of labour is resisted. As one city official in Tongxiang told me,

> The old people, they don't think they'll find jobs so they want to hold onto some land to farm. They can't see that they can get a pension and stay home to look after the grandkids, or work as day labourers … we have to spend a lot of time talking them into it. If they are sensible it might only take us a few days, but others, sometimes we visit them day after day for more than a year, and they still refuse to give up the land.[11]

Whereas wealthier 'old people' become self-funded (and thus respectable) retirees, most become welfare recipients who eke out their meager pensions by taking on casual work as street sweepers and watchmen, tilling wasteland and recycling rubbish. Land expropriation disrupts these people's socially-expected life course, capacity to contribute to their families, social and familial standing and self-respect. Their excision from the population that is valued as being economically productive is another facet of violence as urban development.

These dimensions of violence as urban development can be found all across China. Consider the following three 'most different' cases in Guangdong and Zhejiang. In only one of these cases, Wukan, did land taking involve collective conflicts that resulted in the sorts of physical brutality listed in Table 1. In Guangzhou's 'urban villages', villagers voluntarily participated in and were enriched by income from piecemeal, unplanned land developments. Villagers in both Guangdong locations remained in situ. The Zhejiang case of Tongxiang involves large-scale, integrated planning of both rural and urban land use, government-coordinated relocations that are primarily funded by villagers and below-market value compensation for expropriated land.

What these 'most different' cases have in common is that violence is by no means solely instrumental or epiphenomenal to the development of China's urban areas. Instead, violence is immanent in the urban improvement of people, places, livelihoods, and ways of being and doing. Second, these cases illustrate the long temporal duration and spatial and social extent of violence as development. They show that violence persists long after the land income has been distributed and the excavators have departed. The harms persist in memories and stories about where and how things once were, what wrongs were done and what impacts were felt by various people, and are renewed by news of more expropriation events. Third, in combination with the examples I have drawn from

---

[11]Authors' interview TX 22 April 2013.

Baileqiao, these cases demonstrate that the concept of violence as development opens up fruitful new methods for studying expropriation, including reinterpreting existing literature (as I do with the two cases from Guangdong) and the use of longitudinal studies and qualitative research techniques (as I have used in Zhejiang).

Wukan

Protests against land taking in Wukan became headline news at the end of 2011. Yet neither the land taking nor the protests were new. Since 1993, the village Party Secretary's participation in lawful expropriations and unauthorized land sales had reduced the area collectively owned by villagers by more than one third, till only 600 hectares of farmland remained (Qu 2012). From the several hundred million *yuan* paid for land on which to build the factories, hotels and power plant that eventually surrounded Wukan, villagers received annual subsidies of just a few hundred *yuan*. Villagers repeatedly demanded restitution of their land, petitioning Lufeng county government five times and Guangdong provincial government seven times. In 2009, young villagers formed a group to oppose the land sales (Pomfret 2013). But it was only after the villagers expelled the Party Secretary and demanded new elections, only after they had been surrounded by police and one of their members killed, that their complaints attracted public attention. Following an investigation by a provincial work committee, Wukan villagers were permitted to elect a new representative committee, publicize village accounts and begin negotiating the return of their land. Emboldened by the concessions won at Wukan, expropriation violence erupted in other Guangdong villages, including Zuotan, Xincuozhai and Liantang. 'Across China', wrote John Pomfret (2013), 'people began buzzing about a "Wukan Spring"'.

It was a brief spring. After the elections, violence persisted. To the frustration of villagers, businesses' land leases were protected by law, commercial contracts, official connections and the police. The few plots that were returned to the village became a source of conflict between villagers who wanted the land redistributed, and the newly elected committee that planned to use it to construct state-mandated New Socialist Countryside development projects. Committee members who had been elected because of their honest reputations became the subjects of suspicion, resentment and harassment. By early 2013, some had resigned or taken leave because, one explained, 'the village bristles daily with criticism, and the village committee's authority is weakening' (S. Yang 2013), and the new Party Secretary installed security cameras to protect his house and family.

We would learn little about the violence in Wukan if we were to interpret it simply as a bargaining strategy to capture profit from land development. Nor is it instructive to view it as symptomatic of a political ecological gradient, or a catalyst of villagers' politicization. Over almost two decades, the violence involved, on the one hand, what in China's ideology of urbanism are depicted as urban improvements, including the dramatic expansion of Wukan's built environment, employment and income growth, governments' assignment and enforcement of exclusive property rights, and the introduction of elections, transparent accounting and deliberative practices. On the other hand, it led villagers to become profoundly disillusioned, divided and fearful. The concept of violence as development helps account for precisely these changes.

Guangzhou's urban villages ('chengzhongcun')

What Chung and Unger (2012) dub the 'Guangdong model' of urbanization has a similarly long history. In the early 1990s, village collectives transferred their property to shareholding companies and distributed shares to all village collective members, with premiums given for older age and male sex. Company revenues funded village welfare schemes and shareholders' dividends. When cities such as Guangzhou expropriated village land, contrary to the predictions of game theorists, they elicited villagers' cooperation by paying compensation to their shareholding companies and leaving the companies to control and profit from some development sites. Like Hsing, Chung and Unger view urban villages positively for, although more of the land in Guangzhou is now owned by the state than by native villagers, they argue that 'the shift of all the landed assets into a private shareholding company perpetuates the native villagers' exclusive ownership and control' (Chung and Unger 2013, 34). By the mid-2000s, Guangzhou encompassed 139 such 'urban villages' (Hsing 2010, 145).

This positive image of urban villagers' secure private property and political autonomy is the antithesis of the stereotype in much of the Chinese literature, where 'urban villages' are described as islands of backwardness, disconnected from urban transport and sewerage infrastructure, lacking the civilizing effects of state government and education, and populated by an unsavory mix of slum landlords, the unemployed, outsiders and criminals.

The concept of violence as urban development prompts us to recognise how both these interpretations ideologically endorse urbanization. The negative stereotype justifies government interventions to suppress 'feudal' gender discrimination, root out corruption in shareholding companies, shift jurisdictional boundaries and remove officials opposed to expropriation and land rezoning, and ultimately incorporate these communities into urban construction plans and governing structures. The ideology of urbanism combines with new corporate forms of governance to challenge villagers' valuation of their homes and understandings of community. In Chung and Unger's case study of Xinxiang, for example, residents were 'living in a disorderly, unhygienic and not entirely safe environment' (38). Yet those who opposed another high-rise redevelopment because they distrusted redevelopment plans were sued by the company in which they held shares for damaging 'the community's interest'. In other urban villages, disputes have erupted over lack of transparency and disparities in share dividends and welfare. The concept of violence as urban development also illuminates the creation of new sources of insecurity in urban villages: urban residents must be carefully guarded from the immigrant rural hordes outside their gates, whilst the migrant workers they fear are themselves disproportionately represented among the victims of urban crime, industrial accidents, environmental pollution, slum housing and chronic disease. As Chung and Unger write:

> [T]he Chen community no longer lives in the twisting lanes of small, old, dingy houses that they used to occupy. They rent these out to migrant workers. The shareholding company provided a free block of nearby land on which the village government organized the construction of a fancy gated residential estate. Behind high fences, it contains ten high-rise apartment buildings, playgrounds, a large social-club building, and a giant swimming pool, with two stories of underground parking … . Only Chens are allowed to live in the housing complex. They have sealed themselves off in their privileged lifestyle from the tens of thousands of immigrants who today work and live in Chen-owned and -controlled territory. (2013, 36)

Tongxiang

Compared to Guangdong's Pearl River Delta, less farmland has been used for urbanization in northern Zhejiang's Yangtze River Delta. Part of the reason for this is that since 2000, governments in cities like Tongxiang have been trading quotas for farmland conservation and urban development. Villagers in dispersed hamlets are encouraged to relinquish their house sites and purchase either villas in government-subsidized, planned, centralized communities, or low-cost urban apartments. By turning the old house sites into cultivable land, governments add to their farmland conservation quota. This quota can be used to offset their expropriation of land for development in higher-value locations, or sold to other expanding urban jurisdictions. For each village household that relocates, Tongxiang government estimates that it gains 0.4–0.6 mu of farmland.[12] For each mu of farmland gained, local authorities are awarded substantial bonuses by the city. By 2023, Tongxiang aims to relocate around half of its 500,000 villagers into centralized communities (Tongxiang shi, Tongxiang dangwei 2009).

In addition to relocating, villagers are encouraged to sign their farmland over to shareholding cooperatives for the remainder of the contract term, for an annual rent that is pegged to average national grain prices. The cooperatives merge the farmland and sub-lease it to specialist farming households and agribusinesses.

These moves substantially reduce the cost to government of subsequent expropriations, because villagers already have paid for their own relocation and relinquished their contract farmland. In one instance of expropriation in 2012, for example, villagers received only 3000 yuan per mu compensation, roughly one tenth the amount paid for a similar site in a neighbouring city.

As Huang and Wang explain, the relocations also transform villagers' ways of living:

> From scattered housing, villagers move into row housing and apartments. The old confused functions, backward assemblages and unsafe buildings are transformed by intelligent design into rational, safe, high quality urban housing, and villagers' production and lifestyles are completely transformed … . From having been involved in independent small peasant farming, they become industrial workers with new attitudes, and assimilate into urban life, completing their transformation from peasants to urban citizens. (Huang and Wang 2010, 76)

Behind this rosy façade of government-coordinated urbanization, there is a great deal of pressure and dislocation. Although the swaps are supposedly voluntary, villagers who cannot afford or do not want to relocate are browbeaten by local authorities, disconnected from services and access roads, and are preyed upon by thugs and thieves. Some have returned from work to find their houses destroyed (Cao 2011). When an agreement to move is signed, old houses are demolished and construction on the new home begins. In the interim, many families have difficulty finding affordable rental accommodation. Because only the wealthy can afford to purchase villas in the centralized communities, the countryside is becoming gentrified. Villa owners commute to work in towns. Conversely, poor farmers who have moved into apartments but not relinquished their contract land must commute to work in the fields, and get criticised by urban officials and neighbours for storing seed and tools around apartment blocks. Those who have given up their contract land often end up tilling the same fields as day labourers employed by agribusinesses, earning far less than they did as independent farmers. In May 2012, tensions over these diverse issues prompted people from different communities to block Highway 302, between Shanghai, Jiaxing and Hangzhou.

The theoretical frameworks we use to explain expropriation violence should be able to account for the sorts of non-instrumental motives, and spatially dispersed and socially complex harms, occurring in Tongxiang, Guangzhou and Wukan. The concept of violence

---

[12]Author's interview 17 September 2012. One mu equals 0.0667 hectares.

as development provides us with a tool for thinking about how, even in such 'most different' cases, an iterated, cumulative process of forced urban improvement is changing rural property, governance, livelihoods and ways of being a community'.

Conclusion

I have sought to show that one of the problems with explaining expropriation violence as gaming, or a spatial reinforcement of political economic differentiation, or a catalyst of villagers' politicization, is that those explanations overlook violence as development. Violence as development stretches well beyond the duration of any particular bargaining process, the site of each inner-city, peri-urban and remote rural development project, and the impulse to organize collectively among abstract groups of land-losing villagers. It takes the form of harms that accumulate over, and trouble, entire lifetimes. What the concept of violence as development calls for in the study of land politics in China is greater attention to people's participation in political-economic processes that span long periods of time, space and social and organizational scale.

A second thing to be learned from applying this concept of violence as development is that we need to be more cautious in imputing common concerns, strategic rationality and organizational unity to those caught up in expropriation violence. People come into contentious politics with very different histories, as well as shared grievances. As perceived wrongs and suspicions of deception accumulate, their complaints, demands and strategies mutate. So, violence as development is both inclusive and transformative. It can touch vast numbers of people, and disrupt their ideas of what is in the nation's, collective's, community's and individual's interest. Violence as development involves dispossession, as well as the definition and assignment of exclusive property rights, the centralization and extended reach of authoritarian government, the reclassification of people's utility that results in some losing their livelihoods and social, familial and self-respect. And it leaves people with new sources of security and insecurity. Where does this leave analytical approaches that are based on the presumed antagonistic, rational interests and competitive strategies of governments, village leaders and village collectives? It requires us to animate our theorization of land expropriation politics with much more diverse, dynamic actors.

References

Amnesty International 2012. China Human Rights Briefing 18-23 April. Available from http://www2.amnesty.org.uk/blogs/countdown-china/china-human-rights-briefing-april-18-23-2012 [Accessed on 7 May 2012].

Amnesty International 2012. *Standing their ground: thousands face violent evictions in China.* London: Amnesty International.

Amnesty International 2012. China Human Rights Briefing 6-12 September.

Areddy, J. 2011. China explosions kill two. The Wall Street Journal, 27 May. Available from http://online.wsj.com/news/articles/SB10001424052702304520804576346603464813660 [Accessed 14 June 2012].

Arendt, H. 1970. *On violence.* New York: Harcourt, Brace & World.

AsiaNews, 2011. Inner Mongolia: Fresh protests and clashes with police. Dozens injured. AsiaNews 26 July. Available from http://www.asianews.it/news-en/Inner-Mongolia:-fresh-protests-and-clashes-with-police,-dozens-injured-22191.html [Accessed 17 October 2012].

AsiaNews, 2011. China: three brothers set themselves on fire to protest land expropriation. AsiaNews, 8 November. Available from http://asianews.it/view4print.php?1=en&art=23112 [Accessed 1 July 2012].

AsiaNews, 2012. Forced expropriation and home demolitions continue Communist Party abuses. AsiaNews, 13 April. Available from http://www.asianews.it/view4print.php?l=en&art=24495 [Accessed 1 July 2012].

Australian Financial Review. 2011. Land grabs in China cause major unrest. *Australian Financial Review*, 16 December. Available from http://www.afr.com/p/home/land_grabs_in_china_cause_major_GOLNgbnTmYbvIuTqG8MtCN [Accessed 24 August 2012].

Bai, X.Y. 2012. Difang zhengfu tufa quntixing shijian de yingji guanli yanjiu [Research on local governments' emergency management of the eruption of mass incidents], *Lingdao yu guanli*, 3, 96–98.

Bernstein, T. 1998. Instability in rural China. *In:* D. Shambaugh, ed. *Is China unstable? Assessing the factors*. Washington, DC: Sigur Centre for Asian Studies, pp. 93–110.

Blomley, N. 2008. Law, property and the geography of violence: the frontier, the survey and the grid. *Annals of the Association of American Geographers*, 93(1), 121–41.

Caijing 2012. Guangdong fan zhengdi baodong 2 qian cunmin dazhan cunguan jingcha [2 thousand villagers battle village officials and police in Guangdong land expropriation riot. Available from http://club.china.com/data/thread/1011/2742/81/30/6_1.html [Accessed 7 October 2012].

Cao, H. 2011. Fada diqu tugai liang nan kun jing [Two difficulties in the land reform environment in developed areas]. *Dongfang zaobao*, 4 January, A34.

Chen, Z. 2011. Heyi lu jinbuzhi qiang zhan gengdi shijian [Why do we not prohibit repeated incidents of land grabbing?]. Yunnan zhengxie bao, 25 July.

Cheng, X., C.F. Wang and Y. Chai. 2009. Guanyu nongcun tudi zhengyong buchangkuan fenpei ruo gan wenti de tansuo [An exploration of problems relating to the distribution of land expropriation compensation funds]. *Fazhi yu shehui* 10, 284–6.

China Daily 2010. Rural Land Disputes Lead Unrest in China [online]. China Daily, 6 November. Available from http://www.chinadaily.com.cn/cndy/2010-11/06/content_11510530.htm

China Daily 2011. Save arable land from shrinking [online]. *China Daily*, 12 January. Available from: http://www.chinadaily.com.cn/usa/life/2011-01/12/content_11848807.htm [Accessed 18 April 2011].

Chung, H. and J. Unger. 2013. The Guangdong model: collective village land, urbanization and the making of a new middle class. Paper presented at conference on Urbanization of Rural China, Zhongdian, China 15–17 July. Published 2013. The Guangdong model of urbanisation. *China Perspectives*, 3.

Demick, B. 2012. Chinese woman facing home's demolition blows up herself, two others. *LA Times*, 10 May. Available from http://latimesblogs.latimes.com/world_now/2012/05/chinese-suicide-bomber-kills-herself-two-others.html [Accessed 15 May 2012].

Escobar, A, 2004. Development, violence and the new imperial order. *Development*, 47(1), 15–21.

Fan, P. 2011. Woguo dangqian de nongcun shehui xingshi he nongmin jieceng [Report on the social situation and stratification of China's villagers]. *In:* Chinese Academy of Social Sciences, ed. *2011 nian Zhongguo shehui xingshi fenxi yu yuce [2011 Chinese Society Analysis and Forecast]*. Beijing: Shehui kexue wenzhai chubanshe, pp. 261–71.

Galtung, J. 1969. Violence, peace and peace research. *Journal of Peace Research*, 6(3), 167–91.

Han, Z. 2010. Qiang chai beiju shilu [Revelations about forcible demolition tragedies]. Yunnan zhengxie bao, 15 November 2010.

He, X.F. 2010. Nongcun tudi de zhengzhixue. [The politics of rural land]. *Xuexi yu Tansuo*, 2, 70–5.

Ho, P. 2001. Who owns the land? Property rights and deliberate institutional ambiguity. *The China Quarterly*, 166, 394–421.

Hsing, Y.T. 2010. *The great urban transformation: politics of land and property in China*. Oxford: Oxford University Press.

Huang, Z.H. and P. Wang. 2010. Zhejiang Jiaxing: 'liang fen, liang huan' zhidu liandong gaige de changshi. [Zhejiang Jiaxing: attempting reform through the 'two distributes, two swaps' system]. Minshang, 10 September, 75–8.

Hui, C. 2010. Forced demolition an inevitable pain in China's urbanization. *Global Times*, 18 October. Available from http://www.globaltimes.cn/opinion/commentary/2010-10/582829.html [Accessed 7 December 2012].

Jiang, S.H. 2012. Guangdong 'Wukan shijian' shenceng ci fenxi yu duice yanjiu [A deep analysis of Guangdong's 'Wukan incident' and research into counter-measures]. *Shehui Zongheng*, 27(5), 61–4.

Kalyvas, S.N. 2003. The ontology of 'political violence': action and identity in civil wars. *Perspectives on Politics*, 1(3), 475–94.

Landesa 2012. Research report: summary of 2011 17-province survey's findings [online]. Available from: www.landesa.org [Accessed 4 May 2012].

Lasseter, T. 2011. Unable to stop land grab, Chinese farmer set self afire. McClatchy 27 May. Available from http://www.mcclatchydc.com/2011/05/27/114908/unable-to-stop-land-grab-chinese.html [Accessed 7 June 2013].

Lasswell, H. 1950. *Politics; who gets what, when, how*. New York : P. Smith.

Li, C. 2007a. Woguo fazhan xiandaihua nongye mianlin de zhuyao wenti he zhengce xuanze [Important issues and policy choices confronting China's development of modern agriculture]. *Xuexi yu tansuo*, 4, 123.

Li, C. 2007b. Zhongguo nongcun tudi zhidu gaige de san bu qu [A three step dance for reforming China's rural land system] [online]. *Xuexi shibao*, 7 August. Available from: http://news.ifeng.com/opinion/200708/0807_23_179670_1.shtml [Accessed 17 March 2008].

Li, F. 2011. Cong shehuixue shijiao toushi tudi zhengyong zhong de chongtu [Land expropriation conflicts from a sociological perspective]. *Nongcun jingji yu keji*, 22(1), 66–8.

Li, H.Z. and Q.C. Zhang. 2011. *Zhengdi liyi lun [Theorizing interests in land expropriation]*. Shanghai: Fudan daxue chubanshe.

Li, T. 2007. *The will to improve: governmentality, development and the practice of politics*. Durham, NC: Duke University Press.

Lin, G.C.S. 2009. Developing China: land, politics and social conditions. New York: Routledge.

Lu, X. 1997. The politics of peasant burden in reform China. *Journal of Peasant Studies* 25(1), 113–38.

Ma, J.B. 2013. *The land development game in China*. Lanham: Lexington Books.

Nixon, R. 2011. *Slow violence and the environmentalism of the poor*. Cambridge: Harvard University Press.

Parsons, C. 2012. Chinese protester opposing government takeover of village land is crushed to death by state-owned road flattening truck. *Mail Online*, 25 September. Available from http://www.dailymail.co.uk/news/article-2208507/Chinese-protester-opposing-government-takeover-village-land-crushed-death-state-controlled-road-flattening-truck.html [Accessed 18 September 2012].

Pierson, P. 2004. *Politics in time: history, institutions and social analysis*. Princeton: Princeton University Press.

Pomfret, J. 2013. Special report: freedom fizzles out in China's rebel town of Wukan [online]. *Reuters*. Available from: http://www.reuters.com/article/2013/02/28/us-china-wukan-idUSBRE91R1J020130228 [Accessed 22 March 2013].

Qu, Y. 2012. An insider's account of the Wukan protest [online]. *Caixin*, 19 March. Available from: http://english.caixin.com/2012-03-19/100369893.html [Accessed 26 March 2012].

Radio Free Asia, 2011. *China: Authorities detain protesters*, 12 July. Available from http://www.refworld.org/docid/4e3904d01a.html [Accessed 29 November 2013].

Radio Free Asia, 2012. *Evictee self-immolates*. Radio Free Asia 16 October. Available from http://www.ecoi.net/local_link/251191/362202_en.html [Accessed 22 October 2012].

Radio Free Asia, 2012. *Man dies in Hubei land dispute*. Radio Free Asia, 26 September. Available from http://www.refworld.org/docid/5069a8ee8.html [Accessed 22 October 2012].

Rong, Z. 2010. *Tudi tiaokong zhong de zhongyang yu difang boyi: zhengce bianqian de zhengzhi jingjixue fenxi [Central local gaming in the regulation and control of land: a political economic analysis of policy change]*. Beijing: Zhongguo shehui kexue chubanshe.

Sargeson, S. 2004. Full circle? Rural land reforms in globalizing China, *Critical Asian Studies*, 36(4), 637–56.

Sargeson, S. 2008. Women's property, women's agency in China's 'new enclosure movement': evidence from Zhejiang. *Development and Change*, 39(4), 641–65.

Sargeson, S. 2012. Victims, villains, and aspiring proprietors: framing 'land-losing villagers' in China's strategies of accumulation. *Journal of Contemporary China*, 21(77), 757–77.

Sargeson, S. and Y. Song. 2010. Land expropriation and the gender politics of citizenship in the urban frontier. *The China Journal*, 64, 19–45.

Sun, H.T. 2010. *Zhengdi jiufen de zhengzhixue fenxi: yi Y shi, Z qu chengjiao cun wei lie [Political analysis of land expropriation conflicts: an example from peri-urban villages in Z district, Y city]*. Beijing: Zhishi chanquan chubanshe.

RURAL POLITICS IN CONTEMPORARY CHINA

Tang, Y., R. Mason and P. Sun. 2012. Interest distribution in the process of coordination of urban and rural construction land in China. *Habitat International*, 36, 388–95.

Tongxiang shi, Tongxiang dangwei. 2009. Guanyu tuijin cunzhuang jiju, jiakuai nongcun xin shequ jianshe ruogan yijian [Opinion on the promotion of central villages, speeding up the construction of rural communities]. Document No. 6. Tongxiang city.

2009 nian Zhongguo tudi churang jin da 15000 yi yuan: Hangzhou quanguo di yi [China's income from land sales in 2009 exceeds 1.5 trillion yuan: Hangzhou tops the country] [online]. 2010. Available from: http://www.fdc.soufun.com/news/2010-01-07/3013338.htm [Accessed 18 April 2011].

Wang, X.Q. 2013. 'Tudi guanli fa' boyi wunian zhengdi buchang tiaokuan duzi chuangguan ['Land Administration Law': a sole breakthrough in five years of gaming over the compensation clause]. *Nanfang zhoumo*, 17 January, 17, 19.

Wang, C. and Y. Xing. 2013. Revenue from five major real estate taxes has soared [online]. *Caixin*, 20 March. Available from: http://english.caixin.com/2013-03-20/100504151.html [Accessed 22 March 2013].

Wang, H.J., S.Y.Yang, H. Zhou and H.Q. Huang. 2011. *Chengshihua jinchengzhong de nongmin wenti [Villagers' problems in the course of urbanization]*. Beijing: Zhongguo nongye chubanshe.

Watts, J. 2010. China's new netizens voice suspicions over death of village chief. *The Guardian*, 30 December. Available from http://www.theguardian.com/world/2010/dec/29/china-netizens-qian-yunhi-death [Access 29 November].

Wei, Y. 2011. Chai wei weihe yiding yaodao chaiqian shi [Why demolition violations are inevitable in forcible demolitions]. Fazhi ribao, 15 November.

Wong, C. 2013. Paying for urbanization in China: challenges of municipal finance in the 21st century. Paper prepared for R. Bahl, J. Linn and D. Wetzel, eds. *Metropolitan government finances in developing countries*. Cambridge: Lincoln Institute for Land Policy, Forthcoming.

World Bank and China Development Research Centre. 2012. China 20130: Building a Modern, Harmonious and Creative High-Income Society. Washington: World Bank.

Xinhua. 2003. ''Sannong' redian toushi: shidi shiye nongmin zhuangkuang diaocha' [Perspectives on the 'three rural problems' hotspot: an investigation into the situation of landless, unemployed villagers] [online]. *Xinhua*, 10 July. Available from: http://www.agri.gov.cn/jjps/t20030710_99079.htm [Accessed 16 July 2003].

Xu, J. and A. Yeh. 2009. Decoding urban land governance: state reconstruction in contemporary Chinese cities. *Urban Studies*, 46(3), 559–81.

Yan, J. B. 2010. Mass disturbances in east China over land acquisition spread. Caijing, 22 July. Available from http://english.caijing.com.cn/ajax/ensprint.html [Accessed 25 June 2012].

Yang, D.J. 2003. Qian fada diqu shidi nongmin de chulu zenmeyang jiejue [How to resolve the way out for landless farmers in less developed areas]. *Nongcun jingji daokan*, 2, 22–24 [online]. Reissued on the Ministry of Agriculture website, *Nongye bu xinxi zhongxin*, 19 March 2003. Available from: http://www.agri.gov.cn/llzy/t20030319_66967.htm [Accessed 16 July 2003].

Yang, S. 2013. My thoughts one year on from my election as Deputy Chairman of the Village Committee [online]. *China Media Project*. Available from: http://cmp.hku.hk/2013/03/13/31750/ [Accessed 26 March 2013].

Yep, R. 2013. Containing land grabs: a misguided response to rural conflicts over land. *Journal of Contemporary China*, 22(80), 273–91.

Yew, C.P. 2012. Pseudo–urbanization? Competitive government behavior and urban sprawl in China. *Journal of Contemporary China*, 21(74), 281–98.

Yu, J. 2010a. *Kangzhengxing zhengzhi: Zhongguo zhengzhi shehuixue jiben wenti [Contentious politics: fundamental issues in Chinese political sociology]*. Beijing: Renmin chubanshe.

Yu, J. 2010b. Nongdi chongtu jiang yingxiang Zhongguo shehui de fazhan [Rural land conflicts will influence China's social development]. *In*: J. Yu, ed. *Diceng lichang [My perspective on the bottom stratum]*. Shanghai: Sanlian shudian, pp. 45–8. [Reprinted from *Dangdai Zhongguo*, 3, 2009.]

Zhai, M. 2011. 'Fujian Longquan tudi douzheng quan jilu' [A complete record of the Fujian Longquan land struggle]. 2011. *China Left Review* 2, 1.

Zhao, D.Y. 2009. 'Tudi zhenyong guochengzhong nongmin, difang zhengfu yu guojia de guanxi hudong' [The interactive relationship between villagers, local government and the state in the course of land expropriation]. *Shehuixue yanjiu*, 2, 93–129.

Zhao, X., S. Zhang and C. Ran. 2007. Shilun chengzhenhua jinchengzhong shidi nongmin wenti de zhengfu zeren [Preliminary theorization of government responsibility in the problem of villagers' landlessness during urbanization]. *Nongcun jingji*, 2, 90.

Zhao, Y. and C. J. Webster. 2011. Land dispossession and the enrichment of Chinese urban villages. *Urban Studies*, 48(3), 529–51.

Zhengdi zhidu gaige yanjiu ketizu. 2002. *Yanjiu zhengdi wenti, tansuo gaige zhi lu [Research on land expropriation problems, exploration of paths for reform]*. Beijing: Zhongguo dadi chubanshe.

Zhou, D.G. 2012. Tisheng lingdao ganbu ying dui quntixing shijian nengli de sikao' [Elevating leaders' capacities with regard to mass incidents]. *Xuexi yu sikao*, 4, 9–11.

Zhou, T. 2004. 'Zengjia nongmin shouru guanjian zai nali' [What is the key to increasing villagers' income?] [online]. *Zhengce guancha*, 3. Available from: http://www.cnki.net [Accessed 12 June 2004].

Sally Sargeson, a Fellow at the College of Asia and the Pacific, The Australian National University, is currently researching gender and substantive representation in villages, and the politics of land development. Her most recent books are *Contemporary China: society and social change*, co-authored with Tamara Jacka and Andrew Kipnis (Cambridge University Press, 2013) and *Women, gender and rural development in China* co-edited with Tamara Jacka (Edward Elgar, 2011).

In defense of endogenous, spontaneously ordered development: institutional functionalism and Chinese property rights

Peter Ho

Neo-liberal observers have frequently raised the red alert over insecure property rights in developing and emerging economies. Development would be at a crossroads: either institutional structure needs changing or it risks a full-fledged collapse. Yet, instead of focusing on the enigma between economic growth versus 'perverse' institutions, this contribution posits a functionalist argument that the persistence of institutions points to their credibility. In other words, once institutions persist they fulfill a function for actors. Chinese institutions have been frequently criticized for lack of security, formality and transparency, yet paradoxically, these apparently 'perverse', inefficient institutions have sustained since the late 1970s throughout the entire economic boom. Key to understanding this might be the realization that institutional constellation stems from an endogenous, spontaneously ordered development in which the state is merely one of many actors that ultimately shape institutions into a highly complicated and intertwined whole. The argument is substantiated by reviewing the case of China's rural-urban land rights structure with particular reference to its markets, history and rights of ownership and use.

1. Introduction

In the debates over property rights, social scientists have been heavily divided over what one could term the dilemma of the 'institutional chicken or egg' (P. Ho and Spoor 2006): do institutions affect economy (and *en passant* with it, society, polity and environment) or do market forces (read: relative changes in factor prices and technology) determine institutional structure? As Libecap (1989, 6–7) phrased it, the neo-liberal literature posits that

> [m]arket forces are argued to erode property rights institutions that are poorly suited for responding to new economic opportunities. (...) History, however, shows that property institutions are not mere respondents to broad economic pressures, but that they in turn, shape the path of economic progress.

One might wonder why the scholarly discussion over institutional chickens and eggs plays out so vehemently.

The writing of this contribution has been made possible through the RECOLAND (Rethinking China's Collapse over Land: Development and Institutional Credibility) Project generously funded by the European Research Council (ERC). The author would like to express his sincere gratitude for the detailed and constructive comments by Rafael Wittek and Herman Hoen that helped shape this paper. Special thanks also to go the anonymous reviewers of this journal.

Perhaps a major reason why is because either view leads to an uncomfortable truth. The axiom that relative prices and technology determine institutional structure appears nihilistic because it implies that institutions do not matter, leaving little room for government, policy and regulation. The opposing axiom is equally unsettling, as it has led us on a quest to 'get institutions right', which, whatever way one looks at it, has turned out to be an endless endeavor. Put differently, if institutions determine the economy, then why do the 'right' institutions not result in the desired effects? As we will see in the theoretical discussion below, the mutually contradicting axioms identified here have spilled over into discussions over endogeneity, spontaneous order and credibility.

However, in all our theoretical and empirical considerations, China's emerging economy stands out as a mind-boggling enigma in popular and scholarly perception: an economic powerhouse pushing forward decades of sustained, double-digit growth, yet furnished with all the wrong institutions – authoritarian, non-transparent, unclear, ambiguous and insecure (e.g. Palomar 2002, Pei 2006, Shirk 2007, Prosterman *et al.* 2009). This contribution maintains it could be more illuminating to turn the argument around – rather than focusing on the paradox between growth vis-à-vis 'perverse' (Furubotn 1989, 25) institutions, it is argued that China's institutions, as they exist and persist, are *credible* and the spontaneous cause *and* effect of development. In other words, Chinese institutional structure is *not* the result of intentional design by which institutions can be 'wrongly' or 'rightly' engineered. To those to whom the Chinese state appears to be a classical strong 'developmental state' (e.g. Evans 1995, Leftwich 1995) wielding substantive leverage over the economy and boasting great organizational muscle power, such a postulate might seem odd or even unacceptable. However, forgoing the premise of intention does not exclude intentional action *per se*, but implies that an actor will not see its intentions materialize as these water down into something else through the protracted bargaining with other actors and economic agents. The outcome is a complex, multi-layered, contradictory and, at times, downright *un*intended institutional constellation, that in its bare existence could never have resulted from conscious human design. Or as Adam Ferguson, when describing social structures in his classical essay, eloquently phrased it: 'the result of human action, but not the execution of any human design' (Ferguson 1782, 1).[1] That principle – the result of human action, but not of human design – has theoretically also been conceptualized as *endogenous, spontaneously ordered* development.

Following this argumentation, the paper examines China's land tenure, with particular reference to rural land-based institutions. Where relevant, it will also digress on urban property rights and land markets. There might be various reasons why examining China's institutional structure around land property rights could be important for our understanding of institutional change. First, instead of privatizing land as neo-liberal economists have advocated, the Chinese state firmly holds on to land ownership. Moreover, in many regions land has *not* entered formal registers, which implies that ownership, use and boundaries are unknown, or at least open for continuous government or corporate intervention. Second, despite the lack of formality, security and transparency, land has been and still *is* a core driver of China's capital accumulation and development over the past decades (Hoogerwerf 2002, Flannery 2003, Lin 2009, Hsing 2010).[2] In addition, the widely criticized insecure rural tenure – and the agricultural land lease system in particular – is actually regarded as

---

[1] The precise reference on Ferguson's page 1 is Part III, Section 2.
[2] The rapid land development is also closely related to the highly inflated real estate sector, which in the past few years has witnessed price rises of over 200 percent in major cities over 2000–2010 (Wu *et al.* 2012). For more info, see also Shen *et al.* (2005) and Hou (2009).

having facilitated China's successful transfer from an agrarian to an industrialized, urbanized society (e.g. Svejnar and Woo 1990, Peng 1995). Finally, although land has been identified as a critical source of conflict and socio-economic instability (Editorial Liaowang 2003, Griffiths 2005), China has to date not met similar institutional disintegration as, for instance, befell the former Soviet Union or the Middle Eastern regimes toppled by the Arab Spring.

In fact, despite repeated predictions of imminent transitional collapse (S. Wang and Hu 1999, Chang 2001, Pei 2006, Shirk 2007), China has been relatively stable since the start of the reforms more than 35 years ago. At the same time, this point drives home the importance of recognizing that institutional credibility, and credible land tenure in particular, are not tantamount to a situation of no conflict. For an appropriate assessment of institutional credibility, social conflict should be measured against more variables than its level alone; for instance, the incidence, frequency, nature and geographical distribution might be equally important to consider.[3]

Apart from the introduction and conclusion, this paper is organized into a theoretical part and an empirical part. The theoretical part provides a review of the scholarly dilemmas and debates on property rights and institutional change with particular reference to concepts of endogeneity and spontaneous order. It will be demonstrated how the discussions on these concepts could coalesce into notions of credibility and institutional functionalism as a way to solve the paradox why socio-economically inefficient institutions (read: insecure, opaque and informal) can exist and persist. Against this backdrop, it is posited that the state cannot determine the form of institutions through land titling or privatization, as outer appearance is determined by institutional function as the resultant of a long, arduous and autonomous process of bargaining. In the following, empirical part of the paper, the theoretical considerations put forward in the preceding section will be applied to land – as one of the means of production, apart from labor and capital. The analysis will be done in terms of the rural and urban land markets, their history since 1949, and the rights of ownership and use. It will be shown that the institutional amalgam that has grown from endogenous, spontaneously ordered development around land is a highly intricate, paradoxical institutional whole that, albeit conflictual in nature, has simultaneously persisted during China's economic boom and its profound rural-urban transformation.

2. Revisiting institutional 'hot potatoes': three postulates of neo-liberal thought

> One of the least controversial principles in the economics of land markets is the notion that the more clearly defined the property rights, the greater the land market efficiency.
>
> Micelli *et al.* (2000, 370)

The quotation above is illuminating as it succinctly sums up the main postulates of what is known as the neo-liberal – or neo-classical for that matter – view on property rights and institutional change. This section does not suggest that there is a single, consistent body of literature that represents *the* 'neo-liberal' or *the* 'neo-classical' theory, as that would be making a straw man out of something that in essence consists of various strands and viewpoints, which sometimes concur, and sometimes contradict each other. However, it *is* maintained that there are certain neo-liberal postulates or basic assumptions around which scholarly debate *and* empirical validation take place, which simultaneously wield great influence over development policy and intervention. These neo-liberal postulates

---

[3]For more information, see P. Ho, 2013.

are: (1) institutions affect the economy, (2) institutions can be designed by intention, and (3) secure, private and formal tenure are imperative for stable, economic growth (see e.g. Coase 1960, Alchian and Demsetz 1973, North and Thomas 1973).[4]

By contrast, it will be argued here that the institutional change of China's rural-urban land markets structure critically diverges from these postulates; i.e. derives from *endogenous, spontaneously ordered* development. *Ergo*, the form of institutions is not in our hands;[5] in fact, one might argue, it is of no real significance at all. In order to understand these conceptual distinctions, some additional explanation on each of the postulates might be helpful.

2.1. On 'institutional chickens and eggs': cause and effect

The first neo-liberal postulate that institutions affect the economy (or vice versa) might be given in by a positivistic, econometric requirement that dependent variables need to be strictly separated from independent variables to meaningfully, thus quantitatively, study economic and social phenomena. The vexing problem in this approach is that socio-economic phenomena are rarely a straightforward matter of cause and effect, yet are in reality the result of mutual interaction.

Methodological rigor has frequently forced scholars to struggle with the conceptualization of institutional cause and effect. As Commons described the confusion: 'Sometimes an institution seems to be analogous to a building, a sort of framework of laws and regulations, within which individuals act like inmates. Sometimes it seems to mean the "behavior" of the inmates themselves' (Commons 1961, 69). Moreover, whereas North and Thomas (1973, 1, 23) start their seminal work with the statement that '[e]fficient organization entails the establishment of institutional arrangements and property rights that create an incentive to channel individual economic effort', they paradoxically write a few pages later that 'the pressure to change property rights emerges only as a resource becomes increasingly scarce relative to society's wants'. Vandenberg (2002, 227) also noted this inconsistency as he wrote that North 'implicitly accepts that institutions affect prices', but simultaneously noted that his analysis is also 'based on changes in factor scarcities, which result in changes in relative prices and which, in turn, prompt changes in institutions'.[6]

Some social scientists have criticized the uneasy premise that institutions are regarded as merely cause and not effect.[7] Others, on the other hand, have tried to crack the scholarly

---

[4]Having said this, one should recognize that the crucial element in neo-classical thought is competition and not property rights *per se*. Of course, because of the greater and greater control over firm performance, neo-classical thought has a certain preference for private, decentralized property rights, but this is subordinate to the necessity of competition. For this reason, neo-classical thought has often neglected the debate on property rights, and concentrated on the notion that, regardless the allocation of property rights, they should at least be well defined (so responsibilities are clear, and competition can be given full rein; Herman Hoen, personal communication, Groningen, 12 December 2012).

[5]Meaning: it is not in our hands, regardless, whether we talk about actors from state, civil society and peasant organizations, business, social movements, academia, media and the like.

[6]Vandenberg also thought that this conceptual tension might be inherent to North and Thomas' economic historical analysis, as they left behind hard-core, econometric methodology. He cautiously quotes: 'Clark suggests that North's increasing deviation from the hard and fast parameters of neoclassical analysis has led him to assemble "a noisy rabble of all the pet concepts and theories of a variety of disciplines and sub-disciplines"' (Clark quoted in Vandenberg 2002, 232).

[7]As, for example, Inglehart (1997, 206) wrote: 'institutions *do* help shape their society's culture – along with many other factors. But the plausibility of the interpretation that institutional determinism is the major explanation is severely undermined by the finds (...) that economic development leads to democracy, but democracy does not bring economic development'. (Emphasis in original document).

puzzle by defying the methodological premise for the separation of dependent and independent variables, and maintaining that institutions can simultaneously be dependent *and* independent variable. As, for instance, Putnam (1993, 8) introduced his method: 'Taking institutions as an independent variable, we explore empirically how institutional change affects (...) actors. Later, taking institutions as a dependent variable, we explore how institutional performance is conditioned by history'. Although merging dependent and independent variables in a single analytical go is equal to capital statistical offense, it also shows the reach of quantitative analysis, for which reason some have made the case for mixed methods – the combination of quantitative and qualitative methodologies (Cresswell 2003, Brannen 2005).[8]

2.2. *Endogeneneity and spontaneous order*

How can it be that institutions which serve the common welfare and are extremely significant for its development come into being without a common will directed toward establishing them?

Menger (1883, 146)

Perhaps the prime reason for the heated discussions over the causality between institutions and economy (or society and polity, for that matter) is its inextricable connection to the second postulate of neo-liberal thought: the notion that institutions can be designed by intention.[9] Examples of the influence of this line of reasoning on land policy and administration abound (e.g. De Soto 2000, 47, Micelli *et al.* 2000, Palomar 2002, 1). However, the haunting issue here is, if the form of institutions is the result of human design, why is it then so difficult to 'get institutions right' (e.g. see the cases described by Amsden 1997, Cornia and Popov 2001, Kuran 2012)?[10]

Against the backdrop above, the principle of endogeneity[11] or endogenous development attempts to chart a way out by positing that institutions and property rights are the resultant of social actors' and economic agents' *interaction*.[12] In this view, institutions are not shaped and enforced by a single, outside agent, but instead through the mutual

---

[8]One could imagine, for instance, a study on the causal effect of institutions on economic growth through qualitative historical analysis, while the opposite effect of economic growth on institutions could be assessed through quantitative regression or factor analysis.

[9] The typical example of this is the so-called 'Washington Consensus', a set of policy prescriptions which included privatization of state-owned enterprises and the provision of legal security for property rights (Williamson 1989). The 'Washington Consensus' became the guiding principle for many social engineering programs financed by the World Bank and the International Monetary Fund in the former Soviet Union and other Eastern bloc countries.

[10]Or, as Aoki (2007, 2) rightfully asked: '[I]f institutions are nothing more than codified laws, fiats, organizations, and other such deliberate human devices, why can't badly performing economies design (emulate) "good" institutions and implement them?'

[11]The term 'endogeneity' in a statistical sense refers to circular causality between the independent and dependent variables; in other words, variable X affects Y, which in turn affects X again. Conceptualized in terms of a supply and demand model it would imply that when predicting the demand level at equilibrium, the price would be endogenous because producers change their price in response to demand and consumers change their demand in response to price. Contrarily, a change in consumer tastes or preferences would be an exogenous change on the demand curve.

[12]The notion of interaction could also be termed 'contracting', 'negotiation', or 'bargaining'. See for instance Libecap (1989, 7).

interaction of that agent with others. The endogeneity principle therefore precludes an external agency that can shape institutions, as any actor is involved in the 'game',[13] albeit institutions may be *perceived* as externally shaped. As an eminent student of institutional change aptly phrased it: 'an institution thus conceptualized is essentially *endogenous*, but appears to be an *exogenous* constraint to the individual agents' (Aoki 2007, 3, italics added). The preclusion of externally, and thus also intentionally shaped, established and enforced institutions inevitably leads to the question of autonomy, for which reason social scientists tabled the ultimate question: do actors have a choice at all in the design of institutions, or does it ultimately escape human intentionality? Here one might also recall the late Duesenberry, who noted with a tinge of irony and humor: 'Economics is all about how people make choices; sociology is all about how they don't have any choices to make' (1960, 233).

It is at this point that the endogeneity principle intersects with the notion of spontaneous order, for which some of the recent proponents include Schotter (1981), Knight (1998) and of course Hayek (1976).[14] The latter viewed the economy as 'a more efficient allocation of societal resources than any design could achieve' (Hayek cited in Petsoulas 2001, 2), and described it as 'the order brought about by the mutual adjustment of many individual economies in a market' (Hayek 1978, 101).[15] Although reasoning from a neo-liberal viewpoint, Hayek's conceptualization of markets, prices and the economy as brought forward by spontaneous order is particularly relevant, not only economically speaking, but also for our understanding of social, political and cultural institutions.[16]

2.3. Tenure security and evolution

The third and final neo-liberal postulate reviewed here posits that tenure security is a precondition for sustained growth (e.g. Gordon 1954, Coase 1960, Cheung 1970, Alchian and Demsetz 1973). This postulate is entwined with the idea that the institutional change of property rights follows a given evolutionary trajectory; simply put, from traditional, informal, opaque, insecure and common property rights to modern, formal, accountable, secure and privatized property rights (e.g. North and Thomas 1973, North 1994).[17] Through competition, different institutional forms will be engaged in a

---

[13]This is different from the widely used definition by North, who assumed externality in his view of institutions as 'the rules of the game in a society or, more formally, ... the humanly devised constraints that shape human interaction' (North 1990, 3).

[14]Aoki, for instance, employed the term 'self-sustaining' to reject the principle of intentionality, and subsequently defined institutions as 'self-sustaining, salient patterns of social interactions, as represented by meaningful rules that every agent knows and are incorporated as agents' shared beliefs about how the game is played and to be played' (2007, 7).

[15]Hayek's notion of spontaneous order was inspired by Adam Ferguson, who described social structures in his essay as 'the result of human action, but not the execution of any human design' (Ferguson 1782, 1, Part III, Section 2). Ferguson's most famous contemporary is, of course, Adam Smith (1776), whose notion of the 'invisible hand' has become the textbook example of this line of reasoning. Carl Menger (1871, 1883) can be seen as the nineteenth-century representative of the thought on spontaneous order.

[16]It is relevant, not in defense of free trade, privatization and deregulated markets, but, paradoxically, as opposed to the neo-liberal postulate that institutions can be designed by intention.

[17]A very useful overview of the application of evolutionary principles in economic theory is provided in Bergh and Stagl (2003).

process of selection by which the most efficient institution eventually prevails over others (see e.g. Alchian 1950, Hawley 1968, Hannan and Freeman 1977).[18]

However, whereas evolutionary theory – in its original conception and current readings – consciously steers away from any teleological assumptions,[19] in the neo-liberal view, each stage not only follows from the other, but is also regarded as more advanced (read: preferred) than the preceding stage. Following neo-liberalist theory's second postulate that institutions can be intentionally engineered, it is not a far step to move to a teleological interpretation of institutional evolution; in sum, the belief in what could be labeled a Comtean 'Law of Institutional Progress'.[20]

It is such belief that propels neo-liberal strong interests in the form of institutions and, more specifically, in the ability to predict institutional form over time. The scholarly stakes of institutional prediction are high: should one be able to predict form as a function of time, one could ultimately design and guide societal development and economic growth. Hence the attempts at classifying and staging institutional forms, as well as at creating preconditions for economic 'take-off' and 'high mass consumption'.[21] Hence also the strong interest in institutional convergence or, as the political scientist Inglehart (1997, 17) worded it: 'In a given economic and technological environment, certain trajectories *are* more probable than others: it is clear that in the course of history, numerous patterns of social organization have been tried and discarded, while other patterns eventually became dominant.' (Emphasis in original).

2.4. *Confusion over convergence*

The idea of convergence or the teleological principle of an ultimate final form has incited considerable debate and scholarly confusion (Radice 2000, Streeck and Thelen 2005),[22] not least because empirical study has yielded a dazzling variety of institutional forms. As a matter of fact, the common principle seems: the more study, the more forms, the more confounding matters become. Some scholars found evidence for institutional convergence;

---

[18]The ideas expounded here are no novelty and, in this regard, it might be helpful to recall that new theories often derive from the adage 'originality is the product of a faulty memory' (Louis Wirth quoted in Becker and Richards 1986, 136). Institutional theory clearly draws upon Herbert Spencer's principles of social darwinism. In his 'System of Synthetic Philosophy', Spencer (1897) describes societal evolution, in which societies best adapted to their environment will prevail through a process of the 'survival of the fittest'. Shortly after Spencer's book was published, Veblen (1898) followed suit and coined the term 'evolutionary economics'. The idea of institutional evolution through selection has since then played a crucial role in neo-liberal thought. For instance, Friedman (1953) proposed that markets as one type of institution act as major mechanisms of natural selection. As firms compete, unsuccessful rivals fail to capture an appropriate market share, go bankrupt and have to exit.

[19]Thus, contrary to what is popularly believed, Spencer's notion of 'survival of the fittest' does not refer to physical fitness, but implies that a society is most suited to its direct environment. It is the reason why evolutionary biologists prefer the more neutrally worded 'natural selection' over the popular, yet discriminatory, 'survival of the fittest'. See also Mayr (1992) and Ruse (1998, 93–98).

[20]One of the earliest proponents of the teleological development of humanity is, of course, the Enlightenment philosopher Auguste Comte (1798–1857), who in *The law of human progress* set forward his theory of three succeeding stages in the development of humanity and human knowledge. In his view, each stage was a necessary precondition for the next (Lenzer 1997, Pickering 2006).

[21]Rostow identified various fixed stages in economic development, moving from traditional, static society with limited technology to the age of high mass consumption. See Rostow (1960).

[22]The works by Radice (2000) and Streeck and Thelen (2005) attempt to make sense of the conceptual confusion by providing substantive overviews of the different models of institutional change in relation to convergence.

others saw divergence *and* convergence, while there are those who reject the notion altogether. As an example of the first view, Long (2006, 321), in scrutinizing the Chinese economic transition, noted that its 'good economic performance since 1978 can be attributed to the convergence of China's economic institutions with the economic institutions of modern capitalist economies, particularly the East Asian capitalist economies'. By contrast, when examining political institutions of the European Union, Botcheva and Martin (2001) observed divergence *and* convergence.[23] Lastly, Campbell (2004), for instance, rejects convergence and, in opposition of it, put forward the idea of 'bricolage' which consists of a dual mechanism of the use of locally available elements and an institutional translation to local contexts.

The paradox that has for long puzzled economists, sociologists and political scientists alike, is – if we accept convergence as a result of competition and efficiency – why is there such abundant empirical evidence for institutional *divergence* through the persistence of 'perverse', inefficient and opaque institutions? Put bluntly, why do traditional, informal, insecure and common institutions and property rights continue to persist, while they are blatantly inefficient?[24] Theorists – even within the neo-classical, neo-liberal schools – have tried to account for the paradox by pointing to the role of power in the formation of institutions.[25] For example, North (1994, 360–1) wrote: 'Institutions are not necessarily or even usually created to be socially efficient; rather they or at least formal rules, are created to serve the interests of those with the bargaining power to create new rules'. Due to differences in power, certain institutions can be forced upon others, which in turn lead to their persistence despite their inefficiency (Acemoglu and Robinson 2006). Some have tried to explain the puzzle by drawing attention to organizational bureaucratization and standardization (Meyer and Rowan 1977, DiMaggio and Powell 1983),[26] while again others argued that it is the contracting (bargaining or negotiation) that ultimately explains the existence of inefficient institutions: 'to better understand the observed variety of property rights institutions and the associated levels of economic performance, attention must be directed to the political bargaining underlying the creation and modification of property rules and laws' (Libecap 1989, 7). From bargaining as an explanatory factor, it is but a small step to the idea of social actors' interaction, which means we have come full circle, and are back to Theoretical Square One.

---

[23]Botcheva and Martin's (2001) article may have added to the confusion as it inadvertently reasons from the neo-liberal axiom, in which institutions externally affect human (or in their case: statal) behavior, yet without explicitly stating so.

[24]As Furubotn rephrased Libecap's perspective, the question is 'why seemingly "irrational" or "perverse" structures of property rights are adopted, and are able to persist in society' (Furubotn 1989, 25). An example of such inefficient institutions is described in Kuran, who identified Islamic law in the failure of producing sustained and credible limitations on government takings (2012, 1088).

[25]In this regard, Vandenberg noted that 'North is cognisant of the importance of power and how power affects the design and operation of institutions' (Vandenberg 2002, 227).

[26]This relates to the argumentation around the notion of 'institutional isomorphism'. In this view, it is maintained that, different from institutional convergence, isopmorphism does not occur as a result of efficiency gains but due to external factors, such as bureaucratization, professionalization and standardization (or myths and ceremonies). As Meyer and Rowan wrote: '[O]rganizational success depends on factors other than efficient coordination and control of productive activities' (1977, 352), while DiMaggio and Powell (1983, 147) noted: '[B]ureaucratization and other forms of organizational change occur as the result of processes that make organizations more similar without necessarily making them more efficient'.

Perhaps we should readjust the paradigmatic lens through which we are viewing the problem. If certain institutions have emerged, and more importantly, *persist* as a result of spontaneous, endogenous development, they are likely to fulfill a certain function, and apparently are perceived as credible.[27] It then no longer matters what form these institutions or property rights have assumed, and whether that form is regarded as 'economically inefficient' (i.e. traditional, backward, insecure, informal, undemocratic or common). As Beckert (2010, 151) stated in an illuminating article: 'The question is not whether homogenization or divergence is more important' as the same mechanisms that cause convergence can also cause divergence.[28] An institution that performs a function for the overall survival of an economy, society, polity or a group of actors will persist; or, vice versa, becomes an 'empty institution'[29] that sooner or later will vanish (as might happen during armed conflict, social movements and revolutions). Or, it will evolve into a different institution in accordance with the new function it needs to fulfill. This line of thought flows from a functionalist inspired approach in institutional theory (Aron 2000, 128, Rodrik 2002, 5, H.-J. Chang 2007, 19–20),[30] which has recently revolved around the notion of credibility (e.g. Grabel 2000, P. Ho 2005a, P. Ho and Spoor 2006).

It should be noted that the discussion around credibility as a function of the perceived social support for endogenously shaped institutions is radically different from a neo-classical reading about 'credible commitment' by which credibility is regarded as a measure of the external, intentional commitment by the state to engineer and enforce tenure security through formalization, privatization and titling (see e.g. Haber *et al.* 2003, Frye 2004). In this sense, it is also different from the notion of functional equivalents (Rodrik 2002, 2007), albeit coming from a functionalist approach. That notion still presupposes the principle of intentional and external design, as evident in this quotation:

> [T]he Chinese leadership *devised* highly effective institutional shortcuts. The Household Responsibility System, Township and Village Enterprises, Special Economic Zones, and Two-Tier Pricing, among many other innovations, *enabled* the Chinese government to *stimulate* incentives for production and investment without a wholesale restructuring of the existing legal, social, and political regime. (Rodrik 2002, 8, italics added)

By contrast, the credibility thesis forgoes any principle of intentional, institutional design.

---

[27]Grabel (2000, 11) is probably one of the first scholars to use the principle of endogeneity in a reconceptualization of credibility.

[28]Similarly, H.-J. Chang (2007, 20) noted that 'institutional forms may not matter that much, as the same function can be performed by different institutional forms'.

[29]In other words, an institution that exists on paper alone, and that has no significant effect on social actors' behavior, or might even be socially contested. See also the discussion of the 'empty institution' in P. Ho (2005a, p. 73.).

[30]Aron (2000, 128), for instance, argues that in relation between institutions and economic development, we should describe institutions by certain 'performance or quality measures' (i.e. function variables), rather than variables that 'merely describe the characteristics or attributes' of institutions (i.e. form variables). Also, in Rodrik's later work, the notion of 'functional equivalents' is used. The principles of functionalism go back to the theory of evolution developed by Lamarck, and published in seven volumes as *Histoire naturelle des animaux sans vertèbres* from 1815 to 1822, as well as the ideas of functionalism represented by Émile Durkheim (1858–1917).

3. Credibility as lens for institutional functionalism

The notion of credibility, or what could perhaps be termed the 'credibility thesis',[31] posits that even when rejecting the neo-liberal reading, we might in fact still be examining the question within the same paradigm, as we reason from the importance of form over function. Instead of focusing on what out of paradigmatic necessity must be considered an 'empirical anomaly' through the lens of neo-liberal institutional teleology and evolution, we had better first focus on the question of how institutions function, or fail to function, at a given time and place. Then, and only then, can we – through meticulous description of the 'rules of the game' that constitute the institution – establish what its form is. That scholarly endeavor – the meticulous description of institutions – is done far too little, yet is absolutely critical to understand the distinction between form and function. Why so? Because what we are bound to see, and that is the reason why dichotomies and institutional categorization will never work, will be a bewildering variation and hybridization in institutional shades of gray, because spontaneously ordered and endogenously formed institutions are as numerous as the multitude of times and spaces in which they take shape.

The credibility thesis' premise that institutional form is subordinate to function might have several contributions to make in the scholarly debates described above: (1) 'inefficient' or 'perverse' institutions – be they insecure, fuzzy, undemocratic or traditionally backward – have no relevance in this view, as any institution fulfills a function once it *persists*; (2) following from this, divergence and convergence, too, have no significance for our understanding of institutions – what counts is whether institutions are perceived as credible by the actors through whose endogenous, spontaneously ordered interaction they sprang forward, and finally (3) the credibility thesis might divert us from the overly optimistic and positivist belief in a 'Law of Institutional Progress' and institutional engineering, as it is detached from any theoretical teleology, normative conviction or political doctrine.

Credibility's three critical dimensions identified here beg the 'what' question: *what* is the institutional structure in a given time and space-determined context, and what levels of credibility does it command? The 'what' of institutions implies that if we want to understand why institutions – as endogenous, spontaneously ordered 'rules of the game' – are socially accepted by those whom they govern, one must know their structure over time *and* space. At the same time, when institutions are non-credible and heavily contested, or merely exist on paper as an 'empty institution', we need to know their structural make-up as well. As a first step in dissecting the question of credibility,[32] we need to chart to the minutest detail what rules exist longitudinally and geographically. That is the aim of what can be called 'institutional functionalism', and is the aim of this writing as well. We will do so by looking at land as a means of production with reference to its markets, institutional origins and rights of ownership and use.

4. Land property in China: markets, historical origins and rights

4.1. The land market

In neo-liberal theory, markets are seen as a prime institutional mechanism through which firms compete in their struggle for greater efficiency and survival (Alchian 1950, Friedman 1953). Against this background, we will here zoom in on one of China's major factor

---

[31]See also the discussion in Ho (2013).
[32]Thus, the second question – what levels of credibility does a certain institution command? – will not be taken up in this writing.

markets: the land market. Resulting from the continuous bargaining between actors, the rural-urban land markets form a complex, intertwined whole with certain intended and as many *un*intended outcomes, yet adapted to its time and space-specific context there where institutional arrangements persist.

On 1 December 1987, China auctioned its first piece of land (note: urban *not* rural land) since the Communist government had come to power almost 40 years earlier. This revolutionary breakthrough marked the beginning of China's land market (Walker 1991). When traveling to China's bustling and high-rising cities it becomes clear that the land market is most developed near and in the urban areas. The socialist past makes research on the Chinese land market a daunting undertaking as things are seldom what they appear. First, in a pragmatic marriage between ideology and market needs, certain terms are avoided in the Chinese context. For instance, the word 'contract' (*chengbao*) is the politically correct term for 'lease', while 'valued land use' (*youchang shiyong*) is a euphemism for 'land market' or 'commercial use'. Second, although the Constitution stipulates that rural land is owned by the collective unless stipulated otherwise, there are widely diverging interpretations by which the collective is actually represented: the township, administrative village or villagers' group.[33] Thus, rural land rights are built on legal quicksand which on the one hand results in externalities, such as forced evictions, rent-seeking and land disputes (e.g. Guo 2001, Griffiths 2005), but which, on the other hand, has also provided the 'institutional lubricant' which makes the system tick (P. Ho 2005a. p. 21). Third, China's urban and rural land markets are *very* much in transition and know certain procedures and institutional arrangements (some formal, many informal) that may be uncommon or purposively kept in a legal twilight zone (S.P.S. Ho and Lin 2003, Lin 2009).

4.1.1. *Rural land market*

In China, one cannot speak of a 'true' rural land market in the neo-classical sense of the word, as there is no freely negotiated price for agricultural land determined through demand and supply. Despite repeated calls for reform both within and outside government circles, the central government has been wary to bring market forces into the countryside out of fear for the loss of institutional credibility, and, thus, the eruption of socio-economic instability (read: the rise of landless peasants). As a result, since the introduction of the Household Contract Responsibility System (*jiating chengbao zerenzhi*), the 'rural land market' has been divided into two levels: a primary (or *yiji shichang*) and a secondary market (*erji shichang*).

As we can see from Figure 1 below, at the primary market, the owner of rural land (in practice represented by the administrative village and *not* the original owner, the villagers' group; see explanation in the historical section below) is allowed to 'contract' or lease land to farmers for 30 years. In the past, land tax was paid to the collective based on the acreage of the contract land. However, since 2006, the government has exempted all agricultural taxes in an effort to reduce farmers' financial burden (X. Li 2004, Xing 2005).[34] From the onset of the rural lease system, farmers were allowed to 'circulate' (*liuzhuan*, euphemism for trade) their lease to other farmers *within* the same village at the 'secondary

---

[33]By which 'local' refers to the province, prefecture, municipality, city or county. For more information on this, also see P. Ho (2000).
[34]For more info on the various taxes that existed in the past, see X. Li (2004). Also see Kennedy (2013) as well as Day (2013).

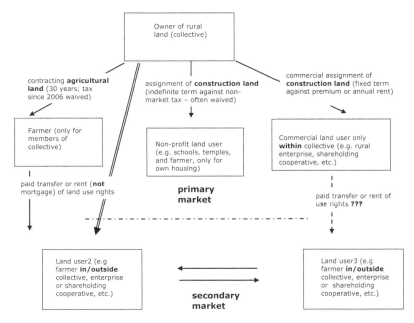

Figure 1. Overview of the rural land market.
Source: drawn by the author.

market'.[35] Simultaneously, however, the central government initially refrained from stipulating whether this could be a 'valued use'.[36] The possibility of paid or commercial transfer of land lease rights was thus intentionally left undefined. (Depicted as a broken arrow in Figure 1).

A significant change in the transfer of lease was made during the revision of the Land Administration Law in 1998. As a result, the right to transfer at the secondary market was also extended to individuals and legal persons *outside* the village collective. At the same time, the law strengthened the role of the village collective by stipulating that such transfers are subject to a two-thirds majority vote by the villagers' representatives meeting (in practice: two-thirds of the village population).[37] Although Chinese law contained no regulations for paid lease transfer, this was already common practice in the urbanized regions.[38] Commercial lease of agricultural land happens in two ways: when the farmer (after migration to the city) subleases land to someone else, or when the collective leases directly to a farmer or agricultural enterprise (shown as a thick arrow in Figure 1).[39] Not until the proclamation of

---

[35]Allowed under the 1986 Land Administration Law, article 2, which states: 'the right to the land owned (…) by the collectives may be lawfully transferred'.

[36]For a detailed review of the internal debates by the Legal Committee of the National People's Congress on the exact formulation of the Land Administration Law in relation to the balance between state and market forces at the countryside, see also P. Ho (2005a pp. 38–41).

[37]See Huang, 2001.

[38]For example, in Jianli County (Hubei Province), the farmers working outside agriculture numbered 220,000 in 2001, which is 49 percent of the total rural labour force. They left behind 520,000 mu (35,000 ha of arable land – one third of the county's total area of arable land – which was subsequently sub-leased. In most cases the original lessee has to pay the new tenant a fee to work his land, as agriculture is less profitable. This fee could amount to 300 Renminbi (Rmb) per mu (G. Huang 2001, 1).

[39]This is common in situations where farmers have migrated to the city permanently and have returned their contract land to the collective.

the Rural Contracting Law in March 2003, some 20 years after the establishment of the Contract Responsibility System, were the principles for commercial lease finally defined:

- It must be a voluntary, consultative and paid lease while no individual or organization may impose or frustrate the transfer;
- The ownership and agricultural land use cannot be changed;
- The term of lease cannot exceed the remaining time of the original term (30 years);
- The new lessee must have agricultural management capacity;
- Members of the same village enjoy priority in obtaining lease rights.[40]

In addition to agricultural land, at the primary rural land market the village collective can also assign *non*-agricultural land – i.e. rural construction land – to users. Commercial users, such as village industries, shareholding cooperatives and private enterprises, can lease rural construction land for 30 years against a fixed premium or an annual land rent. Chinese law has imposed several restrictions on this type of lease:

(1) Only industries, cooperatives and enterprises within the collective are entitled to rural construction land;
(2) It is forbidden to transfer or rent the lease right of rural construction land for non-agricultural construction;[41]
(3) Construction land, in fact *all* rural land – be it agricultural or non-agricultural land – cannot be mortgaged.[42]

The lease of rural construction for *non-commercial* purposes refers to the use of land for housing, schools and public works, such as roads, bridges and irrigation systems. Farmers are only allowed one plot of land for housing, which is provided to them as a type of rural 'social housing'. As no land rent is asked for construction land granted to non-commercial users, no lease term is specified (although this cannot exceed 30 years), and the rural housing land (in Chinese: *zhaijidi*) is generally not titled, the lessee often erroneously (or intentionally) assumes that the land on which his house (or school, temple or mosque) is built, is also his ownership. During Maoist times, such beliefs seldom posed problems.

---

[40]As stipulated in the 2002 Rural Contracting Law, Chapter 5, Article 33.

[41]The National People's Congress justifies this measure by stating that: 'The past years "real estate is hot" (…) as a result of which much land has been left idle. If collective land is allowed to enter the market, large quantities of collective land will be converted into construction land, and even more land will be left idle. This will frustrate the reform of the state-owned land use system'. 1998 Land Administration Law, Article 43 (the Chinese term *'ben'* or 'own' is crucial regarding the first restriction). The prohibition on rent or transfer of rural construction land is further explained in the official legal interpretation of this law (National People's Congress Legal Work Committee 1998, 176).

[42]During the revision of the Land Administration Law, Shanghai and Henan provinces requested to include the following stipulation: 'Under the condition that the ownership and land use will not be altered, the use right to collective land can be legally transferred, rented out, or mortgaged'. But this suggestion was not adopted by the National People's Congress (National People's Congress Legal Work Committee 1998, 381). In July 2010, China started with the first pilot for the mortgage of rural land. Provinces such as Guizhou followed soon after. At the national level, the first change was signaled by Article 20 of the Communiqué of the Communist Party Congress of 12 November 2013. This stipulated: '[F]armers should be given the rights to occupy, use, benefit, and transfer their contract land, as well as the rights to *mortgage and hypothec* the contract management right' (CCP, 2013). How this Party stipulation will work out legally and in actual practice is unknown at the time of writing." (Editorial 2010, Y. Huang 2011).

However, with the commodification and urbanization since the economic reforms, non-commercially-acquired rural construction land is increasingly used for commercial purposes.

For instance, despite the fact that it is strictly prohibited to sell or rent out farm houses and subsequently apply for new plots of housing land,[43] farmers increasingly engaged in the sale of their property. Over the years, clandestine yet extremely lucrative markets for 'small property rights housing' (*xiaochanquan fang*, i.e. without ownership and clear rights on the term of the lease) have emerged, particularly in the peri-urban areas. Chinese scholars and politicians have long been divided over the issue. A renowned law professor and advisor to the Minister of Land and Resources remarked that regarding China's land markets anything might be possible, because 'if it is not explicitly forbidden, it is allowed (*wu jinzhi, ji xuke*)'.[44] Other Chinese academics have argued that farmers should be allowed to marketize and mortgage their property, as it is the only capital asset they have, and would potentially lift them out of relative poverty (Q. Zhou 2004, Gao and Liu 2007).[45] The central government, on the other hand, has stressed that rural housing has been provided by the collective to farmers as free social housing, and should therefore not be sold. By 2012, the central government officially declared its intention to crack down on the illegal sale of rural 'small property rights housing' (Ministry of Land and Resources 2012).

From an institutional credibility perspective, the interesting aspect about small property rights housing is its high market demand, despite the limited, ill-defined property rights – no ownership, no land permit and no clearly defined lease term. The market demand is mainly driven by migrant workers and low-income families, who cannot afford to buy commercial housing in the cities. A study by the Ministry of Land and Resources (C. Wang and Zhang 2010, 116) found that in 62 percent of 1083 sample villages, housing land had been sold, with a higher rate of sale in the more affluent, developed regions. Rural housing with unclear and insecure property rights obviously fulfils a certain need and, thus, a function for social actors.

4.1.2. Urban land market

To have a good sense of the endogenous nature of land-based institutions in China, how they function and what tensions they generate, it is equally important to describe the urban land market as well. The urban sprawl has caused substantive losses in arable land[46] and a sharp rise in social conflict as expropriated farmers are inadequately

---

[43] As is stipulated in the 1998 Land Administration Law, Article 62. The illegal sale is also made possible due to the unclear responsibility over land and housing in the countryside. Instead of being issued by the relevant government housing and land administration departments, the township government or village committee issues the house and land permits.

[44] Personal communication, Wang Weiguo, Beijing, 16 November 2005.

[45] As Gao and Liu (2007, 31) argue: 'the current system overemphasizes the role of administrative approval. As the land use right for farmers' housing has the nature of a usufruct right, the rules of the original acquirement should be restructured according to the principles of the property law. The role of rural land owners must be stipulated'.

[46] For instance, official figures mention a decrease of four percent in the total arable area over 1978–1996: an annual loss of 218,000 ha. See Ash and Edmonds (1998). More information is also available in S.P.S. Ho and Lin (2003) and J. Wang *et al.* (2012).

compensated.[47] This is not to say that there is no conflict over urban land. Contrarily, urban redevelopment,[48] popularly known as *chaiqian* (literally: tear down and relocate), has also led to protracted disputes and civil protests as ex-home owners are poorly compensated (Kahn 2006, Zhu and Ho 2008). This issue, however, falls outside the scope of this contribution.

China has a strongly centralized land planning system under which land use must strictly follow the comprehensive land use plans (*tudi liyong zongti guihua*) and annual land use plans (*tudi liyong niandu jihua*). Under this system, land use plans and any change therein must be approved by the designated government unit, and subsequently reported to a higher level in the administrative hierarchy. However, due to the high potential economic gains in land and real estate development, clandestine deals and 'land theft' frequently occur, despite the strictness of the planning system (Pieke 2005). It is an open secret in China that there are four basic avenues for rent-seeking, a major one of which is over land,[49] and more specifically, the expropriation and subsequent development of rural land.

Ironically, China is losing its most productive land first as urbanization is geographically concentrated in the coastal and southern provinces where land is most fertile.[50] On a more positive note, the cities also provide alternative employment for the farmers. As farmers start to work elsewhere, contract land is directly sub-leased to others within the village, or returned to the village collective, which sub-leases the land to fewer (outside) farmers or agricultural companies.[51] As a result, a farming operation emerges with larger economies of scale and longer-term lease rights.

Similar to the rural market, the urban land market is divided into two levels (see Figure 2 below). At the primary market, the use rights to urban land (already owned by the state, or acquired through expropriation from rural collectives) are assigned to users in a commercial and non-commercial way. In the non-commercial manner, land use rights are 'allocated' (*huabo*) to state or non-profit organizations (e.g. schools, parks and temples) with no time limits to the lease term and at non-market costs (in certain cases the land tax is even waived). In the commercial manner, land use rights are 'conveyed' (*churang*) to commercial users (land developers) for a fixed term (40 years for commercial land, 50 years for industrial land, and 70 years for residential land) (Ma 1991, 93, 183, 891, 988, Walker 1991). This is done against a lump-sum premium that is determined through mutual negotiation, tender or auction. Other variations of 'valued use' at the primary market include an

---

[47]In 1993, China experienced only 8709 incidents related to peasants' and workers' unrest. By 1999, these conflicts had quadrupled to over 32,000. Six years later, they had again tripled to an excess of 87,000 (Chinese Academy of Social Sciences 2005).

[48]A well-known example of this is the conflict over the redevelopment of the Dongbakuai District in Shanghai (Chinaview 2006).

[49]The other ways of rent-seeking are through: (1) government purchase of materials (*zhengfu caigou*), when for instance a corrupt official would buy a government car at a higher price than the market price, with the difference divided between him and the car dealer; (2) infrastructure development (named: *jiben jianshe*), by which similar arrangements exist as with government purchase, or by which the construction company sends gifts to senior state officials to win a tender, and (3) appointments (*dang guan*) when gifts are generally given to the decision-makers in order to obtain a wanted position [personal communication, section head of municipal auditing agency (*shi shenjiju*), 5 June 2008]. For privacy reasons, the official's name and the city to which the auditing agency belongs have not been mentioned.

[50]For instance, in an attempt to halt indiscriminate expropriations and urban sprawl, the governor of Guangdong province decreed that 'collective construction land which rights are disputed is forbidden to enter the market' (literally: 'to be transferred') (Guangdong Provincial Government 2005, 1).

[51]An example of this is provided in Appendix J in P. Ho (2005a).

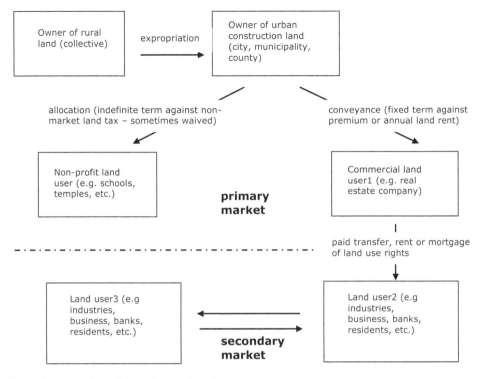

Figure 2.   Overview of the urban land market.
Source: drawn by the author.

annual land rent or, in exceptional cases, the exchange of enterprise shares for the rights to use state land.

At the secondary market, those who have acquired land use rights through commercial state conveyance may transfer (*zhuanrang*) or rent (*churang*)[52] these rights to others, while they may also be used as a collateral (which, as we have seen, is legally still not allowed for rural land). The mortgage of land is considered of vital importance for China's economy which has a relatively undeveloped capital market. At the secondary market, real estate corporations can sell their land use rights to businesses, industries and urban residents. Non-profit organizations that have acquired land use rights through non-commercial state allocation are excluded from participation at the secondary market, unless they compensate the state for the difference between the allocation price and the conveyance price (J. Chen and Wills 1999, S.P.S. Ho and Lin 2004).

The economic reforms have also led to the demise of the urban collective sector in China. During the socialist era, citizens – as employees of a 'work unit' (*danwei*) – were entitled to collective pension and disability schemes, medical insurance and housing. During the late 1990s, two developments forever changed the face of China's urban real estate market. First, a great number of work units sold the housing they owned to their employees at below-market prices in a once-only offer. Second, at roughly the same

---

[52]In the latter case, a new contract and (possible) new conditions are stipulated. In the former case, the original contract is transferred under the same conditions.

time, real estate companies started to develop commercial housing catering to the rising middle class that was increasingly able to afford to buy apartments on their own accord. The urban market has developed at such a quick pace, leading to a boom in the prices of real estate. Over the period from 2005 to 2009, average housing prices in the country have more than tripled (Chovanec 2009).[53] In reaction to what by many has been labeled a 'real estate bubble', the government repeatedly called for a 'cooling down' in the urban housing market, which culminated in 2010 in strict limits on the number of houses that can be owned by one household (in effect in 20 major Chinese cities) (China Daily 2005, 2011). However, by the end of 2012, it was clear that house prices in major cities, such as Beijing, Shanghai, Shenzhen and Guangzhou, had again increased (Anjuke 2012a, 2012b, 2012c, 2012d).[54]

4.2. Historical sketch of Chinese land property

In the neo-liberal view, history is often regarded as the cumulative amalgam of past ideas, beliefs, and decisions that have led to sub-optimal or inefficient economic allocations and present market failures. History, therefore, materializes into an institutional blockage that can impede or 'lock in' efficient performance: the notion of path dependence or (sarcastically) also termed the 'historical hangover' (Arthur 1994, Liebowitz and Margolis 1995, Garrouste and Ioannides 2001). The idea of endogenous, spontaneously ordered development would acknowledge the causal effects between history and institutions, yet simultaneously concur with the historian's view that social actors' interactions are context-specific, and therefore hard to emulate.[55] Some argued that due to history's uniqueness, institutional arrangements, too, are unique, and that as a result, inter-societal comparison is difficult or even beyond our reach (Ben-Dor 1975). Against the latter perspective, this contribution reviews the recent institutional history of Chinese land property rights.

In the years following the communist takeover in 1949, the Chinese government launched a Land Reform (or *tudi gaige*) through which land was seized from 'capitalist rich farmers' and 'landlords' and redistributed to landless and small peasants.[56] However, contrary to widely accepted beliefs, urban land was *not* nationalized during the 1950s land reform. This decision was partially given in by government concerns to avoid alienating the middle and upper class from the Communist cause. However, during the Great Proletarian Cultural Revolution (1966–1976) much of the privately-owned

---

[53]Before this, the average real estate price rose by 14.4 percent over 2003–2004, and by 12.5 percent in the first quarter of 2005 (Lan and Shun 2005).

[54]Per unit (m^2) asking price for Beijing had increased with 19.4 percent from 25166 to 30054 RMB over January to December 2012; Shanghai 3.1 percent from 22665 to 23356 RMB; Shenzhen 10.4 percent from 17428 to 19247 RMB; and Guangzhou with 11. 6 percent from 14762 to 16476 RMB over the same period.

[55]In other words, as opposed to the popular adage *'l'histoire se repète'*.

[56]However, historical research has proven that the rate of tenancy was different from what the Chinese communist regime tried to make people believe. In fact, tenancy showed huge regional differences, with owners generally distributed in North China and tenant farms more abundant in South China. From the famous 1930 study by Buck of a large number of farms, 'owners average 63 per cent for all of the 2866 farms studied' (Buck 1930, 145). Another problem of the land reform movement was the classification of farmers into poor, middle, rich farmers and landlords. Madsen rightly notes that 'the official criteria for determining who was a landlord or a rich peasant, and so forth, were themselves imprecise' (Madsen 1991, 626).

urban land (and real estate) was forcefully confiscated by the Red Guards, and redistributed to factory workers or used for public purposes (such as for schools and government offices).[57]

These shifts in urban land property were, however, never formally included in laws or regulations. Interestingly, with the start of the economic reforms and perhaps also under pressure from descendants of previously expropriated capitalist entrepreneurs, the Chinese state tried to revert, or at least account for, the complex patchwork of private, public and confiscated property in the cities. In March 1982, the State Agency for Urban Construction (*guojia chengshi jianshe zongju*) proclaimed that:

> [T]here are various property systems for urban housing in China [*sic!*]. We should strengthen the titling and administration of housing and land property. The urban land and real estate administrations should according to law, determine property rights, differentiate between the different situations in land ownership and use rights, conduct land surveys, and establish titling documentation and management structures. (State Agency for Urban Construction 1982)

However, shortly after this surprising notice, the National People's Congress adopted the Revised Constitution which quickly smothered what could have become a major ideological debate, and unequivocally, irrevocably stipulated that: 'urban land is state owned' (Revised Constitution 1982, Article 1, Section 1). With certain exceptions,[58] this has become the basic land ownership structure for China today: state-owned in the cities and, as we will see below, collectively owned in the villages.

The legitimacy of Communist rule in the countryside was mainly accomplished through 'land-to-the-tiller' programs by which peasants were won over to the revolutionary cause with the promise of a plot of land of their own. However, this brief interlude when private land ownership was a dream come true for China's peasantry was soon ended. In the late 1950s, the central authorities stepped up their control over land by setting up rural collectives. During the Great Leap Forward in 1958 – Mao's ambition to catapult China into the modern age – the means of production were brought under the control of large-scale collective entities, known as the People's Communes.[59] The result of the sudden, large-scale collectivization during the Great Leap was disastrous: millions of farmers died from starvation as agricultural production plummeted due to a combination of collective inefficiency, bad management and natural disasters. In reaction, the state tried to restructure the ownership of the means of production of the People's Communes into a 'three-level ownership' divided over the commune, the brigade and the team. Yet only after three years of intense debate amongst reformist and conservative factions in the Party did the new ownership structure take effect. In 1962, the Chinese Communist Party decreed in party regulations – but never in law – that land ownership from then on

---

[57]For a detailed description of the history of urban China under Mao, see MacFarquhar and Fairbank (1991).

[58]The main exceptions being: forest, grassland, wasteland, infrastructural, mining and military land in the countryside.

[59]Collectivization was executed in phases with a gradual pooling of the means of production – labor, capital and land. Before the People's Communes were set up in 1958, they were preceded by the Mutual Aid Teams, the Lower Agricultural Production and the Higher Agricultural Production Cooperatives. The official end to private land ownership in China came in the year 1956, when the ownership of agricultural land was vested in the Higher Agricultural Production Cooperatives. See Madsen (1991).

was to be vested in the lowest administrative level, the production team.[60] This remained China's basic ownership structure for rural land until the economic reforms in 1979.[61]

In response to the state of collectivist lethargy in which Chinese agriculture had been for many years, various localities experimented with new institutional arrangements for rural land use (C.-H. Chen 2002). By the mid 1980s, one of these experiments – the one in Xiaogang Village in Anhui Province – had evolved into a national model for more private land tenure. After over 20 years of planned, collectivized agricultural production, the use rights to land – when, what and how much can be produced – were once more returned to the Chinese farmer (Eyferth *et al.* 2004, 1–2). This new model entered the history books as the 'Household Contract Responsibility System', as farm households were allowed to 'contract' or lease land from the village collective for a fixed period of time. Although not at all intended, the practice of household contracts eventually led to the complete disbandment of the communes (C.-H. Chen 2002). These and their subordinate units – the production brigade and the production team – were subsequently replaced by new administrative units, respectively the township, the administrative village and the villagers' group (also known as the natural village).

China is frequently depicted as a country with 'fuzzy' or ambiguous property rights. However, legally speaking, the urban property rights structure is fairly straightforward as urban land is by definition state-owned. Furthermore, to overcome 'local protectionism', the law specifically stipulates that ownership is exercised by the central and *not* the local state.[62] Therefore, the ambiguity of the Chinese property rights structure is more apparent through the rural ownership and use rights, which will be discussed below.

4.3. Land ownership and use

4.3.1. The ownership right

The first level at which the institutional ambiguity of property rights in China manifests itself is that of ownership. Despite the explosive development of 'a market economy with Chinese characteristics', the government has time and again stressed that there is no going back on the ideological foundations of state and collective land ownership.[63] Yet

---

[60]Article 17 of the earlier 1961 draft of the Sixty Articles stipulated that 'all land ... within the limits of the production brigade is owned by the production brigade'. However, by the time the revised draft of the Sixty Articles was issued by the CCP Central Party Committee on 27 September 1962, the team had been given land ownership (CCP Central Party Commmittee 1961, 454). As the final version of the Sixty Articles decreed: 'All land within the limits of the production team is owned by the production team' (CCP Central Party Committee 1962, 141–2).

[61]The Plenum of the 11th Central Committee of the Communist Party in December 1978 marked the beginning of the economic reforms and the official end of collectivism.

[62]The Land Administration Law states that 'the ownership right of state-owned land is exercised by the State Council as representative of the state'. In practice, state ownership is exercised by the Ministry of Land Resources: 'The responsible department for land administration of the State Council [the Ministry of Land Resources] is uniformly charged with the management and supervision of the nation's land'. See Articles 2 and 5, Revised Land Administration Law (Fang 1998, 207). Moreover, the National People's Congress Legal Affairs Work Committee formally issued the legal interpretation that 'the various levels of local government are not the representative for the ownership of state-owned land. They have no right to deal with state-owned land without authorization ... ' It also added that 'the right to profit from state-owned land belongs to the central people's government' (Renda Fazhi Gonguo Weiyuanhui 1998, 37).

[63]A principle also enshrined in the successive revisions of the Chinese Constitution.

there is more to this issue than meets the eye. Communist ideology is nowadays perhaps of less importance in the decision against private ownership than pragmatic considerations. Contrary to China, states-in-transition such as the former East German Democratic Republic (DDR) chose to privatize land after the fall of the Berlin Wall. This decision led to a large-scale eruption of land-related grievances (Kerres 2004).[64] China, on the other hand, has avoided the question of ex-ownership. Instead, the Chinese government has privatized use rights while leaving ownership in the hands of the state: an institutional compromise that over time could incrementally evolve towards a property rights structure not unlike the British system of Crown land to which individual tenants can be entitled to a 'freehold' (Simpson 1976, 28).[65]

Although land use rights had been returned to individual farmers at the time of decollectivization during the mid 1980s, nothing was stipulated regarding ownership. As we have seen above, the ownership of agricultural land during Maoist times was vested in the lowest collective level: the production team. When the teams were disbanded, nothing would have been more logical than to name their successor – the villagers' group – the new owner. But the Chinese government refrained from doing so as it feared to open a Pandora's Box of ex-ownership conflicts.[66] On the other hand, the parliamentary debates about the owner of collective land – the administrative village or the villagers' group – have in practice already been decided in favor of the former. A survey commissioned by the Chinese Communist Party found that the overall majority of the agricultural land was *de facto* already leased out by the administrative village (Wang 1998, 56).[67]

Today, the village collective is a relatively weak institution when it comes to safeguarding and representing peasants' interests over land. The institutional ambiguity over collective land ownership is an important factor in this. Yet its weakness also stems from the fact that the Chinese government has been wary to allow farmers to organize themselves. Whereas in the cities a limited but vibrant civil society has emerged, the countryside

---

[64]One only needs to think of the descendents of Jewish families expropriated during the Holocaust, or the heirs of the former German gentry driven away from their estates when the Russian Red Army defeated the Nazi regime. Ex-ownership issues are notoriously complex: claimants often lack titles as these were destroyed during wars and political upheaval, while new owners can not simply be evicted as they have invested in the land for many decades. This is shown, for example, in a case at the European Court for Human Rights in Strasbourg. In the final days of the German Democratic Republic, Prime Minister Hans Modrow ordained that full ownership be granted to the farmers of the former collectives (LPG) and their heirs. The German government of Helmut Kohl, however, reversed this measure in 1992. In response, around 70,000 former LPG members and their descendants filed a case at the European Court for Human Rights, which in January 2004 ruled that the decision made by the Kohl administration was unlawful. In addition, the German government has been ordered to provide suitable financial redress to the victims of expropriation (Kerres 2004, 5).

[65]As Simpson described: 'Freehold originally meant that the land was held by services of a free nature and not that it was free from all rent and conditions, as its name seems to imply. This indeed is what it has come to mean, for it can now be regarded as synonymous with absolute ownership' (Simpson 1976, 28). In the North of The Netherlands, there is a similar right called 'Beklemrecht', a right originating from the seventeenth century under which the tenant ('Beklemde Meier') had full rights over the land against a fixed, yet relatively low, rent.

[66]There are many claimants to collective land, ranging from individual farmers to schools and temples. For instance, in an administrative village in Inner Mongolia, land ownership title was granted to a primary school. See P. Ho (2005a. Appendices C and Appendix D, 212–25). Many farmers also believe that they, rather than the collective, own the land on which their houses are built (Sargeson 2002).

[67]The villagers' group accounted for only 32.3 percent, the township for 1.1 percent, and other categories for 3 percent (Wang 1998, 56).

generally lacks voluntary peasant organizations.[68] This might be less problematic in the inland regions, yet where commercialization and the urban sprawl have transformed China's countryside at breakneck speed, the collective finds itself in an increasingly undermined position to protect its members' interests over land. It is thus also at this locus where new institutional arrangements might emerge.

4.3.2. The use right

The central state has attempted on several occasions – yet by and large without much success – to strengthen tenure security of farmers' use right.[69] Initially, the Chinese state experimented with a lease period of only five years. To ensure farmers' economic incentives to invest, the lease term was later extended to 15 years, in 1984. This meant that the majority of contracts would expire towards the end of the 1990s. By 1993, the government stipulated that the term should be extended with another 30 years on top of the original contract period.[70] This measure is popularly known as the 'second round of lease' (di'erlun chengbao) as opposed to the first round of lease following decollectivization in the mid 1980s.[71]

The second round of lease was explicitly intended to guarantee a secure tenure of 30 years for farmers, free from any outside (government) intervention. For this purpose, local authorities were instructed to issue farmers individual, standardized and notarized contracts. Reality proved unruly as contracts were frequently kept by the village collective (administrative village or its subordinate villagers' group), instead of being issued to farmers.[72] In addition, changes in property rights were more often than not recorded informally, or based on oral agreement instead of written down in a contract. More importantly, the term of lease – 30 years – is often little more than a paper agreement. In practice, the village collective can appropriate and redistribute contract land whenever deemed necessary. Research has shown that contract land was often redistributed once every four to five years. Strangely enough, at least from a neo-liberal perspective, the land readjustments are actually supported by the majority of farmers (Kung and Liu 1997, Wang 1998, Kung 2000).

---

[68]See also the introduction in P. Ho and Edmonds (2008). The government has organized elections for the cadres of the administrative village since the mid 1980s. Although these have changed the role of the village collective, they have not significantly strengthened it either – certainly not regarding farmers' interests over land. See Z. Wang (1998).

[69]The strengthening of tenure security was done through large-scale titling programs. A detailed overview of the efforts of the Chinese state in this direction is given in P. Ho (2012).

[70]Cheng and Tsang, (1996, p.44) The term of lease is different for other types of land – forest, grassland and wasteland. The 2002 Rural Land Contracting Law defined that 'the lease term for grassland can be 30 to 50 years, for forest land 30 to 70 years, and for forest land with special trees and shrubs the lease term can be extended after approval by the forest administrative departments of the State Council'. Hu 2002 p.174). The term for urban land according to the Land Administration Law is 70 years.

[71]The 1986 Land Administration Law was revised in 1998 to provide the legal basis for the 'second round of lease' (di'erlun chengbao). See article 20 which stipulates that the lease term for agricultural land (not forest, wasteland, and grassland) is 30 years.

[72]Local custom also affected the state's effort to title collective land ownership. As a government notice warned: 'Land permits must be issued to the hands of the rights holder; it is strictly prohibited to withhold these under the pretext of a uniform administration' (Ministry of Land and Resources et al. 2011).

The underlying reason for this apparent contradiction lies in the absence of alternative risk-avoiding arrangements and sufficient employment outside agriculture. In short, the redistributions are a sheer necessity to ensure that everyone in the village has equal access to land. Suppose somebody in a household dies, while a newborn can be welcomed in another; the collective faces substantial social pressure to redistribute land among these two households.[73] In effect, the agricultural lease system thus functions as a social welfare net for the vast surplus of China's rural labour.[74] The readjustments have resulted in land fragmentation, which arguably inhibits a farming operation with greater economies of scale.[75] However, China is no exception, as land fragmentation elsewhere in the world is a logical outcome of farmers' risk-avoiding strategies when alternative social welfare mechanisms are unavailable or more costly (Charlesworth 1983, Ilbery 1984). The rural land lease system's role as an institution for the distribution of social security services, rather than for the distribution of land as a commodity, explains what is described elsewhere as the 'institutional credibility' of property rights.[76]

5. Cursory thoughts on China's relevance for understanding property rights and institutions

China poses a valuable case for the study of property rights and institutional change. In many observers' eyes, the country features an authoritarian, opaque polity with imperfect markets, in which property rights – regardless of whether they relate to immovables, corporate capital or intellectual property – are ill-protected by the state. According to neo-liberal thought, there are only two trajectories out of this developmental deadlock: institutional structure must change, or will collapse. Certain journalists and China-watchers believe that the Chinese nation-state is already heading towards an institutional collapse. They point to the widening gap between rich and poor, and the rise in social conflict (Chinese Academy of Social Sciences 2005, Young 2007). For instance, Pei (2006, 10, 132) speaks of a 'trapped transition' caused by a 'decentralized, predatory state'. Rampant corruption, local protectionism and political rent-seeking will herald the end of the Chinese developmental miracle.[77] What will happen? Some speculate that another Tian'-anmen movement, ethnic separatism or a peasants' uprising will force the Communist Party out of power. Others believe that the People's Republic might implode in a similar vein as the former Soviet Union and Yugoslavia (Gordon 2001, Shirk 2007).

However, despite the predictions of an imminent collapse for many years,[78] Chinese institutional structure has not broken down, but appears to tenaciously persist without

---

[73]In this regard, Kung (2000, 702) made the important observation that: 'reallocations are partial in nature, involving only households that have experienced a change in membership. Villages well endowed with land resources or with abundant off-farm income opportunities have tended to reallocate land less frequently, and involved fewer households and less land'.

[74]As Guhan (1994, 40, 41) notes: 'In the agricultural economy, land is the primary asset from a subsistence point of view: it provides food security, enables utilization of family labour, and reduces vulnerability to labour and food markets. (...) The example of China shows how access to land can provide the fundamental basis for social security in an agrarian economy'.

[75]The average area of cultivable land is less than 0.10 ha per capita (National Bureau of Statistics 2003, 6, 97).

[76]See Chapter 1 in P. Ho (2005a), and P. Ho and Spoor (2006). Note that this is different from the neo-classical explanation of credibility as described in e.g. Diermeyer et al. (1997, 20).

[77]As Pei (2006, 166) asserts: 'Doubtless ... China's state capacity will continue to erode and sustainable development will be put at risk'.

[78]Perhaps the earliest influential report in that direction is the study by S. Wang and Hu (1999).

significant change. This seemingly enigmatic yet tantalizing paradox that China poses to certain students of institutional change sets it apart from the main postulates of neo-liberal thought reviewed in this paper: (1) that there is a direct causal relationship between institutions and the economy (or vice versa); (2) that institutions can be intentionally designed, and (3) that secure, private and formal tenure is a pre-condition for sustained economic growth. In trying to account for this divergence, it was argued that there is no way of 'getting institutions right' because institutional change derives from endogenous, spontaneously ordered development. It implies that institutions that persist at a given time and space perform a function amongst the actors that use and depend on them, as they have come into being through actors' internal, autonomously driven interaction. The outcome is a highly complicated, paradoxical and, generally, unintended institutional amalgam, that could never have been the resultant of human design, yet does spring forth from human action. China's rural-urban property rights structure is a testimony to this.

Land – as one of the means of production – is evidently situated at the crux between institutional function and actors' interaction therewith. For one thing, although Chinese agricultural tenure is regarded as insecure due to its informal, untitled character and frequent land reallocations by the collective, there is ample evidence that it is deemed credible by the overall majority of the rural populace (Kung and Liu 1997, Wang 1998, Kung 2000). Similarly, although the law makes no stipulations about the extension of the urban land lease[79] after its expiration, and lease permits for the land on which homes are built are often lacking (Jia 2005, Qi 2009), Chinese citizens still buy new housing *en masse*,[80] and even settle for so-called 'small property rights housing' (illegal housing with no ownership rights, located at the rural-urban fringe).[81] The above hardly seems to point to a low institutional credibility of land-based property rights in China. Having said this, it would be a misconception to think that credible land tenure can be equated with a situation of no conflict. In any interaction between social actors there will be power divergences, and as a result, frictions and distributional conflict over resources (Libecap 1989, Knight 1998).[82]

If Chinese social actors and economic agents perceive land property rights as credible, is there also evidence that they originate from endogenous, spontaneously ordered development? What we have seen in this contribution is that property rights cannot be externally designed. The efforts and intentions of the Chinese state – albeit perceived as a strong, centralized state with substantial organizational muscle power – are shaped and limited in its

---

[79]As a commentator stated about the Property Law: 'What will happen to the house after 70 years? It is an issue that is now closely watched by society' (Lu 2005, 3).

[80]Note that long before the housing boom and the bubble in real estate even materialized (Shen *et al.* 2005, Hou 2009), a poll carried out in 2003 amongst 23,000 urban citizens of several major Chinese cities (Beijing, Guangzhou, Shanghai, Wuhan, Xi'an and Shenyang) found that close to a quarter of the respondents (24.6 percent) had already bought new housing (at the secondary urban market), while another 73.8 percent expected to buy a house in the next year (Institute for Marketing Information 2003, 135).

[81]There are basically two possible situations: (1) the house is owned by a farming household, which illegally sold its ownership right to a third party; (2) the house has been newly built on rural land that has not been officially expropriated, and is thus not formally designated as urban construction land. For more information, see also J. Cai (2011) and Ministry of Land and Resources (2012).

[82]In this regard, the 'credibility thesis' concurs with Libecap's view that distributional conflicts are 'inherent in any property rights arrangement, even those with important efficiency implications' (Libecap 1989, 2).

endogenous interaction with other actors.[83] Furthermore, in contrast with the (neo-liberally inspired) popular and scholarly views, Chinese property rights and land-based institutions are not static and caught in an institutional deadlock, but change incrementally as the state 'muddles through' (Lindblom 1959) in its attempts to liberalize land markets.

At times, the state forges through with new laws, regulations and policies, as was the case in 1987 when the Chinese government auctioned its first piece of land, which heralded the end to Communist ideology on the control over land as a means of production. In other instances, the central state is forced backwards and needs to adopt a 'hands-off' approach, which involves toning down earlier intentions, such as its land titling programs through the issue of notarized, standardized contracts, or even leaving major issues undefined, such as rural collective ownership. The results of these dynamics are complicated and opaque land markets, which on paper are strictly divided into rural versus urban, and 'primary' versus 'secondary markets', yet which in practice feature extremely fluid institutional arrangements. Many observers remain concerned about the insecurity, opacity, and inconsistency of Chinese landed property rights. Yet China's property rights over land are not cut in stone.

The great mistake in the study of institutional change is to take a snapshot of that which is in flux. Institutional change implies a shift in the endogenous, spontaneously ordered 'rules of the game' over time and space. For instance, we have seen how, from the onset of decollectivization in the mid 1980s, farmers were allowed to 'circulate' or transfer agricultural land use rights. At the same time, however, it was unclear whether such 'circulation' could be done commercially (or as 'valued use', as it is euphemistically called). It was not until over two decades later that the principles of commercial rural lease were finally codified in law, although full-fledged trade in rural land use rights had been going on long before that (Deng 2001). Similarly, although clear laws and regulations are missing, farmers have been freely selling their homes and housing land to urban dwellers, leading to a lively but illegal market in 'small property rights housing'. The trade was recently met with a total ban by the central authorities (Ministry of Land and Resources 2012). It remains to be seen, however, if the ban can be effectively implemented and what will be done with the rural property that has already been sold. For one, politicians, scientists and non-governmental organization (NGO) representatives have become increasingly concerned over the growing rural-urban divide. To close the gap, it has been argued that farmers should be allowed to marketize and mortgage their homes (Q. Zhou 2004, Gao and Liu 2007). Furthermore, as the rural land market is increasingly integrated in the urban land market due to the ongoing urbanization, it will become more and more difficult to separate one from the other.

If we ever want to understand the Chinese property rights puzzle, we should recognize that it is not institutional form that counts, but function. It might well be that the heavily criticized insecurity, opacity and inconsistency of China's property rights are exactly situated at that functional core. Put bluntly, the flexibility and opacity of the land markets are perhaps the very drivers of China's capitalist development. Evidence already supports such a view: over approximately the same period as the institutional structure was in place that is described here, a large-scale conversion of rural land into urban construction land has been taking place (Ash and Edmonds 1998, Lin and Ho 2003, Lin 2009, J. Wang *et al.* 2012),

---

[83]For all analytical clarity, in fact, they are also shaped and limited in the interaction with itself – the different factions and vested interests from within the state, as well as between the different levels of the state (central versus local).

which in turn has propelled substantial profits in the real estate sector.[84] For years on end, real estate has been in the top echelons of most profitable industries.[85] Moreover, arable land conversion has also been crucial in powering the regional economy. The local government heavily relies on revenues from the conveyance of rural land use rights to commercial users at the (primary) urban market, which is partly due to the absence of a property tax[86] and the specifics of the promotion system for officials (Y. Cai 2003, L. Zhou 2007, Kung *et al.* 2009, Hsing 2010). It is estimated that up to 40 percent of local government budgets originates from the sale of land use rights. In 2010, local governments earned close to US $441 billion on land sales, while the annual increase from land sale revenues by local government exploded from just nine percent in 2006 to 116 percent four years later (White 2011, Chovanec 2012, Wu *et al.* 2012).

In China's present stage of development, a large-scale codification and formalization of land rights might not only result in the establishment of non-credible and empty institutions, it might also be socially highly disruptive. This is not to say that at a certain time and place in the trajectory of development, more formalized and secure property rights might not emerge as credible institutions (in fact, we might already see such developments in the wealthier, urbanized areas). However, it *is* definitely to say that credible institutions and property rights are not ours to design or engineer, but appear through the interaction of social actors and economic agents bound together in an endogenous, spontaneously ordered development.

6. Glossary of key Chinese terms

Chaiqian 拆迁, literally: tear down and relocate
Chengbao 承包: To 'contract', politically correct term for 'lease'
churang 出让: Commercial land conveyance to industries and companies
di'erlun Chengbao 第二轮承包: Second round of lease, i.e. extension of lease in the 1990s
erji shichang 二级市场: Secondary land market: land rights transferred/traded between users
huabo 划拨: Non-commercial land allocation to state or non-profit organizations
liuzhuan 流转: Literally 'circulation', i.e. trade of lease contracts
yiji shichang 一级市场: Primary market: land rights assigned to users by state or collective

---

[84]The average profit is 15 percent per transaction of real estate (approximately three times the world average). However, even higher profits can be made, as the following example illustrates: when land from a village near the southern section of Beijing's Fourth Ring Road was expropriated, the farmers received only 177 RMB per square meter. The same land was sold two years later for almost 38 times the original price (China Times 2006, 4). It is therefore no surprise that among the 100 wealthiest Chinese business leaders, more than 40 were in real estate. Even after some of these tycoons were arrested for economic crimes, 35 real estate developers still remained on *Forbes'* 2003 list (Hoogerwerf 2002, Flannery 2003).

[85]From 2001–2004, real estate ended in first place, and in 2005, real estate ended in third place (China Economic Net 2006).

[86]By the end of 2012, there were pilots with property taxation in only two Chinese cities: Shanghai and Chongqing. In the US, property tax accounted for 73.9 percent of all revenues for the local government in 2009, a figure based on the 2009 Comprehensive Annual Fiscal Report (CAFR) and presented by Prof. John L. Mikesell, mikesell@indiana.edu, International Symposium on China's Urban Development and Land Policy, Peking University, Shouren International Conference Center, 14–15 July 2012.

youchang shiyong 有偿使用: 'Valued land use', i.e. commercial land use or land market
zhuanrang 转让: Transfer of original lease contract under same conditions

References

Acemoglu, D. and J.A. Robinson. 2006. De facto political power and institutional persistence. *The American Economic Review*, 96(2), 325–30.

Alchian, A.A. 1950. Uncertainty, evolution and economic theory. *Journal of Political Economy*, 58 (3), 211–21.

Alchian, A.A. and H. Demsetz. 1973. Property rights paradigm. *Journal of Economic History*, 33(1), 16–27.

Amsden, Alice H. 1997. Bringing production back in: understanding government's economic role in later industrialization. *World Development*, 25(4), 469–80.

Anjuke, 2012a. *2012 Beijing fangjia zoushi* (The 2012 Beijing house price movements). Available from: http://beijing.anjuke.com/market/#mode=1&hm=0&period=12 [Accessed 16 December 2012].

Anjuke, 2012b. *2012 Shanghai fangjia zoushi* (The 2012 Shanghai house price movements). Available from: http://shanghai.anjuke.com/market/W0QQcZ1#mode=1&hm=0&period=12 [Accessed 16 December 2012].

Anjuke, 2012c. *2012 Guangzhou fangjia zoushi* (The 2012 Guangzhou house price movements). Available from: http://guangzhou.anjuke.com/market/W0QQcZ1#mode=1&hm=0&period=12 [Accessed 16 December 2012].

Anjuke, 2012d. *2012 Shenzhen fangjia zoushi* (The 2012 Shenzhen house price movements). Available from: http://shenzhen.anjuke.com/market/W0QQcZ1#mode=1&hm=0&period=12 [Accessed 16 December 2012].

Aoki, M. 2007. Endogenizing institutions and institutional changes. *Journal of Institutional Economics*, 3(1), 1–31.

Aron, J. 2000. Growth and institutions: a review of the evidence. *The World Bank Research Observer*, 15(1), 99–135.

Arthur, B.W. 1994. *Increasing returns and path dependence in the economy*. Ann Arbor: University of Michigan Press.

Ash, R. and R.L. Edmonds. 1998. China's land resources, environment and agricultural production. *The China Quarterly*, 156, 836–79.

Becker, H.S. and P. Richards. 1986. *Writing for social scientists: how to start and finish your thesis, book, or article*, 2nd ed. Chicago, IL: University of Chicago Press.

Beckert, J. 2010. Institutional isomorphism revisited: convergence and divergence in institutional change. *Sociological Theory*, 28(2), 150–66.

Ben-Dor, G. 1975. Institutionalization and political development: a conceptual and theoretical analysis. *Comparative Studies in Society and History*, 17(3), 309–25.

Bergh, J.C.J.M. van den and S. Stagl. 2003. Coevolution of economic behaviour and institutions: towards a theory of institutional change. *Journal of Evolutionary Economics*, 13, 289–317.

Botcheva, L. and L.L. Martin. 2001. Institutional effects on state behavior: convergence and divergence. *International Studies Quarterly*, 45(1), 1–26.

Brannen, J. 2005. Mixing methods: the entry of qualitative and quantitative approaches into the research process. *International Journal of Social Research Methodology*, 8, 173–84.

Buck, J.L. 1930. *Chinese farm economy: a study of 2866 farms in seventeen localities and seven provinces in China*. Shanghai: The Commercial Press.

Cai, J.M. 2011. *Guanyu tuoshan jiejue xiao chanquan fang wenti de jianyi* (Suggestions how to suitably solve the problem of small property rights' housing). Available from: http://blog.workercn.cn/?8738 [Accessed 10 October 2012].

Cai, Yongshun. 2003. Collective ownership or cadres' ownership? The non-agricultural use of farmland in China. *The China Quarterly*, 175, 662–80.

Campbell, J.L. 2004. *Institutional change and globalization*. Princeton: Princeton University Press.

CCP Central Party Committee. 1961. *Nongcun Renmin Gongshe Gongzuo Tiaoli Cao'an* [Draft of the Work Regulations for the Rural People's Communes]. March 1961. *In*: Zhongguo Renmin Jiefangjun Guofang Daxue Dangshi Yanjiushi (ed.), *Reference and Educational Material on the CCP*, Vol. 23, p. 454. Beijing: Guofang Daxue Chubanshe.

CCP. 1962. *Nongcun Renmin Gongshe Gongzuo Tiaoli Xiuzheng Cao'an* [Revised Draft of the Work Regulations of the Rural People's Communes]. 27 September 1962, in Zhongguo Renmin Jiefangjun Guofang Daxue Dangshi Yanjiushi (ed.), *Zhonggong Dangshi Jiaoxue Cankao Ziliao* [Reference and Educational Material on the History of the CCP], Vol. 23, Beijing: Guofang Daxue Chubanshe, 1986.

CCP, 2013. *Zhonggong zhongyang guanyu quanmian shenhua gaige ruogan zhongda wenti de jueding* [decision by the CPC Central Committee on several major issues on the overall deepening of the reforms), adopted on November 12, 2013 at the 3rd Plenary Session of the 18th Central Committee of the Chinese Communist Party], available at http://www.ce.cn/xwzx/gnsz/szyw/201311/18/t20131118_1767104.shtml [Accessed 29 November 2013]

Chang, G. 2001. *The coming collapse of China.* New York: Random House.

Chang, H.-J. 2007. *Institutional change and economic development.* Tokyo, New York, Paris: United Nations University Press.

Charlesworth, N. 1983. The origins of fragmentation of land holdings in British India: a comparative examination. *In*: P. Robb, ed. *Rural India: land, power and society under British rule.* London: Curzon Press Ltd. pp. 181–215.

Chen, C.-H. 2002. Property rights and rural development in China's transitional economy. *Economics of Planning*, 35(4), 349–63.

Chen, J. and D. Wills. 1999. *The impact of China's economic reforms upon land property and construction.* Aldershot: Ashgate.

Cheng, Y. and S. Tsang, Agricultural Land Reform in a Mixed System: The Chinese experience of 1984–1994, *China Information*, Vol. 10, No. 4 (Spring 1996), 44–74.

Cheung, S.N.S. 1970. The structure of a contract and the theory of a non-exclusive resource. *Journal of Law and Economics*, 13(1), 49–70.

China Daily. 2011. Beijing issues new rules to limit house purchase [online]. *China Daily*, 16 February. Available from: www.chinadaily.com.cn/china/2011-02/16/content_12028324.htm [Accessed 7 July 2012].

China Economic Net. 2006. *2005 nian Zhongguo shi da baoli hangye* [The 10 most profitable industries in China in 2005] [online]. *Zhongguo Jingjiwang.* Available from: www.chinaeconomy.cn (now www.ce.cn) [Accessed 28 March 2006].

China Times. 2006. *Beijing dijia baozhang 50 duo bei de mimi: zhengdi 'jihu ling chengben'* [The secret behind the 50-fold rise in Beijing land prices: land requisitions at 'almost no cost']. *Huaxia Shibao*, June 4, 4.

Chinaview. 2006. Shanghai: house arrest of lawyer, mass detention of petitioners [online]. Chinaview. Available from: http://chinaview.wordpress.com/2006/10/24/shanghai-house-arrest-of-lawyer-mass-detention-of-petitioners/ [Accessed 10 July 2007].

Chinese Academy of Social Sciences, ed. 2005. *Blue book of China's society: analysis and forecast on China's social development.* Beijing: Social Sciences Academic Press.

Chovanec, P. 2009. China's real estate riddle [online]. *Far Eastern Economic Review*, 8 June. Available from: www.feer.com/economics/2009/june53/Chinas-Real-estate [Accessed 10 June 2009].

Chovanec, P. 2012b. Further Thoughts on Real Estate's Impact on GDP. Patrick Chovanec: An American Perspective from China. January 20. at: http://chovanec.wordpress.com/2012/01/20/further-thoughts-on-real-estates-impact-on-gdp/

Coase, R.H. 1960. The problem of social cost. *Journal of Law and Economics*, 3, 1–44.

Commons, J.R. 1961. *Institutional economics.* Madison: University of Wisconsin Press.

Cornia, A.G. and V. Popov. 2001. *Transitions and institutions: the experience of gradual and late reformers.* Oxford: Oxford University Press.

Creswell, J.W. 2003. Research design: qualitative, quantitative, and mixed methods approaches, 2nd ed.. Thousand Oaks, CA: Sage Publications.

Deng, K. 2001. What can land guarantee for the farmer? [Tudi Neng Baozhang Nongmin Shenme?] *Nanfang Zhoumo*, 14 June, p. 3.

De Soto, H. 2000. *The mystery of capital: why capitalism triumphs in the West and fails everywhere else.* New York: Basic Books.

Diermeyer, D., J.M. Ericson, T. Frye and S. Lewis. 1997. Credible commitment and property rights: the role of strategic interaction between political and economic actors. *In:* D.L. Weimer, ed. *The political economy of property rights: institutional change and credibility in the reform of centrally planned economies.* Cambridge: Cambridge University Press, pp. 20–42.

DiMaggio, P.J. and W.W. Powell. 1983. The iron cage revisited: institutional isomorphism and collective rationality in organizational fields. *American Sociological Review*, 48(2), 147–60.

Duesenberry, J.S. 1960. Comment on 'An Economic Analysis of Fertility'. *In: Demographic and economic change in developed countries*. Edited by National Bureau of Economic Research, Princeton, NJ: Princeton University Press. pp. 231–34.

Editorial. 2010. *Zhongguo jiang shidian nongcun tudi quanyi diya daikuan yewu* [China will have pilots for services for mortgage of rural land rights]. Available from: http://money.163.com/10/0728/17/6CMP0UDN00253B0H.html [Accessed 29 July 2012].

Editorial Liaowang. 2003. China suffers a surge in petitions in 2003. New leaders face a serious test . *Liaowang Dongfang Zhoukan*, 8 December.

Evans, P. 1995. *Embedded autonomy: states and industrial transformation*. Princeton: Princeton University Press

Eyferth, J., P. Ho and E.B. Vermeer. 2004. The opening-up of China's countryside. *In*: P. Ho, J. Eyferth and E.B. Vermeer, eds. *Rural development in transitional China: the new agriculture*. London: Routledge, pp. 1–2.

FangW., ed. 1998. *Zhonghua Renmin Gongheguo Tudi Guanlifa' Shiyong Jianghua [A practical discussion of the 'Land Administration Law of the People's Republic of China']*. Beijing: Zhongguo Minzhu fazhi chubanshe.

Ferguson, A. 1782. *An essay on the history of civil society* [1767], 5th ed. London: T. Cadell.

Flannery, R. 2003. China's 100 richest 2003 [online]. Forbes. Available from: http://www.forbes.com/2003/10/29/chinaland.html [Accessed 2 August 2007].

Friedman, M. 1953. *Essays in positive economics*. Chicago: University of Chicago Press.

Frye, T. 2004. Credible commitment and property rights: evidence from Russia. *American Political Science Review*, 98(3), 453–66.

Furubotn, E.G. 1989. Distributional issues in contracting for property rights – comment. *Journal of Institutional and Theoretical Economics*, 145, 25–31.

Gao, S.P. and Liu, S.Y. 2007. *Zhaijidi shiyongquan chushi qude zhidu yanjiu* [Study on original acquirement of land use right for farmers' housing]. *Zhongguo Tudi Kexue*, 2, 31–37.

GarrousteP. and S. Ioannides, eds. 2001. *Evolution and path dependence in economic ideas: past and present*. Cheltenham, UK: Edward Elgar Publishing.

Gordon, H.S. 1954. The economics of a common property resource: the fishery, *Journal of Political Economy*, 62, 124–142.

Grabel, I. 2000. The political economy of 'policy credibility': the new-classical macroeconomics and the remaking of emerging economies. *Cambridge Journal of Economics*, 24(1), 1–19.

Griffiths, D. 2005. China faces growing land disputes [online]. 2 August, BBC News. Available from: http://news.bbc.co.uk/go/pr/fr/-/2/hi/asia-pacific/4728025.stm [Accessed 25 April 2007].

Guangdong Provincial Government. 2005. *Guangdongsheng Jiti Jianshedi Shiyongquan Liuzhuan Guanli Banfa* [Administrative regulations for the circulation of land use rights of collective construction land of Guangdong Province]. *Guangdongsheng Renmin Zhengfu Ling [Decree of the People's Government of Guangdong Province]*, 100, 23 June, p. 1.

Guhan, S. 1994. Social security options for developing countries. *International Labour Review*, 133 (1), 35–53.

Guo, X. 2001. Land expropriation and rural conflicts in China. *The China Quarterly*, 166, 422–39.

Haber, S., Razo, A. and Maurer, N. 2003. *The politics of property rights. Political instability, credible commitments and economic growth in Mexico, 1876–1929*. Cambridge: Cambridge University Press.

Hannan, M.T. and J. Freeman. 1977. The population ecology of organizations. *American Journal of Sociology*, 82(5), 929–64.

Hawley, A.H. 1968. Human ecology. *In*: D.L. Sills, ed. *International Encyclopedia of the Social Sciences*. New York: Macmillan, pp. 328–37.

Hayek, F.A. 1976. *Law, legislation and liberty, Vol. II: the mirage of social justice*. London: Routledge & Kegan Paul.

Hayek, F.A. 1978. The results of human action but not of human design. *In: New studies in philosophy, politics, economics*. Chicago: University of Chicago Press, pp. 96–105.

Ho, P. 2000. The clash over state and collective property: the making of the rangeland law. *The China Quarterly*, 161, 227–50.

Ho, P. 2005a. *Institutions in transition: land ownership, property rights and social conflict in China*. Oxford: Oxford University Press.

Ho, P., ed. 2005b. *Developmental dilemmas: land reform and institutional change in China*. London and New York: Routledge.

Ho, P. 2012. Revisiting institutional change in China's development: the myth of titling and tenure security. Paper presented at the International Symposium on China's Urban Development and Land Policy, Peking University, *Shouren* International Conference Center, 14–15 July.

Ho, P. 2013. The 'credibility thesis' and its application to property rights: (In)secure Land Tenure and Social Welfare in China. *Land Use Policy*, http://dx.doi.org/10.1016/j.landusepol.2013.09.019, in press.

Ho, P. and R.L. Edmonds, eds. 2008. *Embedded activism: opportunities and constraints of a social movement in China*. London and New York: Routledge.

Ho, P. and M. Spoor. 2006. Whose land? The political economy of cadastral development in transitional states. *Land Use Policy*, 23(4), 580–7.

Ho, S.P.S. and G.C.S. Lin. 2003. Emerging land markets in rural and urban China. *The China Quarterly*, 175, 681–707.

Ho, S.P.S. and G.C.S. Lin. 2004. Non-agricultural land use in post-reform China. *The China Quarterly*, 179, 758–81.

Hoogewerf, R. 2002. China's 100 richest 2002 [online]. Forbes, 24 October. Available from: www.forbes.com/lists/2002/10/24/chinaland.html [Accessed 2 August 2007].

Hou, Y. 2009. Housing price bubbles in Beijing and Shanghai? A multi-indicator analysis. *International Journal of Housing Markets and Analysis*, 3(1), 17–37.

Hsing Y.T. 2010. *The great urban transformation: politics of land and property in China*. Oxford: Oxford University Press.

Hu K.(ed.), 2002. *Zhonghua Renmin Gongheguo Nongcun Tudi Chengbaofa Shiyi* [Legal Interpretation of the Rural Land Contracting Law of the People's Republic of China] Beijing: Falü Chubanshe.

Huang, G. 2001. *Xin Tudi Geming* [The new land revolution], *Nanfang Zhoumo*, 14 July, p. 1.

Huang, Y. 2011. Guizhou starts pilot for mortgage of use rights housing land and land contract right [*guizhou qidong tudi chengbaoquan he zhaijidi shiyongquan diyadaikuan shidian*] [online]. *Xinhuawang*, 15 August. Available from: http://news.cd.soufun.com/2011-08-15/5662324_all.html [Accessed 29 July 2012].

Ilbery, B.W. 1984. Farm fragmentation in the Vale of Evesham. *Area*, 16, 159–65.

Inglehart, R. 1997. *Modernization and postmodernization: cultural, economic, and political change in 43 societies*. Princeton: Princeton University Press.

Institute for Marketing Information, ed. 2003. *IMI Consumer behaviour and life pattern yearbook 2003–2004, Part III*. Beijing: Zhongguo Wujia Chubanshe.

Jia, X. 2005. *Beijing daduo maifangren meilingguo xiao tudizhen, wuzhen ancang fengxian* [Most house buyers in Beijing did not get a small land permit, no permits hides risk]. Available from: http://news.xinhuanet.com/house/2005-10/11/content_3603998.htm [Accessed 18 December 2012].

Kahn, J. 2006. Shanghai party boss held for corruption. *New York Times*, September 25. Also available at www.nytimes.com/2006/09/25/world/asia/25china.html [Accessed 29 November 2013]

Kerres, M. 2004. Hof voor Mensenrechten: Onteigening in ex-DDR onrechtmatig [Court for human rights: expropriation in former DDR unlawful]. *NCR Handelsblad*, 23 January, 5.

Knight, J. 1998. *Institutions and social conflict*, 4th ed. Cambridge: Cambridge University Press.

Kung, J. 2000. Common property rights and land reallocations in rural China: evidence from a village survey. *World Development*, 28(4), 701–19.

Kung, J., Y. Cai and X. Sun. 2009 Rural cadres and governance in China: incentive, institution and accountability. *The China Journal*, 62, 61–77.

Kung, J. and S. Liu. 1997. Farmers' preferences regarding ownership and land tenure in post-Mao China: unexpected evidence from eight counties. *The China Journal*, 38, 34.

Kuran, T. 2012. The economic roots of political underdevelopment in the Middle East: a historical perspective. *Southern Economic Journal*, 78(4), 1086–95.

Lan, Y.L. and Han. S. Shun. 2005. Residential property management in China: a case study of Enjili, Beijing. *Journal of Property Research*, 17(1), 59–73.

Leftwich, A. 1995. Bringing politics back in: towards a model of the developmental state. *Journal of Development Studies*, 31(3), 400–47.

Lenzer, G. 1997 [1975]. *Auguste Comte: essential writings*. New York: Harper and Row.

Li, X. 2004. Rethinking the peasant burden: evidence from a Chinese village. *In*: P. Ho, J. Eyferth and E.B. Vermeer, eds. *Rural development in transitional China: the new agriculture*. London: Routledge, pp. 45–74.

Libecap, G.D. 1989. Distributional issues in contracting for property rights. *Journal of Institutional and Theoretical Economics*, 145, 6–7.

Liebowitz, S.J. and S.E. Margolis. 1995. Path dependence, lock-in and history. *Journal of Law, Economics, and Organization*, 11, 205–26.

Lin, G.C.S. 2009. *Developing China: land, politics and social conditions*. London and New York: Routledge.

Lin, G.C.S. and S.P.S Ho. 2003. China's land resources and land-use change: insights from the 1996 land survey. *Land Use Policy*, 20(2), 87–107.

Lindblom, C.E. 1959. The science of muddling through. *Public Administration Review*, 19(2), 79–88.

Long, G.Y. 2006. An institutional convergence perspective on China's recent growth experience: a research note. *Papers in Regional Science*, 85(2), 321–30.

Lu, W. 2005. *Wuqunfa' Cao'an Quanwen Gongbu Yezhu Gainian Xieru Cao'an* [Full text of the draft of the 'property law' made public, including house owners' ideas into the draft] [online]. *Jiaodian Fangdichan Wang* [House.focus.cn], 11 July, p. 3. Available from: http://house.focus.cn/news/2005-07-11/116041.html [Accessed 21 November 2006].

Ma, K., ed. 1991. *Tudi Dacidian* [*The great dictionary on land*]. Changchun: Changchun Chubanshe.

MacFarquhar, R. and J.K. Fairbank, eds. 1991. *Cambridge history of China: the People's Republic, Part 2: revolutions within the Chinese Revolution 1966–1982*. Cambridge: Cambridge University Press.

Madsen, R. 1991. The countryside under communism. *In*: R. MacFarquhar and J.K. Fairbank, eds. *Cambridge history of China, the People's Republic, Part 2: revolutions within the Chinese Revolution 1966–1982*. Cambridge: Cambridge University Press, pp. 630–638.

Mayr, E.W. 1992. The idea of teleology. *Journal of the History of Ideas*, 53, 117–35.

Menger, C. 1871. *Principles of economics*, 1981 ed. New York: New York University Press.

Menger, C. 1883. *Investigations into the method of the social sciences with special reference to economics*, 1985 ed. New York: New York University Press.

Meyer, J.W. and B. Rowan. 1977. Institutionalized organizations: Formal structure as myth and ceremony. *American journal of sociology*, 83(2), 340–363.

Micelli, T.J., C.F. Sirmans and G.K. Turnbull. 2000. The dynamic effects of land title systems. *Journal of Urban Economics*, 47, 370–89.

Ministry of Land and Resources. 2012. Most recent policy regarding 'small property housing, a comprehensive clean-up is imminent' [*xiao chanquanfang zuixin zhengce jijiang mianlin bei quanmian qingli*] [online]. 22 February. Available from: http://re.icxo.com/htmlnews/2012/02/22/1406339.htm. [Accessed 22 February 2012]

Ministry of Land and Resources, Ministry of Finance and Ministry of Agriculture. 2011. Notice concerning speeding up and furthering the work on the assessment of land rights, titling and issue of permits [*guanyu jiakuai tuijin nongcun jiti tudi quequan dengji fazheng gongzuo de tongzhi*], 6 May. http://www.mlr.gov.cn/zwgk/zytz/201105/t20110516_865762.htm [Accessed 29 November 2013]

National Bureau of Statistics, ed. 2003. *China statistical yearbook 2003*. Beijing: Zhongguo Tongji Chubanshe.

National People's Congress Legal Work Committee, ed. 1998. *Zhonghua Renmin Gongheguo Tudi Guanlifa Shiyi* [*An interpretation of the land administration law of the People's Republic of China*]. Beijing: Falü Chubanshe.

North, D.C. 1990. *Institutions, institutional change and economic performance*. Cambridge: Cambridge University Press.

North, D.C. 1994. Economic performance through time. *The American Economic Review*, 84(3), 359–68.

North, D.C. and R.P. Thomas. 1973. *The rise of the Western World: a new economic history*. Cambridge: Cambridge University Press.

Palomar, J. 2002. Land tenure security as a market stimulator in China. *Duke Journal of Comparative & International Law*, 12, 7–74.

Pei, M. 2006. *China's trapped transition*. Cambridge, MA: Harvard University Press.

Peng, Y. 1995. China's rural enterprises: effects of agriculture, surplus labor, and human capital. *Institute for Social Science Research Paper Series*: Volume VI. 1994–95 – Biotechnology Studies. Paper 7, pp. 1–15.

Petsoulas, C. 2001. *Hayek's liberalism and its origins: his idea of spontaneous order and the Scottish Enlightenment*. Routledge: London and New York.

Pickering, M. 2006 [1993]. *Auguste Comte: an intellectual biography*. Cambridge: Cambridge University Press.

Pieke, F. 2005. The politics of rural land use planning. *In*: P. Ho, ed. *Developmental dilemmas: land reform and institutional change in China*. London and New York: Routledge, pp. 79–102.

Prosterman, R., *et al*. 2009. *Secure land rights as a foundation for broad-based rural development in China: results and recommendations from a seventeen-province survey*. Special Report Number 18, November, Seattle, Washington: National Bureau of Asian Research.

Putnam, R.D. 1993. *Making democracy work: civic traditions in modern Italy*. Princeton: Princeton University Press.

Qi, L. 2009. *Beijing land agency: Beijingshi guotuju: Beijing mai shangpinfang youwang nadao tudizhen* [Those buying commercial housing in Beijing have hope to get land permit]. Available from: http://news.xinhuanet.com/house/2009-11/11/content_12431238. htm [Accessed 18 December 2012].

Radice, H. 2000. Globalization and national capitalisms: theorizing convergence and differentiation. *Review of International Political Economy*, 7(4), 719–42.

Renda Fazhi Gongzuo Weiyuanhui (RFGW), ed. 1998. *Zhonghua Renmin Gongheguo Tudi Guanlifa Shiyi* [*An interpretation of the land administration law of the People's Republic of China*]. Beijing: Falü Chubanshe.

Revised Constitution 1982. article 1, section 1, adopted by the 5th Meeting of the 5th Session of the National People's Congress on 4 December 1982.

Rodrik, D. 2002. *Feasible globalizations*. Discussion Paper No. 3524. London: Centre for Economic Policy Research.

Rodrik, D. 2007. *One economics, many recipes*. Princeton: Princeton University Press.

Rostow, W.W. 1960. *The stages of economic growth: a non-communist manifesto*. Cambridge: Cambridge University Press.

RuseM, ed. 1998. *Philosophy of biology*. New York: Prometheus Books.

Sargeson, S. 2002. Subduing the rural house-building craze: attitudes towards housing construction land and land use controls in four Zhejiang villages. *The China Quarterly*, 172, 927–55.

Schotter, A. 1981. *The economic theory of social institutions*. Cambridge, UK and New York: Cambridge University Press.

Shen, Y., E.C. Hui and H. Liu. 2005. Housing price bubbles in Beijing and Shanghai. *Management Decision*, 43(4), 611–27.

Shirk, S.L. 2007. *China: fragile superpower*. Oxford: Oxford University Press.

Simpson, R.R. 1976. *Land law and registration*. Cambridge: Cambridge University Press.

Smith, A. 1776. *An inquiry into the nature and causes of the wealth of nations*. 1904 ed. Edwin Cannan, ed. Chicago: The University of Chicago Press.

Spencer, H. 1897. *A system of synthetic philosophy: first principles*. New York: D. Appleton and Company.

State Agency for Urban Construction. 1982. *shi(zhen) budongchan he fangdichan de dengji yu guanli de zanqing tiaoli* [Temporary regulations regarding the titling and administration of municipal (town) property and real estate], 27 March, Notice Number 77.

Streeck, W. and K. Thelen, eds. 2005. *Beyond continuity: institutional change in advanced political economies*. Oxford: Oxford University Press.

Svejnar, J., and J. Woo. 1990. Development patterns in four counties. *In*: W.A. Byrd and Q. Lin, eds. *China's rural industry*. New York: Oxford University Press, pp. 63–80.

Vandenberg, P. 2002. North's institutionalism and the prospect of combining theoretical approaches. *Cambridge Journal of Economics*, 26, 217–35.

Veblen, T.B. 1898. Why is economics not an evolutionary science? *Quarterly Journal of Economics*, 12(3), 373–97.

Walker, A. 1991. *Land, property and construction in the PRC*. Hong Kong: Hong Kong University Press.

Wang, C.M. and Zhang, L.H. 2010. *Woguo zhaijidi shiyongquan liuzhuan de xianzhuang kaocha* [Research on the current status of the transfer of use rights of housing land in China]. *He'nansheng Zhenfa Guanli Ganbu Xueyuan Xuebao* (*Journal of the Henan academy for politics and administration for cadres*), 118(1), 116–123.

Wang, H. 1998. *Dangqian Nongcun Tudi Chengbao Jingying Guanli de Xianzhuang ji Wenti* [The present situation and problems facing the management and administration of rural land lease], *Zhongguo Nongcun Guancha*, 5, 56–7.

Wang, J., *et al.* 2012. Land-use changes and policy dimension driving forces in China: present, trend and future. *Land Use Policy*, 29(4), 737–49.

Wang, S. and A. Hu. 1999. *The political economy of inequality in China*. Armonk, NY: M.E. Sharpe.

Wang, Z. 1998. Village committees: the basis for China's democratization. *In:* Eduard B. Vermeer, Frank N. Pieke and Woei Lien Chong, eds. *Cooperative and development in China's rural collective: between state and private interests*. Armonk, NY: M.E. Sharpe, pp. 239–55.

White, G. 2011. Key Revenue Source For China's Local Governments Is About To Get Slammed, March 15, Business Insider, at http://articles.businessinsider.com/2011-03-15/markets/29967282_1_local-governments-china-land-sales [Accessed 16 August 2013]

Williamson, J. 1989. What Washington means by policy reform. *In:* John Williamson, ed. *Latin American readjustment: how much has happened*. Washington, DC: Institute for International Economics, pp. 35–50.

Wu, J., J. Gyourkoc and Y. Deng. 2012. Evaluating conditions in major Chinese housing markets. *Regional Science and Urban Economics*, 42(3), 531–43.

Xing, Q. 2005. Agricultural tax to be scrapped from 2006 [online]. *China Daily*, 6 March, pp. 1–4. Available from: www.chinadaily.com.cn/english/doc/2005-03/06/content_422126.htm [Accessed 16 January 2007].

Young, N. 2007. How much inequality can China stand? *China Development Brief*, February, 1–35.

Zhou, L.A. 2007. *Zhongguo difang guanyuan de jinsheng jinbiaosai moshi yanjiu* [Governing China's local officials: an analysis of the promotion tournament model]. *Jingji Yanjiu (Economic Research Journal)*, 471(7), 37–51.

Zhou, Q. 2004. *Property rights and institutional change* [Quanshu yu Zhidu Bianqian]. Beijing: Peking University Press.

Zhu, J. and P. Ho. 2008. Not against the state, just protecting residents' interests: an embedded movement in a Shanghai neighborhood. *In*: P. Ho and R.L. Edmonds, eds. *China's embedded activism*. London and New York: Routledge pp. 151–170.

Notes on contributor

Peter Ho is Professor at Minzu University, China and Delft University of Technology, The Netherlands. He published extensively in leading SSCI/SCI-rated journals of development, environment and planning. His books include Institutions in Transition (2005, Oxford University Press), China's Limits to Growth (2006, Blackwell Publishers), Developmental Dilemmas (2005, 2009, Routledge) and China's Embedded Activism (2008, 2011, Routledge). In recognition of his scientific achievements, he was awarded the Independent Research Grant for Consolidators by the European Research Council (ERC).

Internal migration and left-behind populations in China*

Jingzhong Ye, Chunyu Wang, Huifang Wu, Congzhi He and Juan Liu

The astonishing scale of internal migration in China since the 1980s can be compared to only a few cases in world history. This migration gave birth to a vast number of peasant workers with family members left behind in rural communities. Dominant studies on migration mainly address the following: why people migrate, what impacts migration has brought about, and how to cope with these positive or negative effects. This review paper builds on this rich body of literature and engages with critical agrarian studies. A better understanding of rural–urban migration in China can be achieved by analyzing the historical emergence of a new (semi-) proletariat class through a study of the biopolitics of their migration. Based on this and the aforementioned discussion, we point to potential future studies as a conclusion.

Introduction

Unlike displaced peasants who gradually became proletarian workers in the land enclosure movement in Britain, a majority of Chinese peasants have become 'peasant workers' (*nong min gong*) through the process of industrialization and urbanization that began in the early 1980s. The number of peasant workers is huge, and their contribution to national economy is considerable. It is estimated that there were around 230 million peasant workers in 2009 (Liu 2010), meaning that one-third of the Chinese peasantry is on the road. With an average annual income of around 8,000 yuan, one peasant worker creates a surplus value of 17,000 yuan a year (Cao 2005).

Internal migration in China has brought about more than economic benefits (here let us put aside the questions who get these benefits and what they do with what they get). The very term 'peasant worker' implies a semi-proletariat situation (Meng 2011, Liu 2012). Peasant workers have not been completely deprived of their means of production but are still forced to sell their labor in the capitalist market. This creates numerous split families – approximately 58 million children (Li 2009), 47 million wives (Cai 2011) and 45 million elderly have been left behind in rural communities by their migrant family members (Xinhua News Agency 2011).

Such a phenomenon, however, is regarded as a necessary step towards modernization in mainstream economics. It is seen as a typical economic issue of how to transform a traditional, agricultural economy into a modernized one. Under the assumption of *homo economicus*, peasants are believed to be able to weigh costs and benefits of working on the farm and in the factory, and choose the latter over the former because it is more remunera-

*We would like to thank Ding Baoyin, Tu Jing, Xu Siyuan, Ning Xia and Chen Jinghuan for their efforts in the collection of references. We thank Prof. Tony Fuller for helping improve the English language in the paper.

tive. In this sense, movements from traditional agriculture to modern industry are encouraged by the wage differentials between the two sectors. The implicit message here is that migration is a free re-allocation of labor in the market and is conducive to the growth of the national economy.

However, there are many relevant studies at the national level showing that impacts of migration have gone beyond the economic sphere. Questionable impacts of migration in the sending areas include education, health, daily care, gender roles, agricultural production and rural development. Most of the findings are contradictory – both negative and positive impacts can be found. Some corresponding countermeasures and policy suggestions have been put forward in response. However, the problem is, if we are satisfied with the conclusion that migration has different impacts in different contexts, we may be trapped in a complicated and diversified world of micro-studies, ending up with either claiming 'it depends', or proposing a few generalized principles for judging what impacts might occur under what conditions. Such findings in previous literature, although important, need to be complemented by a macro-level analysis in order to involve a broader social, economic and political context and present a clearer picture of this phenomenon.

It is in this light that we deem it necessary to emphasize a political economy perspective in a critical agrarian studies approach to the study of migration, so that power structures, government policies and economic relations can be brought into the analysis. This paper, divided into seven sections, develops such a perspective. After presenting the methodology, the third and fourth parts of the paper review previous literature that analyze causes and impacts of migration, with a particular focus on studies of China. The fifth section briefly summarizes actions and strategies combating negative influences of migration on left-behind populations. This is followed by a reflection on the above problem-solution logic in the sixth part. The reflection moves upwards to the macro level with a biopolitical analysis of migration and social care. To conclude, we point out what can be further explored through a rethinking of migration and left-behind populations.

Methodology

In China, we pioneered the first comprehensive research on left-behind children in 2004, and have expanded the research to left-behind women and left-behind elderly since 2006. Our research has always been community based, and has covered 20 villages in nine provinces in China. We have now published four papers in leading international academic journals and 25 in top national journals. In addition, we have published five books in Chinese.[1] On the basis of our 10 years of empirical research on internal migration and left-behind populations in China, we collected altogether over 500 papers from multiple academic databases, with around 300 originally in Chinese and the rest in English. Thus, this paper is built partly on our own empirical work in China and partly on the survey of the literature.

[1] There are, in total, 26 researchers in the team. The team's research on left-behind populations is of leading significance within the Chinese academic community. There has been in-depth media coverage of the team's research on left-behind populations. The team's research on left-behind populations won the highest prize for social sciences research in China in 2013.

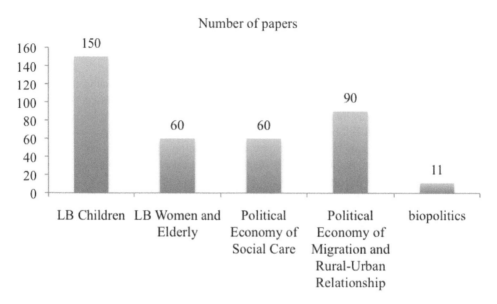

Figure 1 Literatures collected (in Chinese) by themes.

For the latter, we included studies published from 1985 (when migration studies became increasingly prevalent) to 2012. (The advent of literature on the left-behind population came much later, in the 2000s.) Because we were concerned with why migration occurred and with what effects on sending areas, we paid special attention to studies addressing these issues. In doing so, we identified three perspectives and four aspects of impacts that are most frequently adopted, as discussed below. We used key words to locate articles, including 'migration', 'social care', 'left-behind children', 'left-behind women', 'left-behind elderly', 'left-behind population' and 'biopolitics'. Then we screened literature based on frequencies of citation. We also included articles and books referenced by journal articles. Figures 1 and 2 show the breakdown of the literature collected and analyzed.

Because of the impossibility of covering all literature on migration in a single paper, we narrowed the scope of our review to internal migration. Transnational migration is not discussed here. Due to length limitations, not all studies collected are listed in the references.

Causes of migration

This section reviews how mainstream economic, anthropological, and neo-Marxist studies have interpreted causes of migration differently. We argue that migration is triggered by multiple factors at the household-individual scale and in different social–political spheres, with different factors playing more or less important roles at particular times.

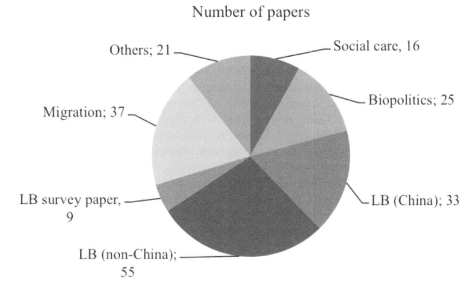

Figure 2 Studies analyzed (in English) by theme.

Three perspectives on migration

The mainstream economic perspective holds that economic factors are the main reason for migration. It seems to be nothing new – population movements occurred in the past and continue at present. People move – from rural to urban, rural to rural, urban to rural, and one country to another – in the hope for better living conditions. Three major theories dominate this perspective: Lewis's unlimited supply of labor, Todaro's expected income, and Stark's relative deprivation. All three suppose that income disparity is the main reason of migration, but focus on different sources of disparity. Lewis (1958) argues that the income difference between urban and rural will inspire the transfer of rural surplus labor from agriculture to industry and that such a movement will not stop until the surplus labor has been entirely absorbed by the advanced industrial sector. Todaro (1981) points out that it is not necessarily the actual income that would encourage a laborer to migrate. The anticipated higher income would also push him or her to move to cities even when the urban unemployment rate might be high in some periods. Stark and Taylor (1991) focus on the income gap within rural communities and argue that the more a laborer feels deprived in the village, the more likely he or she would choose to migrate.

Although powerful and fitting with some everyday observations, the aforementioned theories cannot explain the following phenomena. First, the poorest areas do not have the highest migration rates (Connell et al. 1976, Mallee 1995/96). For instance, the widest income gap in China is that between the western inland and the eastern coastal areas. However, the largest number of migrants does not originate from the western provinces, but from middle China. Six provinces (Jiangxi, Hunan, Anhui, Hubei, Henan and Guangxi) contain 32 percent of the total rural population and 59 percent of the peasant workers;

none of these is in the western areas of China (Yang 2005). Second, the peasants who are most likely to migrate are not the poorest or those who feel most deprived (Du 2000). When examining those who migrate, de Hass argues that migration is 'a socially stratified process in which particular families, ethnic groups or classes participate in and monopolise specific forms of migration' (de Haas 2010, 1601). That is, rural communities and societies are often highly differentiated and migration 'opportunities' are not open to all (ibid.). In extreme cases, some peasants are simply too poor/unhealthy/bound by family responsibilities to migrate. Third, migrant workers, in many cases, are forced to migrate, not for more income, but for survival. Migration implies a historical fact that capital and labor are looking for one another, but not all aspirant laborers were able to find capital (Li 2009). In the colonial period, the dispossessed population could not all be absorbed by plantations, mines and other industries in the local area (Breman 1990). Rather, capitalists preferred to take the trouble to recruit hands from faraway places, making employees easier to manipulate and discipline (Li 2009). In short, the conventional economic perspective draws up a perfect market in which rural laborers migrate out of their own volition and after a careful calculation of costs and benefits. However, in reality, rural–urban migration does not happen mostly in the poorest areas, or among the poorest peasants. It is often not a voluntary 'individual' choice, and not a household's strategy to diversify or specialize their livelihoods. Other factors apart from economic incentives or motivation influence migration.

Anthropologists and sociologists prefer to seek social and cultural reasons behind migration. For instance, in some areas, it is a cultural norm to invest in the first-born boy in the hope that his migration (and later the migration of other family members) might change the fate of the family (Oucho 1996). Unlike those who see migration as a rupture of everyday life, De Haan (1999) disputes this sedentarism and argues that migration should be interpreted as an integral part of societies, and as the norm rather than the exception. In addition, a culture of migration in the community may spur young men to migrate for work, because otherwise they will be depicted as being unambitious and therefore unqualified as future husbands (Kandel and Massey 2002). Migration is then associated with success. With various celebrations of migration in the media, 'people learn to migrate and they learn to desire to migrate' (Ali 2007, 39), as is the case in Guangxi province where modernized ways of consumption and mounting demands have changed the Yao and driven them to migrate (Yang 2012b). In general, according to this perspective, migration could be a demographic response to poverty, but it is never a pure reaction to push and pull factors in simple economic terms.

This perspective also encompasses social network theory, which examines how the migrant population is influenced by an individual's social network, previous migration experiences and other social institutions (De Haan 1999, de Haas 2010). Social network theory can explain quite well why, apart from economic reasons such as potential for higher income, migration usually happens in certain regions or among certain groups of peasants. However, it is more plausible when explaining subsequent or repeat migration. It does not work as well for explaining why migration occurs in the first place.

The third perspective, neo-Marxism, strives to take a broader economic and political context into account in explaining the advent of initial migration (see Safa 1982, Breman 1985, Standing 1985). In this perspective, migration in the contemporary capitalist world is often not a 'choice' for poor inhabitants in rural areas, but the only option for survival under the dull compulsion of economic forces, because their means of subsistence have been largely commercialized (Bernstein 2010). In the global capitalist production chain, migration reflects how labor is channeled to the industrial heartlands of developed countries, as in the case of Mexico (Rubenstein 1992). In this sense, depoliticized economic perspectives cannot explain the forces of migration in periods of primitive capital accumulation, or the more recent migration of peasants who have been deeply absorbed into the capitalist market. In view of this, it is essential to incorporate power and structure into the analysis of migration.

This perspective has a complex interpretation of migration. On the one hand, it laments the social costs that the working classes have to bear during the whole migration process. On the other hand, it respects the necessity of transferring labor from agriculture to industry for the sake of efficiency, productivity and modernization (cf. 'the last peasant' in Yang 2005).

Three perspectives in the Chinese context

Turning more specifically to the Chinese context, we can find plenty of literature explaining migration since the opening up and Reform policy period. In general, this research has gone through shifts in perspectives over time, from neoclassical to neo-Marxist, with social and cultural considerations in between. From the neoclassical perspective in the 1980s and 1990s, researchers argued that rural laborers migrated out for work because of agricultural involution due to the increasing population (Song 1982). A study of 34 villages in Hubei province suggests that the 'pull' factors (mainly urban–rural income differences) mattered more in attracting rural labor to migrate, than 'labor surplus' (Wu 1992). Another study on eight villages in four provinces draws a different conclusion, finding that the relative low returns from agricultural production pushes the peasants out (Huang and Clare 1998). Research in the 1980s and 1990s focused on how rural labor could be better transferred to urban settings, and from agriculture to industry.

With the introduction of social network theory into the Chinese academy in the 1990s, scholars began to pay attention to the role that cultural elements have played in migration. For example, Li (1996) analyzes the social network and social status of peasant workers, pointing out that they rely on family and kinship ties for information and communication, in career seeking and job hopping.

From a class perspective, Ye (2011b) emphasizes that the urban-biased development model in contemporary China has made migration a 'must' instead of an 'option'. Rural areas are increasingly becoming a place where human and other resources have been siphoned off. Pun et al. (2012) pay attention to the migrant workers themselves. They depict the miserable situation that peasant workers are in through studies of construction sites in Peking. Pun's earlier study on women workers in the factory (Pun 2005) show that these are not isolated stories.

Migrant workers across national contexts have long provided an ideal labor force for the capitalist production system: cheap (because their reproduction can be partly complemented by the production in their home villages), easy to control (rootless, without much support from local communities) and able to be used to weaken the bargaining power of local labor (Bernstein 2010). But peasant workers in China are under even harsher exploitative relations because of their double identities (Yang 2005). They are excluded from the social welfare system because they are peasants. They do not rebel easily when fired or unpaid because they still have the last resort of their land. They do not organize themselves because they are family-oriented peasants, not a working class with clear class awareness yet. In addition, this class is also differentiating. A very small proportion become capitalists with relatively larger amounts of capital and hired laborers; some become self-employed individuals; while the majority sell their own labor (Li 1996). Interpreting this situation, both Pun and Yang believe that a proletariat class is in the making, but is hindered by the power structure in China (Yang 2005, Pun et al. 2012). After all, the capitalists prefer to keep workers in a semi-proletariat condition instead of a proletarianized one since it is more conducive to capital accumulation (Wallerstein 1983). The recent popularity of class analysis in migration studies implies that we are confronted with an increasingly conflictual society in which classes are highly differentiated.

In short, it seems that economic, social and cultural factors co-occur in a migration decision for most rural households before migration. Some causal factors are dominant but others peripheral, and vary greatly with different age groups of migrants and diverse households in a specific region or particular time.[2] Young men migrated out to work in the first wave of China's internal migration in the 1980s, followed by middle-aged men and then young women. Now, the tide of migration involves almost all capable laborers in rural communities. As peasants frequently put it, 'all those who can go out to work have gone.'[3] This large-scale and long lasting trend needs to be examined taking micro, meso and macro factors into account.

Impacts of migration

Impacts of migration on the left-behind population are another important issue to be studied. This is not an easy job because of the endogeneity problem (Antman 2012). First, the boundaries of the left-behind population are unclear. For instance, to be called 'left-behind children', should the children be under 16 or 18, or even under 21 years old? How long should peasant workers migrate before their children, spouses, and parents are counted as being 'left behind' – three months or six months or even one year? Meanwhile, at the epistemological level, some scholars believe that the problem of being left behind is general to the entire population in the countryside, which has been left behind socially and economically (Xiang 2007). In India, for example, it has been argued that 'the root of the problems of male outmigration and female suffering lies in the structural inadequacy of the rural society' (Jetley 1987, WS-53). In this sense, the term 'left-behind populations' refers to three different sets of people: (1) the people whose family members have migrated for work; (2) households that have no family members working outside but who are consequently left behind compared to those who do; (3) the entire rural population in many

---

[2] We thank one of our anonymous reviewers for pointing this out.

[3] Our research team conducted another study on left-behind populations at the request of World Bank in 2013. This quoted description was heard repeatedly during the course of the research process.

developing countries who have been left behind in the transition to capitalism. Such different definitions of the very basic term may make it difficult, if not impossible, to compare results of different research studies.

Second, the migrant and non-migrant families may be in different social-economic situations, and comparisons between them may reflect such differences instead of the actual impacts of migration. Third, the longitudinal comparison before and after migration is also questionable since the influencing factors on the living conditions of a left-behind family may involve far more than migration, especially when it covers a long span of time. As Rigg (2007) argues:

> Migration ... may narrow inequalities in source communities, or widen them; it may tighten bonds of reciprocity between migrants and their natal households, or it may serve to loosen or break these bonds; it may help to support agricultural production, or it may be a means to break away from farming altogether. (163)

The following section is drawn from context-specific literature concerning impacts of migration. Generally speaking, impact studies can be classified into two groups: those taking individuals and households as units of analysis and those focusing more widely on the rural community as a whole.

At the household/individual level

Based on frequencies of research content, we selected four aspects to discuss: education, marriage and gender roles, emotional and psychological aspects, and physical health and daily care.

Education

Impacts of migration (and desire for migration) influence educational patterns, performance and duration for both sexes in the sending area. For instance, in India, a girl's chances of being educated are reported to be lower when fathers migrate because they have to take on more domestic duties (Srivastava and Sasikumar 2003). Similar findings are recorded in China (Meyerhoefer and Chen 2011). However, remittances sent back home in El Salvador (Acosta 2011) result in a decrease of child labor and an increase in girls' schooling. In Mexico, a father's migration leads to a reduction of expenditure on boys' education compared to that of girls (Antman 2010). When considering the gender of migrant workers, a mother's migration seems to be more detrimental to her children's education than a father's migration, as shown by studies in the Philippines (Battistella and Conaco 1998, Cortes 2010) and Thailand (Jampaklay 2006). However, Louise Elizabeth Tokarsky-Unda (2005) finds that in Mexico, if mothers migrate with their husbands, it is more likely they will invest in children's education than when fathers migrate alone. Also in China, fathers' migration negatively influences children's school performance because men are usually better educated than women and can help children with their study (Ye and Pan 2011). In Albania and Swaziland, parental or fathers' migration leads often to poor school attendance and performance (Zoller Booth 1995, Giannelli and Mangiavacchi 2010) .

Researchers in Mexico report more negative aspects of migration on children's

Table 1. Impacts of migration on children's education.

	Mothers' migration	Fathers' migration	Parental migration
Impacts on boys	\	Negative (in Mexico)	\
Impacts on girls	\	Negative (in India and China)	\
Impacts on children	More detrimental (in the Philippines and Thailand)	Less detrimental (in the Philippines and Thailand) More detrimental (in China)	Negative (in Albania, China, Mexico, the Philippines, Swaziland, Thailand) Positive (in Bangladesh, Cape Verde, El Salvador) No influence (in China and Mexico)

education than in other regions. Boucher et al. (2005) find no significant impact of international migration on educational investments in Mexico. What's worse, one study on the reduced flow of remittances due to the financial crisis of 2007–2008, finds that remittance reduction is highly related to increases in child labor and reduction in school attendance (Alcaraz et al. 2012). Kandel and Kao (2000), Dreby (2007), and McKenzie and Rapoport (2011) point out that in Mexico, parents' migration might result in children's poor performance at school, reduction in school attendance or drop-outs due to the following: parents' detachment and gang involvement; more housework for girls and migration desires for boys; and that financial benefits of migration discourage above-average schooling.

Research on Thailand reports migration as an alternative to education (Curran et al. 2004). In China, a negative relationship between internal migration opportunities and high school enrolment has been found (deBrauw and Giles 2006). However, there may not be much difference in student interest in studying and cognition of their own school performance, before and after parents migrate (Wu 2004). Likewise, Yeoh and Lam (2006) find that left-behind children are performing poorly in the Philippines, although remittances sent back have been used for children's education, such as sending them to expensive private schools (Adi 2003, Yang 2008). By contrast, in Cape Verde, the possibility of migration encourages people to complete at least intermediate secondary school (Batista et al. 2007). Similarly, children in Bangladesh enjoy more educational opportunities since remittances reduce the amount of household and non-household child labor (Kuhn 2005). These findings about different regions can be summarized as in Table 1.

Table 1 shows how difficult it is to generalize about the impacts of migration on children's education. Although we could say that migration is likely to produce negative influences in more countries, we cannot deduce why this is the case. For instance, remittances may help with education fees, but migration also creates an environment unfavorable for study at home (more workload, feelings of being abandoned, less supervision and guidance, etc.). Experiences and ideas brought back by parents may encourage children to study harder, but are also likely to lure them to drop out and migrate as their parents did. As in the case of China, the influence of migration on children's education might be negative, or there may be no significant differences before and after parents migrate.

Physical health and daily care

It has been argued that the health of 'left-behind' people has been in general improved because they may have better food habits, higher nutritional status and more health-seeking behaviors under the influence or support of migrated family members (Hadi 1999). In addition, the sanitation and health facilities in the sending communities may also be improved with remittances (Taylor 1999). However, the impacts on health in different regions and for different groups vary greatly.

Based on weight-for-height and weight-for-age figures, Anton (2010) reports a positive relationship between remittances and short-term and middle-term nutritional status of left-behind children in Ecuador. Kanaiaupuni and Donato's research points out that infant survival rates may increase in the longer term as a result of remittances sent back home and the institutionalization of migration from Mexico to America, although the infant mortality rate may rise at the beginning (Kanaiaupuni and Donato 1999). Based on evidence from a 2003 Children and Families Survey conducted in the Philippines, Yeoh and Lam reported that children being left behind 'were generally taller, heavier and fell ill less frequently than children whose parents were still with them' (Yeoh and Lam 2006, 127). Differentiating by gender, the non-labor income that Brazilian women obtain has improved the health of girls, but not boys (Duncan 1994). However, the health of left-behind children (especially girls) in China is negatively influenced because they have to undertake more housework, and sometimes also farm work. Their diet is poorer because their guardians have less time to take care of them (Ye and Murray 2006).

Migration of a close family member (without much remittance) has generally negative impacts on the health of the left-behind, with women more vulnerable than men (Duncan 1994). In Nepal and Pakistan, left-behind women who did not receive much in the form of remittances, have less money to obtain health care and food, but instead have a heavier workload (Smith-Estelle and Gruskin 2003, Siegmann 2010). The same is the case in China where left-behind women suffer from higher workloads, heavier psychological pressure and loneliness (Ye and Wu 2008). Moreover, various impacts on left-behind women may be observed even in the same country. For instance, a recent study in India suggests that left-behind women experience higher stress and reproductive morbidity due to sexually transmitted diseases from their migrant husbands (Burazeri et al. 2007). Some Indian and South African women were infected with HIV in this way (Kahn et al. 2003, Archana and Parveen 2005) This finding is also supported by Massey's research (2009) in rural West Bengal, India, and northern Bangladesh. However, another study 10 years earlier suggests that the absence of husbands for a long time may reduce the number of baby deliveries and improve women's health in India (Roy and Nangia 2005).

Long-distance care relations – the quality and accessibility of care – between migrant children and left-behind elderly may vary because of migration stages, historical contexts, and family lifecycles (King et al. 2006). Migration disrupts traditional kinship systems and care structures (Gulati 1993), as reflected especially in the case of Albanian elderly whose illegal migrant children cannot come and go freely when elderly parents are ill (de Haas 2010). They lead a near-starvation life, making soup out of grass and weeds because their migrant children could not (or would not) send back remittances (Vullnetari and King 2008). The Chinese elderly in some areas have been in a miserable situation due to the unwillingness of migrant children to support them and weakened social controls to punish

Table 2. Impacts of migration on physical health and daily care of left-behind populations.

	Positive	Negative
Impacts on children	In Brazil, Ecuador, Mexico, the Philippines	In China
Impacts on women	In India (fewer babies)	In China, India, Nepal, Pakistan
Impacts on elderly	In Cambodia and Thailand (co-residence)	In Albania and China

these children (Baldassar 2007a). Around 80 percent of the left-behind elderly have to do farm work, in comparison to about 56 percent of non-left-behind elderly (Ye and He 2008). Their health has been damaged due to greater workloads after their children migrate (Cui 2007). However, senior citizens in Thailand and Cambodia enjoy a relatively better-off life, co-residing with or living adjacent to one grown child (Ikels and Beall 1993), and as already pointed out by Linda, co-resident members usually provide important emotional and material resources (Linda 1989). The above findings are summarized in Table 2.

Table 2 reveals that children's physical health might be improved because of remittances, but this is not the case in China. Two reports on improved health for left-behind women in India and left-behind elderly in Thailand and Cambodia are not relevant in China, because they are attributed to fewer babies (husbands not at home) and other response strategies (co-residential arrangement in multiple-children families).

Emotional and psychological aspects
The emotional suffering of the left-behind population has been extensively discussed (Zimmer et al. 2007). On the one hand, there are optimistic arguments about the maintenance of emotional support and provision of distant care based on the frequent use of modern communication technologies and means of transport, as evidence from Thailand (Mazzucato and Schans 2011) and Italy (Knodel and Saengtienchai 2007) shows. On the other hand, material benefits, in the form of various gifts, may seek to replace parental presence and convey both love and guilt during parents' migration. However, although a particular idealized childhood acts as a justification for parental migration, it may be in conflict with their children's own vision of an ideal childhood (Baldassar 2007b). Left-behind children generally exhibit greater anxiety and loneliness, and mothers' migration produces a greater impact than fathers' in this regard (Battistella and Conaco 1998). The feelings of being abandoned will linger on and often develop into an estranged relationship with parents (Ye 2011a). Loneliness, feeling insecure, worry, and helplessness are frequently reported among left-behind women in China (Ye and Wu 2008). But the traumatic feelings of being left behind may be fewer in societies where migration is widely accepted and practiced (Suarez-Orozco et al. 2002).

The left-behind elderly suffer similar emotional and psychological shocks as children. Some studies find that adult children's migration is closely related to the poor mental health of their parents, although their physical health might not necessarily be negatively influenced (Adhikari et al. 2011). Depression occurs because of unmet needs for hands-on care, financial support and/or emotional connections, especially in areas where expectations for intergenerational support are high (Krause 2001, Ye and He 2008). Depressive symptoms may be relieved when the left-behind elderly can get support from their daughters-in-law, but is increased when such support comes from their sons, as studies in China show

(Zhen 2008). In Bangladesh, however, elderly peasants left behind are reported to be less depressed (Hadi 1999a). Some research also finds that Chinese left-behind elderly may feel satisfied towards their lives in general, although they are more lonely, because remittances can provide medical care, which is most needed by the elderly, and also because the elderly are relatively young (under the age of 69) (Zhang 2003).

Again, the findings are puzzling. Although most reports claim that the psychological health of left-behind populations may be impacted unfavorably by the migration of family members, there is some evidence to show that a culture of migration and remittance dependency somewhat offsets these pains.

Marriage and gender roles
The separation between husbands and wives due to migration may lead to marital instability and family break-up (Horton 2008), whereas it can also lead to the establishment of stronger emotional affection between spouses in some families if the separation is not too long (Zheng and Xie 2004, Hugo 2005). For instance, in Mexico, both males who migrated and females left behind were more likely to have affairs due to male outmigration, and this often resulted in the collapse of marriage (McEvoy 2008). Marriages might be endangered by different living environments of husbands and wives (Xiang 2006), psychological pressure and sexual suppression (Wang 2007). While a growing number of women have been absorbed into the labor market, leading to perceived threats to masculinity in some areas (Waters 2002, Chee 2003, Huang and Yeoh 2005), in other areas, women's migration sustains traditional gender roles instead of fundamentally altering them (Cleaver 2002). For instance, in Indonesia, the migration of young women has produced a rupture of traditional gender roles and aroused the unexpected and threatening reactions of young males who seek to defend their masculinity (Elmhirst 2007). However, in the Red River Delta, the change of gender roles is not evident because they have been continually reconstructed in a process of negotiation and use of cultural resources and discourses to 'make claims on each other's labor and time' (Resurreccion and Khanh 2007, 213).

As for the change of gender roles, some argue that male migration can lead to a reallocation of agricultural resources and therefore creating a valuable space for rural women in China to develop 'independently' (Li 2003) . However, one could argue that women are allowed to share agricultural resources only because agriculture is already a marginalized sector and undervalued in the market (Gao 1994b). Females left behind in rural villages are not likely to be much empowered if they are still relying on males for resources and information (Elson 1992). Even if their status has been improved, it is more due to general social and cultural change, and migration only helps to accelerate this process (de Haas and van Rooij 2010). This is echoed in Rodenburg's research, which shows that women seek to improve their own power with or without husbands' migration (Rodenburg 2000). The women's secondary role in the family may not be altered much in China since husbands stay in control through cell phones and other communication tools (Zhou 2006). The disadvantaged position of Chinese rural women is maintained through gender discourses, institutions and practices after husbands' migration (Jacka 2012).

In sum, findings about impacts of migration on marriage and gender roles are not only complicated and conflicting, but also difficult to identify. The emotional relations and power structure within the family are influenced by more factors than migration alone.

At the community level

Does the migration of rural laborers promote or worsen the development of the rural community? Some optimists argue that south–north migration might result in greater equality in the long run and therefore should be encouraged (Martin and Taylor 2001). The 'balanced-growth' tradition predicts that the remittances and skills brought back by returned migrants may improve employment opportunities at home and eventually make migration unnecessary (Ghosh 1992).

Some believe that there is a discrepancy between individual households and the rural community in terms of impacts of migration. In other words, migration might be conducive to the development of households, but detrimental to that of the community, especially in the north (de Haas 2010). Situations in the south might be quite different, where the ties between those migrating and those left behind are much stronger. Backflows of resources are more guaranteed (ibid.). This is echoed in a developmental mantra, emphasizing the flow of remittances, skills and able-bodied workers back to the area of origin in a longitudinal perspective (Wise and Covarrubias 2008). For instance, significant remittances might be used for agricultural production (Oberai and Singh 1980). In this way, the 'migration process and remittances modernize the rural sector, both directly and indirectly, through their impact on the production-increasing technological and institutional changes in the agricultural sector' (Lakshmansamy 1990, 479).

However, disparities in regional development may not be narrowed due to remittances from migrant workers, as some scholars have found (Cashin and Sahay 1996). For instance, studies in Passore Province of Burkina Faso and Malawi show that females have already been overburdened in farming and remittances sent back are inadequate to hire labor after the migration of males (David 1995, Findley 1997). Although migration contributes to poverty alleviation in many instances, the differentiation of migrants also occurs and inequality among them increases (De Haan 1999). For example, land could become concentrated in the hands of some returned migrants (Islam 1991).

More often than not, those better educated, with skills and in good health are much more likely to migrate. This has led to a discussion of 'brain drain' and 'brain gain' (Batista et al. 2007). The selectivity of migration in many ways leads to a drain-off of the local stock of human capital. As a study in China shows, those left behind in rural communities are more likely to be women, children and elderly, and their average educational level is comparatively low (Gou 2011).

In addition, exposure to modernized lifestyles and the comparative wealth of the returnees results in changes in rural tastes (Lipton 1980), which, in turn, increases the demand for goods from outside and locally produced goods become less desirable. As de Haas (2010) also suggests, the local economy will be undermined due to the decline of demands for local goods and their dependence on the outside world will be increased. Thus, migration may only lead to the 'development of underdevelopment' (Frank 1966).

Other literatures discuss impacts of migration on agriculture. There is a noticeable trend of feminization of agriculture and gray-hair agriculture. This phenomenon is prevalent in many countries in Asia, Latin America and Africa (see Chiriboga et al. 2006, Lastarria-Cornhiel 2006). As mentioned earlier, a study in China reports that women were allowed to share agricultural resources because it is a marginalized sector with low returns (Gao

1994a). There is no agreement as to whether agriculture has been degraded or not as a result of outmigration of the rural labor force (Fan 2003).

This literature paints a complicated and conflicting picture about the impacts of migration. These complexities result partly from different social, economic and demographic situations of each individual household. Much of the existing literature proposes suggestions for compensating for the social costs of migration, depending on what social costs are found. Here we focus on the case of China and discuss what is missing in this problem-solution logic.

'Love' and 'care' for left-behind populations in China

Since the left-behind population started to gain attention in China in the mid-2000s, a number of social actions have been suggested to share the social costs of migration. Generally speaking, these include efforts to promote the physical and psychological health of left-behind populations, to strengthen the connection between left-behind children and their parents, to maintain the connection between school and migrant parents, and to strengthen the relationship between urban and rural families.

Social care for left-behind children

In 2006, the Work Team for Left-behind Children was established under the guidance of the Agriculture and Industry Office in the State Council of China. This team is led by the National Women's Federation, composed of members from 13 different agencies, including the Ministries of Education, Public Security, and Civil Affairs, the Department of Justice, the Ministry of Finance, and the Ministry of Agriculture. The main objective of the Work Team is to motivate resources and personnel from these agencies to assist left-behind children in terms of education, security, health and so forth. For instance, local governments and the Women's Federation in Shaanxi, Shanghai and Sichuan provinces have paid particular attention to the psychological health of left-behind children. They have organized regular sessions of psychological education and established consultancy rooms in some schools (Ye and Yang 2008). Free lunch has been provided to left-behind children in Hubei, Chongqing, and Jiangxi provinces, from the government, various NGOs and private enterprises (Liang 2011, Jiangxi Charity Federation 2012, Wang 2012b).

Phone booths or phone credit cards are provided in Yunnan, Shandong and Sichuan provinces by local governments so that left-behind children can communicate with their parents more easily. In some schools and administrative villages of Fujian and Zhejiang provinces, records of left-behind children have been set up, in order to maintain necessary contacts with migrant parents when unexpected incidents happen (Xinhua News Agency 2006). In Sichuan, Henan and Anhui provinces, activity centers for left-behind children, parental schools, and seminars on how to communicate with left-behind children have been set up or organized for left-behind families (Ye and Yang 2008). There are also efforts to connect families and children in urban areas with left-behind families and children in rural communities, for instance, hand-in-hand activities have been encouraged and undertaken in Chongqing, Guangdong, Zhejiang, Hubei, Sichuan, Anhui and Guangxi provinces (ibid.).

Pu Wei (2008) categorizes support for left-behind children into four types:

1 sporadic, short-term social support, such as hand-in-hand activities between urban and rural children, usually provided by individual sponsors and NGOs
2 systematic, long-term social support, such as training for guardians of children, usually provided by local governments
3 self-care, such as educating left-behind children to be independent and strong minded, usually provided by the schools
4 the empowerment model, training left-behind children to be capable of expressing their needs and organizing themselves, usually provided by local governments and schools.

According to Pu Wei, these models should complement one another, instead of applying only one type of activity to all left-behind children. Wen (2006) believes that the root of the left-behind children problem is the limited farmland and the urban–rural dual structure, which cannot be solved in a short period of time. He therefore argues for more inputs into rural communities and more actions to provide care to left-behind children while the institutional limits are being removed.

Social care for left-behind women and the elderly

Local governments in Qin Yang County, Gansu Province (Yang 2012a) have organized village leaders, women Communist Party members and women elites to sign long-term agreements with left-behind women, helping this last group with production and self-development. Some experts in education, health, law and technology have been teamed up to provide free training to left-behind women. Psychological consultancy has also been provided to them in order to relieve some of their mental stress.

Apart from these, some special training sessions have been held in order to assist left-behind women to obtain certain skills such as embroidery, so that they can earn additional income. Festivals and other entertainment activities have also been organized. A 'women's home' has been established, in which left-behind women can read books, chat with their husbands online, and play badminton or chess. Activities such as handicraft exhibitions, cooking exhibitions and the like have greatly promoted their enthusiasm and broadened their social networks (ibid.). In Jiangxi Province (Lin 2011), micro-finance has been provided to left-behind women to establish their own small businesses under the support of the Women's Federation. Mutual help teams have also been set up so that left-behind women can exchange labor during busy farming seasons.

As for left-behind elderly, the focus of social support is on their daily care. For instance, the Communist Party in Hunan province (Xiao 2006) launched community activities in Chestnut Tree Township, in which every Communist Party member was required to help two households in which there were left-behind elderly. In River Mouth Town, Communist Party members help the left-behind elderly to deal with various complicated documents and certificates. In Anhui province, some daily consumer goods have been given to the left-behind elderly (Wang 2012a).

In general, these efforts are supposed to share some of the social costs brought about by migration in China. However, they are either temporary or limited in scale, with questionable sustainability and effects. More importantly, these actions see the issue of left-behind children or left-behind elderly as a moral one. That is, this is a way of showing 'love', 'care' and 'charity'. They tend to accept the present institutions and policies that create a care dualism, instead of challenging them. The issue of left-behind populations is seen as 'a problem' on its own and the deeper political economic reasons behind it are left untagged. We feel that instead of seeing this issue as an unfortunate problem and trying to find 'solutions', it is necessary to ask what causes this problem and how peasants are responding to it. We are thereby particularly interested in the levels beyond individuals and households.[4] For this, the biopolitics of migration and the politics of social care need to be closely examined.

Biopolitics of migration and politics of social care

The biopolitics of migration and the politics of social care are two essential aspects of how the government regulates the population and how social care has been transferred between different classes and across borders. Biopolitics is a term that can be traced back to Nietzsche. Foucault later defined it as the state-led management of life, death, and biological being – a form of politics that places human life at the very center of its calculations (Foucault 1980, 2003, 2007). It is concerned 'with the family, with housing, living and working conditions, with what we call "lifestyle", with public health issues, *patterns of migration*, levels of economic growth and the standards of living' (Dean 2010, 119; emphasis added). Migration, left-behind populations and related issues of social care in this sense have become legitimate objects for state manipulation and control.

Biopolitics of migration

Biopolitics falls between liberalism and interventionism (Mo 2011). It arose after the triumph of bourgeois revolutions and developed all through the eighteenth century in modern European society. It gradually 'assumes control over life and, through discipline and regulation technologies, and spreads its control over the simple human body up to the entire population' (Gorski and Beiras 2007, 2), and this comes into being in the name of setting the population free from sovereign power, or the absolute power to take life. In this sense, instead of viewing migration as a threat to social stability, policies should be geared towards regulating to facilitate migration – reducing discrimination against migrants and red tape that hinders the free flow of labor (see also Skeldon 1997, Khan 1998), and it should be accepted that migration is part of life, or 'an essential element in livelihood strategies' (De Haan 1999, 7). For liberalism, 'government is not about "the art of leading men", but about observing those natural and economic laws that provide security and subsistence and beyond this, leaving men free' (Dean 2010, 118).

---

[4] As Bronfenbrenner's theory of social ecological system points out, micro- and macro-gaps can be bridged in five systems, namely, the microsystem (the direct environment we live in), the mesosystem (relationships between the microsystems), the exosystem (links between settings with active and inactive participations), the macrosystem (the actual culture), and the chronosystem (transitions and shifts in lifespans) (Bronfenbrenner 1979).

However, the flow of population has never been really free. A study on the movement of Madeirans from the nineteenth century onwards shows that the Portuguese government sponsored 'engineered migration' of Madeirans in order to colonize the southern Angola plateau. Unable to attract farmers from the mainland to the plateau and afraid of losing the land to the neighboring German, Boer and British groups, the Portuguese government pathologized the mobile population and forbade them to leave the plateau. This is, as the author claims, 'an example of biopolitics of population in the 21st century' (Bastos 2008, 27). This indeed reflects how political actions penetrate, invade and define the whole of life. Biopolitics is supposed to be an orientation to intervene in populations to enhance their health and wellbeing (Li 2009). But Bastos' study reveals that not all subsets of groups have been taken good care of. 'Leaving men free' in many cases means to 'let [them] die' (Li 2009).

The global prevalence of neoliberalism since the 1980s has 'deliberately exposed the rural population of the global South to the full blast of market discipline, while withdrawing social protection' from the state (Davis 2006, 174). For instance, immigrant African women in Italy have been confined to domestic domains and exploited deeply due to international boundaries. This enables Italian women (from both the north and the south) to benefit, although previously the Italian women from the south were denigrated (Merrill 2011). A study of Burmese migrant laborers reflects as well how their precarious status has been evoked and strengthened by racialized governance practices and biopolitical arrangements (Arnold and Pickles 2011). Underdeveloped regions have been compared to a 'necropolis' with resources taken away by developed regions (or 'biopolis'), which reflects the geographical fluidity of accumulation and mandated differences among groups of people (Nast 2011). In certain contexts, the competition between migrant and local labor may not be fierce since the former usually take the dirty, difficult and dangerous jobs that local workers shun (Rigg 2007), although in other contexts, local laid-off workers are competing with migrant workers for employment opportunities (Knight et al. 1999). The biopolitics of migration could therefore move between encouraging and regulating the movement of migrants, as can be observed in the Chinese context.

In China, the examination of the history of rural labor movement from the 1980s onwards shows clearly how rural laborers have been regulated by various means in accordance with different requirements of capital and state. In the early 1980s, the migration of rural labor was strictly controlled through the Hukou registration system. It is argued that with regard to the high unemployment rate in the cities at that time, the government gave priority to urban youngsters who waited for the allocation of an urban job. Peasants were discouraged to come and those who came were sent back to their home villages, with only a few exceptions in certain industries, such as in ports and mines (Shang 1984). Peasants were not allowed to move to cities until 1984, and then they were required to bring food with them when looking for jobs in urban areas. Peasant workers who managed to work in the city usually undertook dangerous and low paid jobs, in short-term and informal contracts. They were encouraged to 'leave the land but not the hometown' (*litu bulixiang*), i.e., doing non-farm jobs in the township or in village enterprises. This 'administrative–legal' exclusion put migrant workers in a 'state of exception'.

The massive bankruptcy of township and village enterprises in the late 1980s cast peasant workers back onto the market and massive scales of migrant workers started to form.

In the early 1990s, the number of migrant workers reached 50 to 60 million, and continued to grow rapidly. Such a large-scale population movement was depicted as a threat to social stability and was related to rising crime rates, labor disputes, and more pressures on social infrastructure and other services (Tong 1995). Although the development of labor-intensive industries in the coastal areas created a huge demand for workers, the state did not loosen the control of labor flows. Temporary residence certificates (to collect personal information) and government camps[5] (to take in those without jobs, village introduction letters and/or temporary residence certificates) were invented to regulate migrants in the city. The controls became more severe in the 1990s when state-owned enterprises laid off large numbers of workers (Cai et al. 2005). The social status of migrant workers was deliberately suppressed and access to better jobs became increasingly more difficult. For almost a decade, the salaries of migrant workers seldom increased. In spite of all these discouragements, the sending areas still took various measures, such as the provision of various types of training, to encourage the outflow of local inhabitants for possible remittances (Zhou 1994, Xu 2003). Yang (2005) believes that this is de facto unfavorable for the working class because, for one thing, the training provided by the sending area ensures that employers do not need to include education expenses into the wages they pay. For another, the government training is neither sustainable nor guaranteed, in that it may not provide peasant workers with the training they need. All of these government policies and measures collectively create a genre of biopolitics in which migrant workers are heavily deprived (doing dangerous, dirty, insecure work with low payments and negligible social care), but allowing the producers and investors to benefit from the low cost of labor. This is what Agamben called 'inclusion by means of exclusion' (Agamben 1998, 216) – recognizing them as workers but denying them a great deal of other rights as human beings.

The most recent decade witnessed the return of the first generation of migrant workers to the countryside. Most come back with aged, injured, and unhealthy bodies. Biopolitics is supposed to increase 'utility and docility' of populations through tools of welfare – bodies, health, subsistence and habitation (Foucault 1980). But more often than not, the manipulation of the population has led to inclusion of some and exclusion of others. Peasant workers have been left out of the welfare system as a result of the selective inclusion of biopolitics.

During the whole process of internal migration, from 1984 to the present, poems, television programs, advertisements and novels about migrant workers and left-behind populations have formed part of the subtle exercise of power. They represent an art of balance through which the contribution of peasant workers is eulogized, on the one hand, while the association of urban crimes with peasant workers is established, on the other. The image of peasant workers has been framed in such a way that senses of both tolerance and rejection are nurtured.[6] Similarly, we have observed policies that encourage migrant workers to stay in the city (providing more slots of schooling to migrant workers' children), and to go back

---

[5] The government camp system empowered the urban government to capture 'tramps' and keep them in custody without legal trials. These people had no idea how long they would be confined or what treatment they would get. This system has long been challenged and criticized, especially when a graduate, Sun Zhigang, was accidently locked up as a homeless and jobless person and died in the cell after 48 hours. The system was finally terminated in 2003. The government camp is an extreme form of biopolitics in China. See Agamben (1998).

[6] A good example of the nurturing of rejection came from Zhong Nanshan, a doctor made famous

home (subsidies and favorable taxation for those who return to rural communities and set up a business there). In both cases, peasant workers are 'foreigners' who do not fit into the normal standard of norms and values, and need to be treated 'specially'. With carrot and stick, certain types of population are to be produced – regulated, docile, and harmless to the 'equilibrium' of society.

As power is everywhere, so too is resistance. Even though the chances of winning are quite slim, there exists a protective biopolitics in unions, social movements, left-leaning political parties and the 'left hand' of the state apparatus (Li 2009). In China, what can be witnessed is the rise of the second or third generation of migrant workers in the 2000s who are much more active politically than the first generation. These groups have different migration pathways than their parents, who relied more on social networks for job opportunities. They are better educated, but usually have no experience in farming. They are unwilling to go back to rural communities, but also are not accepted in the urban environment (Wang 2001). Unlike their fathers' generation who endured unfair social care systems and undesirable working and living conditions, these new generations of peasant workers will be more likely to rise to their feet against expropriation and unfairness – for their unpaid salaries (Cheng 2006) and for human rights (Fu 2006). Organized strikes for fairness and better payments have taken place in recent years (Chan 2010, Pun et al. 2009). These struggles and forms of resistance reflect that they are no longer on the passive receiving end of the biopolitics of migration.

In short, biopolitics, as a new subtle way of exercising power, is widely observed in 'État de population' in recent centuries. It is inseparable from the sovereignty of the state, but more indirect, gentle, and hidden beautifully behind everyday political actions and discourses. It is a power that 'through the implementation of discipline and regulation technologies, is able to cause destruction, exclusion, violence and death among the population' (Gorski and Beiras 2007, 2). Peasants' migration history in contemporary China is interwoven with biopolitics of various forms. The biopolitics of migration determines whether, in which places and for how long they can migrate.

This careful control and policing of population movement has been conducted against the backdrop of the dull compulsion of economic forces. Peasant workers become 'foreigners' in their own country, placed in the state of exception with many rights dismissed. In this context, only a few peasant workers can climb upwards on the class ladder, while the majority are stuck in the same position through their whole lives. It is difficult for the second or even third generation of migrant workers to break through the vicious circle without fundamental macro-level economic, social and political changes.

Politics of social care

The social care that these migrant workers are fighting for has three distinctive features and premises, all of which makes it difficult to analyze with traditional economic thinking (Himmelweit 2005):

for his significant contribution in SARS. He angrily called for the recovery of the government camp system after being robbed in the streets of Guangdong City in broad daylight.

1 Care is inseparable from the person who delivers it because it comes originally along with a certain family relationship.
2 Resources to provide care are not equally distributed in society and the person who has the responsibility to provide care may not have access to necessary resources.
3 Care is highly culturally embedded, and cultural norms determine who needs care, who should fulfill the needs for care and how (ibid.).

For instance, in India, South Africa, Korea and Tanzania, the public discourse preaches ideologies in which care is a woman's responsibility; mothers, wives and daughters are the 'natural' care providers (Lund and Budlender 2009, Meena 2010, Palriwala and Neetha 2010, Peng 2010). Even moving the care work out of the family does not reduce its feminization (Franzoni and Voorend 2011) as most professional caregivers are women.

Migration, split families, and modernized ways of working and living create an upsurge in the care market in which the cost of providing care is increasingly shifted from the care purchaser to the care providers, who also have the responsibility to take care of family members, but may have to leave them behind. Hochschild (2000) proposes a global care chain to explain the phenomenon that women as care providers are also consuming other women's care labor (both paid and unpaid). This chain is forged in global inequality, and is also strengthened by the transfer of care resources from poorer regions to richer ones for consumption (ibid.). Such a globalization of the care 'business' is promoted by both care-labor production in developing countries and by the active recruitment of overseas care labor in developed economies to address the unfolding 'care deficit'(Razavi 2007) or 'care crisis' (Yeates 2010). Importing skilled nurses and domestic workers from poorer countries exports the care crisis (ibid.). This process reproduces class inequality among women and children (Faur 2011). Some might be able to retain family caregiving and supplement it with paid care, while some cannot afford either (Palriwala and Neetha 2011). While some better off people enjoy costly care services provided by the private sector, others are poorly served by an inadequate public sector (Budlender and Lund 2011). This phenomenon is also found in China (Yan 2010). According to Yan's study, intellectual families hire babysitters from rural areas, which reflects a special reconceptualization of intellectual and manual labor, urban and rural, men and women. Care work needs to be understood as transnational biopolitics, in which 'transactions between care workers and their clients are embedded in local, regional and global orders of inequality and difference' (Mackie 2013, 15). We could say, due to its special gender-sensitive nature, social care of and by migrant workers and that of their left-behind families reflects the exploitation not only between classes and across regions, but also between genders.

Jenson and Saint-Martin (2003) use the term 'welfare triangle' to summarize institutional arrangements of social care in contemporary societies in which inputs from the state, the market and the family compose a welfare package. However, the state has increasingly become less reliable in this regard. Some portions of social care originally carried out by the state have gradually been outsourced to non-state entities (Razavi and Staab 2010). For instance, the decentralization and privatization of state-owned enterprises in China have destroyed some charitable mechanisms that could have internalized the costs of reproduction. Women have to balance domestic responsibilities with occupational requirements, which have the largest impacts on females in socially and economically disadvantaged groups (Cook and Dong 2011). Yeates (2011) proposes a model of producer-based care networks, which changes the focus of discussion from the care recipients to givers – mostly (transnational) migrant workers. Their faraway work locations and their remittances could be conceptualized as 'distant care', with monetary and emotional contributions to the population left-behind at home (ibid.). But the effectiveness of this materialized compensation and caregiving far from home awaits discussion. Therefore, with the retreat of the state and the migration of family members, left-behind populations are increasingly dependent on the market for the provision of daily care. The market, however, has been questioned in terms of the provision of good-quality care and decent work conditions (Faur 2011). Such a segregation of the market and differentiated social care access exacerbates their disadvantaged status. The politics of social care in this sense exhibits who is providing for whom and who is losing with diminishing compensation.

From the discussion thus far, the actions and suggestions proposed by local governments, enterprises, NGOs and society are called into question. Peasant workers and their left-behind families in rural China have been deprived structurally of family care, parenting and nuptial affection, and forced to enter deeply exploitative work relations. Capital, the state and the society, which have benefited from such exploitation, then seek to 'help' the peasant workers through all these activities to provide 'love' and 'care'. And here, the logic goes bankrupt.

Rethinking internal migration and left-behind populations in China

Previous studies of left-behind populations in rural China have focused mainly on the impacts of migration on the left-behind population and the strategies they utilized in response, as well as corresponding social actions. We have also been engaged in similar impact studies in dozens of villages around China, with influential research findings and important social actions following. It is important to see the phenomena of migration and the left-behind population in context-contingent ways. But we have become increasingly dissatisfied with merely describing the miserable situations of migrant workers and their left-behind family members, and attaching a few recommendations at the end. We cannot help asking why the majority of peasants have to suffer. Why do they have to migrate and why does migration bring them not wealth but rather aged, injured and unhealthy bodies? We heard frequently from peasants during our fieldwork that 'we will be better off when our children grow up.' However, when we looked both back and forward in time, what we observed was migration as a rite of passage – generations of migration without radical changes in households and rural communities. As a result, we started to sit back and try to go deeper, striving to explain how such institutional arrangements come into being in the first place and what politics are behind them.

The literature review presented in sections 3 and 4 in this paper inform the evolution in our thinking. Impact studies do matter, but depend heavily on context. Comparisons between left-behind and non-left-behind populations as well as comparisons before and after migration can be problematic. It is extremely difficult, if not impossible, to rule out intervening factors and isolate migration as an independent variable. In any case, what are such comparisons trying to achieve? They may add some cases to the discussion on the issue of migration and left-behind populations, but not to the analysis of its essence. Even if we could overcome the endogeneity problem, would it matter much if we found that 40 percent of left-behind children were injured psychologically instead of 45 percent in certain areas? We increasingly realized that migration studies need to go back to the very fundamental questions of 'who wins, who loses?', 'what factors are playing a role in this process?', and, most importantly, 'how do we change it?' In an urban-biased and capital-favoring macro environment, migration in China produces unbearable social costs for left-behind populations compared to meager material compensation. Some social actions and strategies from various actors contribute to the softening of certain negative effects, but not by much, not on a sustainable level, and not to the point.

It is for this reason that we brought the concept of biopolitics into the analysis. This concept has already been popularly deployed in studies of transnational migration, especially in terms of racism, gender, health, refugees and aid. By applying it to internal migration in the history of contemporary China, we find that biopolitics exerts powerful control over and manipulation of the movement of the population, through the Hukou registration system, welfare systems, propaganda, discourses, government camps and so forth. Peasant workers in China are in a state of exception, with their rights compromised or dismissed. This analysis can usefully complement findings at the micro levels of individuals and households.

We do not want to argue for a romantic rural countryside in which local people enjoy family union more than migration for better incomes. Indeed, we have been working for a long time with people who are eager to get rid of their identities as peasants, similar to those in other countries who want to struggle out of subsistence production and lead a life with better food, housing, education and health care (Faur 2011). The problem is that in reality, their wishes are denied when biopolitics in favor of capitalist production regulates their movement. They sell labor in a socially isolated environment, being expropriated routinely while witnessing other people prospering because of such expropriation. In addition, there are numerous others, who do not even have the chance to be expropriated and dispossessed – the 'surplus population' in Tania Li's terms and the left-behind population in this context. In this sense, this paper would argue for a right to move for peasants (which implies also a right not to move, and to lead a life they cherish).[7]

The aim of development is not capital accumulation, but the development of human beings, which includes peasants who should not be merely regarded as a surplus resource to be 'absorbed' into the industrial sector. The research on migration and left-behind populations thus needs to bring impact studies and political economy together. In view of this, studies on care dualism and biopolitics of migration in China might assist us to assess how an organized working class can be formed out of the present cocoon of the semi-proletariat, so that it can change the current power structure and exploitative relations and, like the wings of a butterfly, bring about the wildest storm.

References

Acosta, P. 2011. School attendance, child labour, and remittances from international migration in El Salvador. *Journal of Development Studies*, 47 (6), 913–36.

[7] We thank Professor James Scott for the idea – the right to move also includes the right *not* to move.

Adhikari, R., A. Jampaklay, and A. Chamratrithirong. 2011. Impact of children's migration on health and health care-seeking behavior of elderly left behind. *BMC Public Health*, 11, 143.

Adi, R. 2003. Irregular migration from Indonesia. *In*: G. Battistella and M.M.B. Asis, eds. *Unauthorized migration in Southeast Asia*. Quezon City, Philippines: Scalabrini Migration Center, pp. 128–68.

Agamben, G. 1998. *Homo sacer: Sovereign power and bare life*. Translated by D. Heller-Roazen. Stanford: Stanford University Press.

Alcaraz, C., D. Chiquiar, and A. Salcedo. 2012. Remittances, schooling, and child labor in Mexico. *Journal of Development Economics*, 97 (1), 156–65.

Ali, S. 2007. The culture of migration among Muslims in Hyderabad. *Journal of Ethnic and Migration Studies*, 33 (1), 37–58.

Antman, F. 2010. International migration, spousal control, and gender discrimination in the allocation of household resources. In *Working Paper*. Boulder, CO: Department of Economics.

– – –. 2012. The impact of migration on family left behind. In *Discussion Paper*. Bonn: The Institute for the Study of Labor (IZA).

Anton, J.-I. 2010. The impact of remittances on nutritional status of children in Ecuador. *International Migration Review*, 44 (2), 269–99.

Archana, K.R. and N. Parveen. 2005. Impact of male out-migration on health status of left-behind wives – a study of Bihar, India. http://iussp2005.princeton.edu/papers/51906.

Arnold, D. and J. Pickles. 2011. Global work, surplus labor, and the precarious economies of the border. *Antipode*, 43 (5), 1598–624.

Baldassar, L. 2007a. Transnational families and aged care: the mobility of care and the migrancy of ageing. *Journal of Ethnic and Migration Studies*, 33(2), 275–97.

– – –. 2007b. Transnational families and the provision of moral and emotional support: the relationship between truth and distance. *Identities: Global Studies in Culture and Power*, 14, 385–409.

Bastos, C. 2008. Migrants, settlers and colonists: the biopolitics of displaced bodies. *International Migration*, 46(5), 27–54.

Batista, C., A. Lacuesta, and C.P. Vicente. 2007. Brain drain or brain gain? Micro evidence from an African success story. In *IZA Discussion Paper*. Bonn: Institute for the Study of Labor.

Battistella, G. and M. Conaco. 1998. The impact of labour on the children left behind: a study of elementary school children in the Philippines. *Sojourn*, 13(2), 220–41.

Bernstein, H. 2010. *Class dynamics of agrarian change*. Halifax: Fernwood Publishing.

Boucher, S., O. Stark, and E.J. Taylor. 2005. A gain with a drain? Evidence from rural Mexico on the new economics of the brain drain. In *Davis Working Paper*. Berkeley, CA: University of California.

Breman, J. 1985. *Of peasants, migrants and paupers: rural labour circulation and capitalist production in west India*. Oxford: Oxford University Press.

– – –. 1990. *Labour migration and rural transformation in colonial Asia. Comparative Asian studies*. Amsterdam: Free University Press.

Bronfenbrenner, U. 1979. *The ecology of human development: experiments by nature and design*. Cambridge, MA: Harvard University Press.

Budlender, D. and F. Lund. 2011. South Africa: a legacy of family disruption. *Development and Change*, 42(4), 925–46.

Burazeri, G., A. Goda, N. Tavanxhi, G. Sulo, J. Stefa, and D.J. D Kark. 2007. The health effects of emigration on those who remain at home. *International Journal of Epidemiology*, 36(1265–1272).

Cai, F., Y. Du, and Y. Wang. 2005. A political economy of expelling migrant labor from the city: a case study in Beijing. *In*: *Case studies in China's institutional change*. Beijing: China Financial and Economic Publishing House, pp. 285–314.

Cai, M. 2011. Ministry of civil affairs reported we now have 47 million left-behind wives. Xinhua News Agency. http://news.qq.com/a/20110307/002362.htm [Accessed 20 December 2012].

Cao, L. 2005. How much have peasant workers contributed to China? Xiao Xiang Morning News. http://finance.stockstar.com/SS2005090230262226.shtml [Accessed 8 December 2012].

Cashin, P. and R. Sahay. 1996. Internal migration, center-state grants, and economic growth in the states of India. *IMF Staff Papers*, 43(1).

Chan, C.K. 2010. Class struggle in China: case studies of migrant worker strikes in the Pearl River Delta. *South African Review of Sociology*, 41(3), 61–80.

Chee, M. 2003. Migrating for the children: Taiwanese American women in transnational families. *In*: N. Piper and M. Roces, eds. *Wife of worker? Asian women and migration*. Lanham, MD: Rowman & Littlefield, pp. 137–56.

Cheng, X. 2006. A hard course: the literary reflection of farmer in city since 1980s. *Journal of North China Institute of Water Conservancy and Hydroelectric Power (Social Sciences Edition)*, 3, 73–6.

Chiriboga, M., R. Charnay, and C. Chehab. 2006. Women in agriculture: some results of household surveys data analysis. Background paper for World Bank Report (2008). https://openknowledge.worldbank.org/bitstream/handle/ 10986/9254/WDR2008_008.pdf?sequence=1 [Accessed 17 November 2013].

Cleaver, F. 2002. Men and masculinities: new directions in gender and development. *In*: F. Cleaver, ed. *Masculinities matter! Men, gender and development*. London: Zed Books, pp. 1–18.

Connell, J., B. Dasgupta, R. Laishley, and M. Lipton. 1976. *Migration from rural areas: the evidence from village studies*. Delhi: Oxford University Press.

Cook, S. and X.Y. Dong. 2011. Harsh choices: Chinese women's paid work and unpaid care responsibilities under economic reform. *Development and Change*, 42(4), 947–65.

Cortes, P. 2010. The feminization of international migration and its effects on the children left behind: evidence from the Philippines. In *Working Paper*. Boston: Boston University.

Cui, R. 2007. The living situation of left-behind elderly in impoverished areas – an investigation in Long Feng township, En Shi municipality, Hubei province. *Journal of Hubei Administration Institute*, 2, 63–7.

Curran, S., Y.C. Chang, W. Cadge, and A. Varangrat. 2004. Boys and girls' changing educational opportunities in Thailand: the effects of siblings, migration, and village remoteness. *Research in Sociology of Education*, 14, 59–102.

David, R. 1995. *Changing places: women, resource management and migration in the sahel*. London: SOS Sahel.

Davis, M. 2006. *Planet of slums*. London: Verso.

De Haan, A. 1999. Livelihoods and poverty: the role of migration – a critical review of the migration literature. *Journal of Development Studies*, 36(2), 1–47.

de Haas, H. 2010. Migration and development: a theoretical perspective. *International Migration Review*, 44(1), 227–64.

de Haas, H. and A. van Rooij. 2010. Migration as emancipation? The impact of internal and international migration on the position of women left behind in rural Morocco. *Oxford Development Studies*, 38(1), 43–62.

Dean, M. 2010. *Governmentality: power and rule in modern society*. 2nd edn. London: Sage Publications.

deBrauw, A. and J. Giles. 2006. Migrant opportunity and the educational attainment of youth in rural China. In *IZA Discussion Paper 2326*. Bonn: Institute for the Study of Labor.

Dreby, J. 2007. Children and power in Mexican transnational families. *Journal of Marriage and Family*, 69, 1050–64.

Du, Y. 2000. Rural labor in contemporary China: an analysis of its features and the macro context. *In*: L. West and Y. Zhao, eds. *Rural labor flows in China*. Berkeley, CA: University of California Press, pp. 67–100.

Duncan, T. 1994. Like father like son; like mother like daughter: parental resources and child height. *Journal of Human Resources*, 29(4), 950–88.

Elmhirst, R. 2007. Tigers and gangsters: masculinities and feminised migration in Indonesia. *Population, Space and Place*, 13(3), 225–38.

Elson, D. 1992. From survival strategies to transformation strategies: women's needs and structural adjustment. *In*: L. Benería and S. Feldman, eds. *Unequal burden: economic crises, persistent poverty, and women's work*. Boulder, CO: Westview Press, pp. 26–48.

Fan, C.C. 2003. Rural–urban migration and gender division of labor in transitional China. *International Journal of Urban and Regional Research*, 1, 24–47.

Faur, E. 2011. A widening gap? The political and social organization of childcare in Argentina. *Development and Change*, 42(4), 967–94.

Findley, S. 1997. Migration and family interactions in Africa. *In*: A. Adepoju, ed. *Family, population and development in Africa*. London: Zed Books, pp. 109–38.

Foucault, M. 1980. *The history of sexuality*. Vol. 1. New York: Vintage Books.

———. 2003. *Society must be defended: lectures at the College de France 1975–1976*. New York: Picador.

———. 2007. *Security, territory, population: lectures at the College de France 1977–1978*. New York: Palgrave Macmillan.

Frank, A.G. 1966. The development of underdevelopment. *Monthly Review*, 18.

Franzoni, J.M. and K. Voorend. 2011. Who cares in Nicaragua? A care regime in an exclusionary social policy context. *Development and Change*, 42(4), 995–1022.

Fu, P. 2006. Drifting and struggling: the living situation of young peasant workers. *The World of Survey and Research*, 9, 20–25.

Gao, X. 1994a. Rural labor migration and feminization of agriculture in contemporary China. *Sociological Studies*, 2, 83–90.

———. 1994b. Rural labour transfer and the feminization of agriculture. *Shehuixue Yanjiu (Sociology Research)*, 2, 83–90.

Ghosh, B. 1992. Migration-development linkages: some specific issues and practical policy measures. *International Migration*, 30(3/4), 423–52.

Giannelli, G.C. and L. Mangiavacchi. 2010. Children's schooling and parental migration: empirical evidence on the 'left-behind' generation in Albania. *Labour*, 24, 76–92.

Gorski, S.H. and R.I. Beiras. 2007. Contemporary biopolitics in front of migration flows and prison universe: a reflection on the return of 'camps' in Europe. *In*: C. Holgan and M. Marin-Domine, eds. *The camp: narratives of internment and exclusion*. Newcastle: Cambridge Scholars Publishing, pp. 22–38.

Gou, Y. 2011. Exploring the cultivation of new farmers in Gansu from rural staying population structure. *Journal of Linyi University*, 33(6), 1–4.

Gulati, L. 1993. *In the absence of their men: the impact of male migration on women*. London: Sage Publications.

Hadi, A. 1999a. Overseas migration and the well being of those left behind in rural communities of Bangladesh. *Asia-Pacific Population Journal*, 14, 43–58.

Himmelweit, S. 2005. Can we afford (not) to care: prospects and policy. *In* Genet Working Paper. Milton Keynes: Open University.

Hochschild, A.R. 2000. Global care chains and emotional surplus value. *In*: W. Hutton and A. Giddens, eds. *On the edge: living with global capitalism*. London: Jonathan Cape, pp. 130–46.

Horton, S. 2008. Consuming childhood: 'lost' and 'ideal' childhoods as a motivation for migration. *Anthropological Quarterly*, 81(4), 925–43.

Huang, P. and E. Clare. 1998. Promotion or shock? A village study on peasants' migration for work in China. *Sociological Studies*, 3, 71–82.

Huang, S. and B. Yeoh. 2005. Transnational families and their children's education: China's study mothers in Singapore. *Global Networks*, 5(4), 379–400.

Hugo, G. 2005. Indonesian international domestic workers: contemporary developments and issues. *In*: S. Huang, B. Yeoh, and N.A. Rahman, eds. *Asian women as transnational domestic workers*. Singapore: Marshall Cavendish, pp. 54–91.

Ikels, C., and C. Beall. 1993. Settling accounts: the intergenerational contract in an age of reform. *In*: D. David and S. Harrell, eds. *Chinese families in the post-Mao era*. Berkeley, CA: University of California Press, pp. 307–34.

Islam, M.D. 1991. Labour migration and development: a case study of a rural community in Bangladesh. *Bangladesh Journal of Political Economy*, 11(2B), 570–87.

Jacka, T. 2012. Migration, householding and the well-being of left-behind women in rural Ningxia. *China Journal*, 6(7), 1–21.

Jampaklay, A. 2006. Parental absence and children's school enrolment: evidence from a longitudinal study in Kanchanaburi, Thailand. *Asian Population Studies*, 2(1), 93–110.

Jenson, J. and D. Saint-Martin. 2003. New routes to social cohesion? Citizenship and the social investment state. *Canadian Journal of Sociology*, 28(1), 77–99.

Jetley, S. 1987. Impact of male migration on rural females. *Economic and Political Weekly*, 22(4), WS47–WS53.

Jiangxi Charity Federation. 2012. Left-behind children ate free lunch in Wannian country. Jiangxi Charity Federation. http://www.jxcs.org.cn/system/2012/06/ 25/012021228.shtml [Accessed 12 November 2012].

Kahn, K., M. Collinson, S. Tollman, B. Wolff, M. Garenne, and S. Clark. 2003. Health consequences of migration: Evidence from South Africa's rural northeast. http://time.dufe.edu.cn/wencong/ africanmigration/5kahn.pdf [Accessed 17 November 2013].

Kanaiaupuni, S.M. and K.M. Donato. 1999. Migradollars and mortality: the effects of migration on infant survival in Mexico. *Demography*, 36(3), 339–53.

Kandel, W. and G. Kao. 2000. Shifting orientations: how U.S. labor migration affects children's aspirations in Mexican migrant communities. *Social Science Quarterly*, 81(1), 16–32.

Kandel, W. and D. Massey. 2002. The culture of Mexican migration: a theoretical and empirical analysis. *Social Forces*, 80(3), 981–1004.

Khan, A.R. 1998. Poverty in China in the period of globalization. new evidence on trend and pattern. In *Issues in Development Discussion Paper*. Geneva: ILO.

King, R. and J. Vullnetari. 2006. Orphan pensioners and migrating grandparents: the impact of mass migration on older people in rural Albania. *Ageing & Society*, 26, 783–816.

Knight, J., L. Song, and J. Huaibin. 1999. Chinese rural migrants in urban enterprises. *Journal of Development Studies*, 35(3), 73–104.

Knodel, J. and C. Saengtienchai. 2007. Rural parents with urban children: social and economic implications of migration for the rural elderly in Thailand. *Population, Space and Place*, 13, 193–210.

Krause, N. 2001. Social support. *In*: L.R. Binstock and K.L. George, eds. *Handbook of aging and the social sciences*. San Diego, CA: Academic Press, pp. 273–94.

Kuhn, R. 2005. A longitudinal analysis of health and mortality in a migrant-sending region of Bangladesh. *In*: B. Yeoh, M. Toyota, and S. Jatrana, eds. *Migration and health in Asia*. London: Routledge, pp. 177–208.

Lakshmansamy, T. 1990. Family survival strategy and migration: an analysis of returns to migration. *Indian Journal of Social Work*, 51(3), 473–85.

Lastarria-Cornhiel, S. 2006. Feminization of agriculture: trends and driving forces. http://www. rimisp.org/getdoc.php?docid=6489.

Lewis, W.A. 1958. Economic development with unlimited supplies of labor. *In*: N.A. Agarwala and P.S. Singh, eds. *The economics of underdevelopment*. London: Oxford University Press, pp. 400–49.

Li, F. 2009. National women's federation reported we now have 58 million left-behind children. Xinhua News Agency. http://news.xinhuanet.com/ society/2009-05/26/content_11440077.htm [Accessed 20 December 2012].

Li, J. 2003. An exploratory discussion on the change in the life style of the left- behind rural women. *Journal of Shanxi College for Youth Administrators*, 16(2), 38–40.

Li, T.M. 2009. To make live or let die? Rural dispossession and the protection of surplus populations. *Antipode*, 41(S1), 66–93.

Li, P. 1996. Social network and social status of migrant peasant workers. *Sociological Research*, 4, 42–52.

Liang, S. 2011. Left-behind children ate free lunch in En Shi. http://cjmp.cnhan.com/whwb/ html/2011-12/02/content_4928345.htm [Accessed 12 November 2012].

Lin, L. 2011. Caring for left-behind women, building up a harmonious society. Guangdong Womens College http://feminism2007.blog.163.com/blog/static/ 109262455201131010151 2232/ [Accessed 8 December 2012].

Linda, M. 1989. Living arrangements of the elderly in Fiji, Korea, Malaysia and the Philippines. *Demography*, 26, 627–44.

Lipton, M. 1980. Migration from rural areas of poor countries: the impact on rural productivity and income distribution. *World Development*, 8(1), 1–24.

Liu, J. 2012. Proletarianization: theoretical explanation, historical experiences and its enlightments. *Society*, 2, 51–83.

Liu, Z. 2010. National statistics reported we now have 220 million peasant workers. Xinhua News Agency. http://news.xinhuanet.com/fortune/2010-03/23/ content_13232348.htm [Accessed 20 December 2012].

Louise, E.T.-U. 2005. *The impact of migration: women's voices from Mexican migrant communities*. New York : SUNY.

Lund, F. and D. Budlender. 2009. Paid care providers in South Africa: nurses, domestic workers, and home-based care workers. Geneva: UNRISD.

McEvoy, J.P. 2008. Male outmigration and the women left behind: a case study of a small farming community in southeastern Mexico. *All Graduate Theses and Dissertations*. Paper 179. http://digitalcommons.usu.edu/etd/179 [Accessed on 17 November 2013].

McKenzie, D. and H. Rapoport. 2011. Can migration reduce educational attainment? Evidence from Mexico. *Journal of Population Economics*, 24(4), 1331–58.

Mackie, V. 2013. Japan's biopolitical crisis: care provision in a transnational frame. *International Feminist Journal of Politics*, 37(3), 1–19.

Mallee, H. 1995/96. In defence of migration: recent Chinese studies on rural population mobility. *China Information*, 10(3–4), 108–40.

Martin, P.L. and J.E. Taylor. 2001. Managing migration: the role of economic policies. *In*: A.R. Zolberg and B. P. M., eds. *Global migrants, global refugees: problems and solution*. New York and Oxford: Berghahn, pp. 95–120.

Massey, D. 2009. Staying behind when husbands move: women's experiences in India and Bangladesh. In *Briefing Paper*. London: DFID.

Mazzucato, V. and D. Schans. 2011. Transnational families and the well-being of children: conceptual and methodological challenges. *Journal of Marriage and Family*, 73, 704–12.

Meena, R. 2010. Nurses and home-based caregivers in the united republic of Tanzania: a dis-continuum of care. *International Labour Review*, 149(4) 529–42.

Meng, Q. 2011. Semi-proletarianization,commercialization of labor force and Chinese peasant-workers. *Economics Study of Shanghai School*, 1, 135–53.

Merrill, H. 2011. Migration and surplus populations: race and deindustrialization in northern Italy. *Antipode*, 43(5), 1542–72.

Meyerhoefer, D.C. and C.J. Chen. 2011. The effect of parental labor migration on children's educational progress in rural China. *Review of Economics of the Household*, 9(3), 379–96.

Mo, W. 2011. Police: from body to population – exploring Michel Foucault's thought of police. *Academic Monthly*, 43(7), 45–52.

Nast, H.J. 2011. 'Race' and the bio(necro)polis. *Antipode*, 43(5), 1457–64.

Oberai, A. and H.K. Singh. 1980. Migration, remittances and rural development: findings of a case study in Indian Punjab. *International Labour Review*, 119(2), 229–41.

Oucho, J.O. 1996. *Urban migrants and rural development in Kenya*. Nairobi: Nairobi University Press.

Palriwala, R. and N. Neetha. 2010. Care arrangements and bargains: Anganwadi and paid domestic workers in India. *International Labour Review*, 149(4), 511–27.

———. 2011. Stratified familialism: the care regime in India through the lens of childcare. *Development and Change*, 42(4), 1049–78.

Peng, I. 2010. The expansion of social care and reform: implications for care workers in the republic of Korea. *International Labour Review*, 149(4), 461–76.

Pu, W. 2008. An analysis on rural left-behind child studies, supports and actions. *China Youth Study*, 6, 25–30.

Pun, N. 2005. *Made in China: women factory workers in a global workplace*. Durham, NC, London and Hong Kong: Duke University Press and Hong Kong University Press.

Pun, N., C.K. Chan, and J. Chan. 2009. The role of the state, labour policy and migrant workers' struggles in globalized China. *Global Labour Journal*, 1(1) 132–51.

Pun, N., H. Lu, and H. Zhang. 2012. *The construction site – the existence of peasant workers in the city*. Beijing: Peking University Press.

Razavi, S. 2007. The political and social economy of care in a development context: conceptual issues, research questions and policy options. In *Gender and development programme paper*. Geneva: United Nations Research Institute for Social Development.

Razavi, S. and S. Staab. 2010. Underpaid and overworked: a cross-national perspective on care workers. *International Labour Review*, 149(4), 407–22.

Resurreccion, P.B. and V.H.T. Khanh. 2007. Able to come and go: reproducing gender in female rural–urban migration in the Red River Delta. *Population, Space and Place*, 13, 211–24.

Rigg, J. 2007. Moving lives: migration and livelihoods in the Lao PDR. *Population, Space and Place*, 13(3), 163–78.

Rodenburg, J. 2000. Staying behind: conflict and compromise in Toba Batak migration. *In*: J. Koning, N. Marleen, J. Rodenburg, and R. Saptari, eds. *Women and households in Indonesia: cultural notions and social practices*. London: Routledge Curzon, pp. 235–61.

Roy, A.K. and P. Nangia. 2005. Impact of male out-migration on health status of left behind wives – a study of Bihar, India. http://iussp2005.princeton.edu/papers/51906 [accessed on 21 May, 2014].

Rubenstein, H. 1992. Migration, development and remittances in rural Mexico. *International Migration*, 30(2), 127–53.

Safa, H.I. 1982. *Towards a political economy of urbanization in third world countries*. Delhi: Oxford University Press.

Shang, L. 1984. How to improve economic efficiency of labor in porting, transportation and ship-making industries in Shanghai. *China Labor*, 13, 25.

Siegmann, A.K. 2010 Strengthening whom? The role of international migration for women and men in northwest Pakistan. *Progress in Development Studies*, 10345–61.

Skeldon, R. 1997. Rural–urban migration and its implications for poverty alleviation. *Asia-Pacific Population Journal*, 12(1), 3–16.

Smith-Estelle, A. and S. Gruskin. 2003. Vulnerability to HIV/STIs among rural women from migrant communities in Nepal: a health and human rights framework. *Reproductive Health Matters*, 11(22), 142–51.

Song, L.F. 1982. The surplus rural laborers and their future. *Chinese Social Sciences*, 5, 121–33.

Srivastava, R. and S. Sasikumar. 2003. An overview of migration in India, its impacts and key issues. Country Overview Paper, DFID.

Standing, G. 1985. Circulation and the labour process. *In*: G. Standing, ed. *Labour circulation and the labour process*. London: Croom Helm, pp. 1–45.

Stark, O. and E.J. Taylor. 1991. Migration incentives, migration types: the role of relative deprivation. *Economic Journal*, 101, 1163–78.

Suarez-Orozco, C., I.L.G. Todorova, and J. Louie. 2002. Making up for lost time: the experience of separation and reunification among immigrant families. *Family Process*, 41, 625–43.

Taylor, E.J. 1999. The new economics of labor migration and the role of remittances in the migration process. *International Migration*, 37(1), 63–88.

Todaro, M.P. 1981. *Economic development in the third world*. 2nd edn. New York and London: Longman.

Tokarsky-Unda, L.E. 2006. The impact of migration: women's voices from Mexican migrant communities. Diss. State University of New York at Buffalo.

Tong, X. 1995. The tidal wave of rural migrant workers and urbanization. *Open Times*, 5, 61–5.

Vullnetari, J. and R. King. 2008. Does your granny eat grass? On mass migration, care drain and the fate of older people in rural Albania. *Global Networks*, 8(2), 139–71.

Wallerstein, I. 1983. *Historical capitalism*. London: Verso.

Wang, C. 2001. Social identity of the new generation of rural hobo and merger of urban and rural. *Sociological Studies*, 3, 63–76.

Wang, F. 2007. The left-behind women. *China Society Periodical*, 4, 26–8.

Wang, J. 2012a. Caring for the left-behind elderly. Huai Ning News. http://www.hnnews.cc/system/2012/06/14/006041186.shtml [Accessed 8 December 2012].

Wang, Z. 2012b. 800 left-behind children ate free lunch. ChongQin Business News. http://e.china cqsb.com/html/2012-02/27/content_246573.htm [Accessed 12 November 2012].

Waters, J. 2002. Flexible families? 'Astronaut' households and the experience of lone mothers in Vancouver, British Columbia. *Social and Cultural Geography*, 3, 117–34.

Wen, T. 2006. Solving the problem of left-behind children in three levels. *Henan Education*, 5, 113–19.

Wise, R.D. and H.M. Covarrubias. 2008. Capitalist restructuring, development and labour migration – the Mexico–US case. *Third World Quarterly*, 29(7), 1359–74.

Wu, H. 1992. Why would peasants leave the land – a survey in 34 villages of 4 counties and munici-palities, Hubei province. *Rural Economy and Society*, 2, 59–61.

Wu, N. 2004. A survey report on the education of hometown-remaining children in rural areas. *Educational Research*, 10, 15–18.

Xiang, B. 2007. How far are the left-behind left behind? A preliminary study in rural China. *Population Space Place*, 13(3), 179–91.

Xiang, L. 2006. Left-behind rural women: a disadvantaged group to be cared for. *Guangxi Social Sciences*, 1, 176–80.

Xiao, J. 2006. Who is going to look after the left-behind elderly? Sina News. http://news.sina.com.cn/s/2006-06-12/06529181572s.shtml [Accessed 8 December 2012].

Xinhua News Agency. 2006. Archives to be set up for left-behind children in Fujian. Xinhua News Agency. http://www.fj.xinhuanet.com/news/2006-11/19/ content_8875029.htm [Accessed 12 November 2012].

– – –. 2011. The left-behind elderly reached more than 40 million. Xinhua News Agency. http://news.qq.com/a/20111005/000104.htm [Accessed 20 December 2012].

Xu, P. 2003. From government behavior to market actions: a case study on labor migration in Zhu township. *In*: P. Li, ed. *Peasant workers: an economic and social analysis of migrant workers in the city in China*. Beijing: Social Sciences Academic Press (China), pp. 237–51.

Yan, H. 2010. Class utterance and class transformation: second essay on brain work, physical work, gender, and class. *Open Times*, 6, 121–39.

Yang, D. 2008. International migration, remittances and household investment: evidence from Philippine migrants' exchange rate shocks. *Economic Journal*, 118, 591–630.

Yang, R. 2012a. Three ways to push the care for left-behind women in Qinchen county. The Communist Party Committee in Qinchen Municipality. http://www.qysw.gov.cn/2012/0925/21400.html [Accessed 8 December 2012].

Yang, S. 2005. *A political economy perspective on Chinese peasant workers*. Beijing: Peking University Press.

Yang, X. 2012b. Migration for work and the social change in the poverty-stridden area of minority nationalities – a case study in Bei Long Yao nationality in Lin Yun county Guang Xi province. *Guizhou Social Sciences*, 5, 113–19.

Ye, J. 2011a. Introduction: the issue of left-behind children in the context of China's modernization. *Journal of Peasant Studies*, 38(3), 613–20.

– – –. 2011b. The left-behind and the development encounter. *Journal of China Agriculture University*, 1, 5–12.

Ye, J. and C. He. 2008. *Lonely sunsets: the elderly left behind in rural China*. Beijing: Social Sciences Academic Press.

Ye, J. and J. Murray. 2006. *Left-behind children in rural China: impact study of rural labor migra-tion on left-behind children in mid-west China*. Beijing: Social Sciences Academic Press (China).

Ye, J. and L. Pan. 2011. Differentiated childhoods: impacts of rural labor migration on left-behind children in China. *Journal of Peasant Studies*, 38(2), 355–77.

Ye, J. and H. Wu. 2008. *Dancing solo: Women left behind in rural China*. Beijing: Social Sciences Academic Press.

Ye, J. and Z. Yang. 2008. *Caring for left-behind children: actions and strategies*. Beijing: Social Sciences Academic Press.

Yeates, N. 2010. The globalization of nurse migration: policy issues and responses. *International Labour Review*, 149(4), 423–40.

– – –. 2011. Going global: the transnationalization of care. *Development and Change*, 42(4), 1109–30.

Yeoh, B. and T. Lam. 2006. The costs of (im)mobility: Children left behind and children who migrate with a parent. In *ESCAP Regional Seminar on Strengthening the Capacity of National Machineries for Gender Equality to Shape Migration Policies and Protect Migrant Women*. Bangkok: UNESCAP.

Zhang, X. 2003. Elderly care in the development of rural communities. *Journal of Changsha Social Work College*, 10(4), 17–21.

Zhen, C. 2008. Children's migration and the financial, social, and psychological well-being of older adults in rural China. Faculty of the Graduate School University of Southern California, California.

Zheng, Z.Z. and Z.M. Xie. 2004. *Migration and rural women's development*. Beijing: Social Sciences Press.

Zhou, F.L. 2006. *Research on left-behind families in rural China*. Beijing: China Agricultural Press.

Zhou R. 1994. An investigation on 'waves of migrant workers' in Anhui, countermeasures and suggestions. *Chinese Rural Economy*, 1, 53–7.

Zimmer, Z., K. Korinek, J. Knodel, and N. Chayovan. 2007. Support by migrants to their elderly parents in rural Cambodia and Thailand: a comparative study. *In: Poverty, Gender, and Youth working Papers*. USA: ELDIS.

Zoller Booth, M. 1995. Children of migrant fathers: the effects of father absence on Swazi children's preparedness for school. *Comparative Education Review*, 39(2), 195–210.

Jingzhong Ye is a professor of development studies and deputy dean at the College of Humanities and Development Studies (COHD), China Agricultural University. His research interests include development intervention and rural transformation, rural 'left-behind' population, rural education, land politics, and sociology of agriculture.

Chunyu Wang is an associate professor at the College of Humanities and Development Studies (COHD), China Agricultural University. Her research interests include rural transformation, county governance and planning, rural politics, and land grabbing.

Huifang Wu is an associate professor at the College of Humanities and Development Studies (COHD), China Agricultural University. Her research mainly covers gender studies, agrarian sociology, and the rural left-behind population.

Congzhi He is a senior lecturer at the College of Humanities and Development Studies (COHD), China Agricultural University. Her research mainly covers rural social security issues, rural left-behind elderly, rural development policy, and civil society.

Juan Liu is a researcher at the College of Humanities and Development Studies (COHD), China Agricultural University. Her research interests include rural politics, land politics, and land grabbing.

The politics of industrial pollution in rural China

Bryan Tilt

After more than three decades of extremely rapid industrial growth, China faces an environmental crisis. The rural industrial sector, which includes millions of loosely regulated factories and employs hundreds of millions of workers, is a major focal point of this crisis. This paper provides a critical review of scholarship on industrial pollution in rural China and advances a new framework for thinking about the topic as a political domain with three inter-related parts:

- The politics of knowledge: What do rural citizens know about environmental contamination, and how do they know they know it? What sources of information are available to the public regarding pollution incidents? How does uncertainty about pollution sources and severity, as well as the potential links to health risks, shape rural peoples' experience of pollution?
- The politics of action: What strategies do individuals, communities and civil society organizations use to combat pollution? What outcomes are associated with such strategies?
- The politics of regulation: How are national laws and policies regarding pollution control implemented in rural areas? How do agencies and enforcement officials balance the competing objectives of environmental protection and economic growth?

The paper concludes by considering the implications of this framework for how scholars understand industrial pollution in rural China and briefly discussing a future research agenda for this field.

Introduction

After more than three decades of extremely rapid industrial growth, China faces an environmental crisis. News headlines such as 'Pollution turns China village into cancer cluster' and 'Choking on growth' have become routine in recent times. The rural industrial sector, which sounds paradoxical to many Western ears, includes millions of loosely regulated factories and employs hundreds of millions of workers in the countryside; it is also a major focal point of this environmental crisis. For China's 800 million villagers, the rural

I would like to thank the organizers of this collection on politics in rural China, as well as the anonymous reviewers whose comments helped to focus and refine my analysis. Many colleagues recommended important research to include in the paper or provided critical feedback on earlier drafts, including Chen Ajiang, Jennifer Holdaway, Rachel Stern, Anna Lora-Wainwright, Benjamin Van Rooij and Zhang Lei.

industrial sector represents some of the most far-reaching social, economic and environmental changes in human history.

As natural and social scientists try to make sense of these dramatic changes, their efforts have produced a large and wide-ranging body of knowledge about the problem of rural industrial pollution and its effects on individuals, communities and livelihoods. My aim in this paper is to provide a critical review of scholarship on industrial pollution in rural China and advance a new framework for thinking about the topic as a political domain with three inter-related parts:

- *The politics of knowledge:* What do rural citizens know about environmental contamination, and how do they know they know it? What sources of information are available to the public regarding pollution incidents? How does uncertainty about pollution sources and severity, as well as the potential links to health risks, shape rural peoples' experience of pollution?
- *The politics of action:* What strategies do individuals, communities and civil society organizations use to combat pollution? What outcomes are associated with such strategies?
- *The politics of regulation:* How are national laws and policies regarding pollution control implemented in rural areas? How do agencies and enforcement officials balance the competing objectives of environmental protection and economic growth?

In the process, I survey the breadth and depth of this topic, which includes major contributions from both Western and Chinese scholars, and features interdisciplinary collaboration between disparate fields such as environmental science, public health, anthropology, geography, political science, law and sociology. This new analytical framework encourages us to consider all aspects related to rural industrial pollution as a political domain, from the uneven distribution of environmental hazards, to the ways that individuals and groups perceive these hazards and mobilize themselves against them, to the attempts of state agencies and non-governmental organizations to regulate, mitigate or otherwise address pollution.

Pollution in rural China

Two related processes must be understood in order to grasp the current situation of industrial development in China: the transformation of the Chinese countryside into an industrial space, and the emergence of industrial pollution as a topic of personal and political concern. Since its founding in 1949, the People's Republic of China (PRC) has pursued a strategy of rapid industrial development. In the early years, China followed the Soviet model of development, with industrial production in the hands of the central government and agricultural production controlled by a network of rural collectives. The seeds of rural industry were planted during the Great Leap Forward (1958–1961), a tumultuous period in which China's economic planners, at the behest of Mao Zedong, funneled expertise and resources away from agriculture and into key national industries such as steel production. The results of the Great Leap Forward were catastrophic: rural areas produced steel of such poor quality that it proved useless for most military and industrial purposes, and millions of people starved to death (Becker 1996). Following the Great Leap, the central government continued to advocate for industrial growth in rural areas, albeit on a more modest scale. The so-called 'five small industries' – which produced key inputs for agriculture such as iron and steel, chemical fertilizers, farm machinery, cement and electrical power – were the result.

The era of Reform and Opening (*Gaige Kaifang*), which began in 1978, has ushered in sweeping social and economic changes and altered the lives and livelihoods of one-fifth of humanity (more than 1.3 billion people), as China has seen a return of smallholder agriculture under the Household Responsibility System, the privatization of industry, greater integration into the global economy and the rise of an urban consumer class. The expansion of the rural industrial sector, spurred by central policy, is one of the most important components of the economic reforms and represents arguably the greatest employment shift in human history. At the beginning of the reform era, more than 90 percent of China's massive rural labor force worked in agriculture. By the 1990s, approximately one-third of the nation's 500 million rural laborers had taken up factory jobs (China TVE Yearbook Editorial Committee 2004), heeding the exhortation of economic planners to 'leave the land, but not the countryside' (*li tu, bu li xiang*). These so-called 'township and village enterprises' served two important purposes: they absorbed some of the surplus labor in the countryside, which had been created by more efficient agricultural practices under the Household Responsibility System, and they provided township and village governments with much-needed revenue to carry out local development tasks such as infrastructural improvement and the provision of education.

Today, millions of rural factories remain an important part of the national economy, but their ownership structures and operations have changed quite dramatically in recent years. Over the past two decades, local governments have sold communal factories to private investors, a trend that is in line with economic liberalization throughout China (Oi 2005). In many cases, government officials sold factories to insiders with connections to local government or industry or both, a phenomenon called 'insider privatization' (Li and Rozelle 2003). In other cases, factories have attracted outside investors and workers, reducing the financial benefits to local people (Tilt 2010, Lora-Wainwright *et al.* 2012).

The rise of rural factories has caused environmental degradation on an unprecedented scale; air and water quality ratings in some areas of the countryside rival any city. For example, during field research in Sichuan more than a decade ago, I collected air quality samples and had them analyzed by an organic chemistry laboratory that specializes in atmospheric polycyclic aromatic hydrocarbons (PAHs), a class of chemicals with known carcinogenic effects. As it happened, another scientist was collecting air samples in the city of Guangzhou at roughly the same time, and the laboratory examined both batches of samples at the same time. The laboratory technicians were surprised to learn that the rural samples from Futian showed considerably higher levels of PAHs than the urban samples, suggesting greater health risks from air pollution in this small township of 4000 residents than in the megalopolis of Guangzhou (Murray *et al.* 2007).

Air pollution, water pollution and soil contamination are daily facts of life for most rural villagers. Current estimates suggest that rural factories are responsible for as much as two-thirds of the nation's total air and water pollution burden, and that that pollution-related economic losses drag down the country's overall gross domestic product (GDP) (World Bank 2007). In fact, pollution-related damages – in the form of health-care costs, lost worker productivity, infrastructural damage, threats to food safety and lost agricultural productivity – may well be one of the most intractable 'limits to growth' currently faced by China (Ho and Vermeer 2006).

A number of factors contribute to this grim assessment. Coal, because of its abundance and its state-regulated pricing structure, remains the fuel of choice in rural industry. Small-scale factories often lack access to significant capital, which means they typically employ outdated technology and take few environmental mitigation measures. Institutional factors also play a key role. While national air and water quality standards are set by the Ministry of Environmental Protection (MEP), monitoring and enforcement are the responsibility of county-level Environmental Protection Bureaus (EPBs), which typically lack the funding, expertise and other resources necessary to effectively carry out their jobs.

China's environmental regulation bureaucracy has grown from a fledgling agency at its inception in the 1970s to a ministry-level entity with a broad mandate to draft environmental laws, monitor environmental performance and enforce compliance. Speaking before the Sixth National Congress on Environmental Protection sponsored by the State Council, Premier Wen Jiabao in 2006 stressed what he called the 'three transformations':

(1) From an economy-centered model of development to equal attention on both environmental protection and economic development;
(2) From a mindset of 'develop first, then clean up', to a simultaneous emphasis on environmental protection and economic development, and
(3) From a sole emphasis on administrative measures to control environmental problems (e.g. hierarchical, institutional measures) to a more comprehensive system involving the state, the business sector, and civil society (SEPA 2006, Xue *et al.* 2007).

China's institutional capacity for preventing and controlling pollution is clearly gaining momentum. But Wen's rhetoric about the 'three transformations' highlights the gap between the policy goals and aspirations of political leaders and the current reality in rural China. In what follows, I examine current debates related to the three themes of current scholarship on pollution as a political domain: the politics of knowledge, the politics of action and the politics of regulation.

The politics of knowledge

What do rural citizens know about environmental contamination, and how do they know they know it? What sources of information are available to the public regarding pollution incidents? How does uncertainty about pollution sources and severity, as well as the potential links to health risks, shape rural peoples' experience of pollution? Very little is known about the air and water pollution levels to which most villagers are exposed. Environmental monitoring data related to both air and water quality are extremely difficult to obtain, both for scholars and for the general public. Moreover, the data that are available tend to be aggregated at the county level or above, which makes it difficult to assess pollution exposures at the community and individual levels.

Scholars who study pollution in rural China are making considerable efforts to understand what types of information are available to the public about both major pollution incidents and day-to-day pollution levels. Zhang Lei, for example, undertook a study of information disclosures (*xinxi gongkai*) about pollution incidents from EPBs to their constituents (L. Zhang 2010). Tilt and Xiao (2010) examined a range of media reports on the 2005 benzene spill in the Songhua River, including state-sponsored media outlets such as Xinhua News and China Central Television. Government authorities failed to notify the public for more than a week, shutting down the municipal water system of Harbin for several days under the guise of conducting routine repairs. In that case, however, China Central Television, which is state-owned but enjoys some editorial autonomy, conducted investigative journalism into the cover-up, which ultimately resulted in

MEP Minister Xie Zhenhua's dismissal.[1] But this work is unfortunately still extremely modest in scope, which leaves us with an incomplete understanding of how the Chinese public receives information about pollution-related health risks, and how this information shapes their perceptions and attitudes.

In regards to villagers' perceptions of environmental contamination, one of the major areas of progress in this field over the past several decades is the dispelling of common assumptions about how poor, marginalized people perceive the threat of pollution. Much of the cross-national research on environmental perceptions and values makes the 'postmaterialist' assumption that environmental consciousness is only made possible by the satisfaction of material needs through economic development (Ingelhart 1997). While wealthier nations and individuals, having met their basic economic needs, have the 'luxury' of worrying about air and water pollution, people in developing nations or in poor communities remain so mired in the daily struggle for livelihood that environmental quality is nothing more than an afterthought at best (Inglehart 1995, 1997, Dunlap and Mertig 1997).

This conclusion, however, is supported by tenuous data that suffers from several methodological shortcomings. First, approaching the study of environmental perceptions from the global scale requires the aggregation of data at macro-scales such as the nation or region, while most human-environment interactions occur on a local scale. This ultimately masks a great deal of the nuance and variation in how people perceive pollution in their daily lives. Second, cross-national surveys often frame the questions and issues in ways that are not culturally or politically salient for study participants, asking rural villagers in China and elsewhere in the developing world, for example, whether they would be willing to pay more for 'environmentally friendly' cleaning products, an option that simply doesn't exist.

These problematic conclusions have been questioned, and in many cases directly controverted, through a research approach that should have been obvious much earlier: undertaking in-depth field research in rural villages and townships. After a decade or so of careful research in communities facing environmental contamination, we now have a much fuller understanding of how people perceive, experience and cope with pollution on a day-to-day basis (Tilt 2006, Weller 2006, Chen 2009, Lora-Wainwright 2010, Tilt 2010, Van Rooij *et al.* 2012).

Nevertheless, people's perceptions of pollution are often marked by uncertainty and ambiguity as they struggle to make sense of their changing environments. One obvious source of such uncertainty relates to the availability and quality of environmental monitoring data. These macro-level figures on pollution emissions are easy to find in databases such as the China Statistical Yearbook and the China Environmental Statistical Yearbook (Managi and Kaneko 2009). But data are collected by different agencies with different jurisdictions and different policy mandates. Responsibility for monitoring surface water quality, for example, is divided among the MEP, the Ministry of Water Resources and the Centers for Disease Control (Holdaway 2010, 8), each of which employs different data collection methods and reports to different constituents.

Moreover, when researchers attempt to drill down beyond nationally aggregated figures, information about local environmental quality in rural areas is extremely hard to come by. Although the rural industrial sector is a major contributor to the national pollution burden, systematic air- and water-quality monitoring in the countryside is infrequent,

---

[1]Xie has the dubious distinction of being the highest-ranking political official ever to lose his post over an environmental issue.

sporadic and incomplete. What little monitoring data can be collected at the local level suggests that, in rural areas with high concentrations of factories, residents may be exposed to key pollutants that exceed World Health Organization guidelines by a factor of 10. Ambient air-quality monitoring in rural Sichuan, for example, found levels of respirable particulate matter (PM^{10}) that far exceeded Chinese national standards and World Health Organization standards (Tilt 2010). The MEP sets primary, secondary and tertiary air-quality standards based on geographic zoning. In 2012, this agency initiated a national program for monitoring $PM^{2.5}$, ultra-fine particles that are more closely linked to adverse health outcomes. In some cases, data monitoring and disclosure can become the subjects of international politics. For the past several years, the US Embassy has been releasing its own data on the concentration of $PM^{2.5}$ in Beijing, which officials have collected from monitoring equipment within the embassy compound. These figures have often been at odds with those released by the MEP, and the controversy has received considerable coverage in the press, prompting public discussion and concern (Moore 2011).

Another source of uncertainty relates to a disjuncture between scientific monitoring and people's ways of knowing and experiencing the world around them. People often use sensory details – the sight of effluents discoloring a river, or the smell of coal smoke and noxious gases coming from a factory smokestack – to assess the severity of pollution. For example, when a toxic algal bloom occurred on Lake Tai in 2007, due largely to untreated industrial effluents, local residents could see and smell the problem; many people also lost access to drinking water until alternative sources could be provided (Chen 2009).

How pollution exposure may be linked to adverse health outcomes is another area of significant uncertainty (Holdaway 2010). The sociologist Phil Brown (2007, 1) has defined environmental health problems as 'health effects caused by toxic substances in people's immediate or proximate surroundings (soil, air, water, food, household goods)'. Environmental health problems are often characterized by uncertainty because many environmental health hazards are difficult to detect and because it remains extremely difficult to demonstrate a causal link between environmental contamination and human health outcomes. People often focus on local sources of pollution that are easily detectable to the senses, remaining unaware of less apparent problems.

Researchers, too, struggle to understand causation in the midst of conflicting evidence. One basic challenge is how to deal with different etiologies and lexicons in the lay understanding of environmental health problems. As Lora-Wainwright found in northeast Sichuan, stomach and esophageal cancers can be caused by chronic exposure to farm chemicals, but for villagers, these cancers are often conflated with myriad folk maladies such as 'spitting illness' and 'vomiting illness' (Lora-Wainwright 2010). Similarly, Tilt's research (2010) describes the myriad ways in which rural people perceive the linkages between pollution exposure and adverse health outcomes: a teenage girl with cognitive disabilities and slurred speech, a developmentally delayed toddler, and a new water distribution system that brought piped water into the homes of several villages but also brought the fear of arsenic and mercury contamination, since the well and pump were located just downhill from the local factory compound. Such cases highlight the methodological difficulties scholars face in this arena: linking adverse health outcomes with pollution exposure is an epidemiological task requiring large amounts of data. Meanwhile, the perceptions of people most harmed by pollution are often discounted by scientists and public officials as merely 'anecdotal'.

Furthermore, people may feel ambivalent about pollution sources because of their position in the political economy; they may blame farm chemicals for cancer rates, for example, but they also rely heavily on chemical inputs to produce market-quality vegetables and earn a cash income (Lora-Wainwright 2009). Or they may be migrant workers who have traveled hundreds of kilometers from home in search of employment; disconnected from their home towns and extended family networks, and highly dependent upon wage labor in industry, they may overlook the health risks involved in their jobs (Tilt 2010). Or they may have a sense of fatalism, viewing pollution as an inevitable consequence of modernization, something they must 'learn to live with' (Lora-Wainwright *et al.* 2012). As a consequence, many rural villagers experience what Brown *et al.* (2000) have called 'ontological insecurity', a state of uncertainty about the future and about how to weigh the various risks they face in their day-to-day lives. In such a state of uncertainty, high-profile hazard events play an important role in raising villagers' consciousness of environmental problems. Recent case study research in southwest China found that, after a period of relative quiescence over local pollution, an explosion at a sulfuric acid factory, which killed several people and sickened others, resulted in a spike of environmental consciousness and activism from villagers, who demanded to be relocated, to have access to health monitoring and to be compensated (Van Rooij *et al.* 2012).

Perhaps no single phenomenon better illustrates the uncertainty surrounding environmental illness than the rise of so-called 'cancer villages' (*aizheng cun*), a problem with both sociological and epidemiological dimensions.[2] The sociologist Chen Ajiang has reported monitoring data for heavy metals (including lead, zinc, cadmium and copper) on soil, well water, grain and vegetables in villages near industrial factories and mines in Jiangsu, Jiangxi and Guangdong Provinces. The results of this research show levels of harmful pollutants that exceed national standards by a factor of 10. Moreover, mortality and morbidity rates from cancer in the study areas were far higher than expected. But making the link between pollution emissions, human exposure and adverse health outcomes is exceedingly difficult, requiring consistent monitoring, epidemiological data from large numbers of study participants and a great deal of time and resources (Chen and Cheng 2011).

In this, China and the West are facing similar difficulties. Many recent studies on environmental perceptions have noted the prominent, if polemical, role of science in public debates about causation in environmental health. Such works, grounded in a critical science studies approach, rightly point out the fallacy of looking to science to provide rational, indisputable data to override the 'irrational' fears of the lay public, who are most affected by contamination (Fischer 2003, Jasanoff 2004, Brown 2007, Leach and Scoones 2007).

In a study focused on the Lake Tai area, sociologist Chen Ajiang (2009) found that villagers exposed to pollution, and in some cases the industrial firms emitting pollution, increasingly possess what has been termed 'ecological consciousness' (*shengtai yishi*). As we gain more knowledge about peoples' perceptions and experiences of industrial pollution through community-level research, we are learning that villagers are acutely aware of, and concerned about, the threats they face from pollution – to their health, their families and their livelihoods. Moreover, as we study environmental perceptions, we cannot decontextualize them from other components of social, culture and economic life. For example, in a study in rural Anhui province on local people's perceptions of water quality following emissions from paper mills on the Huai River, anthropologist Robert Weller pointed out

---

[2]Cancer is the leading cause of death in both urban and rural China. See MOH (2008) and Holdaway (2010).

that his research participants 'lack environmental consciousness only in the sense that they are not concerned with the same issues as national and global elites, or as people who write questionnaires about values' (Weller 2006, 157).

The politics of action

If rural villagers do in fact possess 'ecological consciousness' and a general concern about the effects of pollution on their health and well-being, what can they do to address the problems they face? What strategies do individuals, communities and civil society organizations use to combat pollution? What outcomes are associated with such strategies? Recent scholarship shows that collective action aimed at addressing environmental problems is on the rise in China, but that understanding this action requires a set of conceptual tools that differ from those commonly employed in the West.

As the Chinese Communist Party has liberalized the nation's economy over the past three decades, it has also gradually reduced the scope of its administrative power, increasing the space within which nongovernmental organizations (NGOs) and civil society organizations may operate (Weller 2005, O'Brien and Li 2006, Chan 2005). This trend is especially important in the environmental arena, where a new Environmental Impact Assessment (EIA) Law, promulgated in 2002, mandates public hearings for major development projects. Scholars now point out that environmental lawsuits are increasingly common, as are environmental NGOs, which number in the thousands (Yang 2005). Yet little is currently known about how these organizations actually affect environmental decision-making processes. In my field research site in rural Sichuan, villagers' concerns about their health and livelihoods motivated them to get involved in collective acts of environmental advocacy, including petitioning the EPB and alerting the provincial media to the infractions of local factories (Tilt 2010). A coalition of farmers, for example, contacted a news crew at the Sichuan Television Station, who traveled to the township to film an exposé on local factories. This media scrutiny had a powerful effect on officials in the EPB, who acted swiftly to close down factories that violated emissions standards (Tilt 2007).

Other recent studies based primarily in urban China have suggested that the country is experiencing dynamic environmentalism but no coherent 'environmental movement' (Stalley and Yang 2006, 333). But such conclusions rest on a definition of 'movement' espoused by Western scholars that may overlook the subtle, spontaneous, ad hoc actions of people who share common interests regarding environmental problems. While it is difficult to estimate the frequency of environmental protests, or 'mass incidents' (*dazhong shijian*), the term preferred by the central government, data published by the government show a marked increase in citizen complaints to EPBs beginning in the late 1990s and culminating in more than 600,000 per year by the mid-2000s (Mol and Carter 2006, see also French 2005, Van Rooij 2012).[3] Such incidents – which may include a diverse range of tactics such as conducting independent sampling of drinking water from wells, filing public petitions and even blockading factory compounds (Jing 2010, Van Rooij 2012) – undoubtedly exert pressure on environmental regulators and policy-makers.

In this regard, scholarship on environmental action in China is linked to a growing body of literature in political science and related fields dealing with the rise of non-state actors in politics, including both formal organizations such as NGOs and informal popular

---

[3]Article 35 of the Constitution of the PRC guarantees citizens 'freedom of speech, of the press, of assembly, of association, of procession and of demonstration'.

movements. Throughout rural China, popular protests on a range of issues from pollution to land annexation tend to be most successful when people work together toward shared goals, and when they invoke the policies and rhetoric of the state itself (O'Brien 2002, O'Brien and Li 2006, Cai 2010). A case in point is the scholarship of sociologist Chen Ajiang, who draws upon the concept of a 'harmonious society' (*hexie shehui*), a rhetorical line currently in vogue within the central government, to argue that such harmony also requires a balance between humans and the biophysical environment. He suggests that 'harmony between people and water' (*ren shui hexie*) is a focal point of this struggle (Chen 2009). Similarly, Pan Yue, Vice Minister of Environmental Protection, has argued that environmental balance is a fundamental part of a 'harmonious society' (Pan 2006). These self-conscious deployments of state-sanctioned terminology are attempts to frame environmental issues in ways that are more palatable to the government.

Referencing a wide range of mass movements and public protests, Ho and Edmonds (2008) have termed this phenomenon 'embedded activism', since it works within existing power structures and political channels rather than directly challenging them (see also Ho 2001). As public consciousness grows about the rule of law and the reciprocal obligations between citizens and the state, anti-pollution campaigns enjoy a wider discursive space in which to operate. Political scientists Kevin O'Brien and Lianjiang Li have described this phenomenon as 'rightful resistance', which they define as:

> A form of popular contention that operates near the boundary of authorized channels, employs the rhetoric and commitments of the powerful to curb the exercise of power, hinges on locating and exploiting divisions within the state, and relies on mobilizing support from the wider public. In particular, rightful resistance entails the innovative use of laws, policies, and other officially promoted values to defy disloyal political and economic elites. (2006, 2)

What does environmental activism look like, and what tactics does it employ? Michelson (2007), adapting the concept of a dispute pyramid from legal scholarship, outlines what he has termed the Chinese 'dispute pagoda'. Using a dataset of roughly 3000 households collected from surveys in rural areas of six provinces, he finds that, for a wide variety of grievances, a sizeable proportion of villagers take no action at all. For those who do, the most common tactic is bilateral negotiation with the responsible party. Only a very small fraction of villagers with grievances actually file formal complaints with public officials or seek legal redress.

This pattern appears to hold true for environmental grievances. When villagers do take action against polluters, disputes are typically resolved through bilateral negotiation and compensation, rather than actual increases in pollution control and regulation, a process that Van Rooij *et al.* (2012) have called the 'compensation trap'. Compensation may be monetary in nature, or it may consist of land contracts or increased access to various resources. Van Rooij *et al.*'s research, based primarily on case studies in southwest China, documented compensation to village collectives and individuals from factories and mines, most of which consisted of 'agricultural compensation' (*nongye buchang*) for crop damages from pollution. One obvious reason why compensation may constitute a 'trap' is that it routinizes pollution as a problem to be dealt with through redress rather than through improved regulation, oversight and enforcement. In the process, it perpetuates a system of environmental governance that is reactive rather than precautionary.

Outside of official government circles, the proliferation of NGOs and other civil society organizations represents a major part of the anti-pollution movement; as Economy (2004, 131) has noted, 'environmental NGOs in China are at the vanguard of nongovernmental activity'. Yang (2005) categorized environmental NGOs into seven groups on the basis of their formal registration status with the government, including: (1) registered NGOs, (2) non-profit enterprises, (3) unregistered voluntary groups, (4) Internet-based groups, (5) student environmental associations, (6) university-affiliated research centers, and, most paradoxically, (7) government-organized NGOs, or GONGOs.

While environmental NGOs represent the vanguard of non-state activity, they also face a range of problems in China, including the requirement to register and receive sponsorship from a government agency, a limited political and legal framework that requires them to tread carefully or risk being closed down, and, like environmental NGOs everywhere, a critical shortage of funding for their operations (S.Y. Tang and Zhan 2008). NGO representatives deal with these constraints in a variety of ways. One tactic is to seek official state sponsorship, as was done by China's earliest and most well-known environmental NGO, Friends of Nature. But this approach comes with the risk of having one's agenda watered down or co-opted by the state. Another tactic is to operate covertly, concealing the true agenda of the organization and distorting membership roles (see Ho 2001).

Scholars differ in their assessments of the efficacy and significance of environmental NGOs, which have admittedly shown limited success in influencing policy outcomes (S.Y. Tang and Zhan 2008) but have proved to be useful tools for pooling information and resources among activists (Yang 2005). One high-profile success story is the Center for Legal Assistance to Pollution Victims (CLAPV), which is headquartered at Beijing Legal University and sponsored in part by a Spanish government agency. CLAPV brings together scholars, lawyers and scientific experts to support citizens in taking legal action against polluters (see J.J. Zhang 2010).

As grassroots movements, online communities and other forms of social networks play a larger role in anti-pollution campaigns, there is a scholarly push to use the tools of network analysis to understand activism. The study of social networks, particularly in the form of reciprocal exchange relationships (referred to in Chinese as *guanxi*), is fundamental to social science research in China. Social networks, through which people maintain interdependence, trust and reciprocity with one another, provide a means to adapt to change, cooperate and produce mutual advantage (Field 2003, 12). A great deal of environmental activism, including anti-pollution campaigns, now takes place on the Internet through social media sites (Sullivan and Xie 2009), but this type of activism is largely limited to urban, educated people (Yang 2005, 2010). Online environmental activism will likely become increasingly important in the countryside as Internet access is expanded to rural users.

The politics of regulation

The final dimension of pollution politics in rural China relates to regulation, oversight and decision-making within the official organs of state power. How are national laws and policies regarding pollution control implemented in rural areas? How do agencies and enforcement officials balance the competing objectives of environmental protection and economic growth? China's environmental protection bureaucracy has grown from a fledgling agency in the 1970s to a ministry-level entity as of 2008, with a broad mandate to draft environmental laws, monitor environmental performance and enforce compliance. The MEP is responsible for overseeing science and policy on such disparate topics as water pollution, atmospheric pollution, solid waste disposal, nuclear safety and radiation, soil and water conservation, biodiversity and wildlife conservation, ecosystem degradation and desertification. China's Environmental Protection Law, the foundation upon which all other environmental laws and regulations rest, was passed on the eve of economic reforms in 1979 and amended significantly in 1989 (see NPC 1989). Since that time, the legal framework for environmental protection has expanded to include at least 20 major statutes, in addition to lesser regulations passed by the State Council.

Moreover, the MEP and other relevant agencies increasingly take a precautionary view of environmental problems. The EIA Law, passed by the National People's Congress (NPC) in 2002 and implemented in 2003, requires environmental assessment for all new development projects, including industrial factories, likely to have significant environmental impacts. It mandates that government entities base their decisions to approve or reject projects in part on the EIA report (B.S. Tang *et al.* 2008).[4] However, the EIA law itself lacks specificity about who should conduct EIAs and exactly which government agencies should exercise which oversight capacities. This makes the EIA review process subject to influence from a range of parties, including local government agencies that may be dependent upon revenue generated from a given project. Moreover, Article 5 of the law states that 'The government encourages relevant entities, experts and the general public to participate in appropriate ways in the environmental impact assessment process', but meaningful public participation seldom takes place (Beach *et al.* 2006).

The MEP has established air quality standards for key pollutants based on a three-tiered system of 'classes' corresponding to geographic zoning. Water pollution is monitored and regulated on a five-tiered system. Establishing standards for pollutants means that the MEP or another relevant agency has determined, based on varying degrees of scientific evidence, how much of a given pollutant is 'safe' to be exposed to on an ongoing basis. For example, particulate matter standards are based on the lowest levels at which total mortality, and mortality specifically linked to cardiopulmonary problems and lung cancer, have been shown to increase in response to long-term exposure to particulate matter.

Despite clear advancements in environmental law and policy, enforcement remains the weakest link in the environmental oversight chain (Ma and Ortolano 2000). It is clear that many rural counties, townships and villages undergo limited and sporadic environmental monitoring at best. Shortages in financial resources, technology and institutional capacity limit the ability of even the most dedicated professionals to carry out their jobs (Tilt 2007). Moreover, officials in EPBs must prioritize the areas and industries to carry out their sampling. This problem has aptly been described as incomplete enforcement; faced with limited resources and capabilities, EPB officials exercise considerable discretion over which polluters to monitor, and how to enforce compliance (H. Wang *et al.* 2003). In many cases, EPB officials are charged with monitoring emissions and enforcing compliance for a huge number of factories in their jurisdictions.

Pollution enforcement occurs at the confluence of multiple factors, including the fiscal situation in the enforcement area, the institutional capacity of regulators, and even pressure from civil society organizations and the media (Tilt 2007). Pollution enforcement can have serious impacts, both real and perceived, on the financial viability of townships and villages. In a recent case study in rural Sichuan, a former township mayor, reflecting on the closure of local factories for non-compliance with pollution regulations, stated that 'over

---

[4]The EIA law has the potential to dramatically change environmental decision-making in China. One example of the EIA law's potential was the halting of more than 30 major infrastructure projects by Premier Wen Jiabao in 2004. In taking this action, Wen invoked the EIA law and criticized various government organizations and publicly traded corporations for failing to conduct thorough environmental reviews. Like other laws and regulations, however, it is subject to the political considerations, financial constraints, and institutional limits of enforcement and oversight. See Hu *et al.* (2012).

the past few years, there has been more economic development, and more opportunities than ever before. But now our factories are shut down, so we've stopped developing' (Tilt 2010, 127). His fears were validated by township government budgetary records, which showed a debt of millions of yuan and a catastrophic decline in revenue from local factories. This interdependence between industry and development, despite the obvious pollution problems, is echoed in Van Rooij *et al.*'s research, in which a village leader noted that 'if there is no pollution, where will development come from?' (2012, 36–7). Fiscal constraints at the township and village levels appear to play a key role in the decision about how strictly to enforce pollution standards. As one local political official complained about the growing severity of pollution enforcement, 'the upper level of government invites you to dinner, but the local government pays the bill' (Tilt 2007, 972). Within this political economy of enforcement, the stakeholders include regional and local government entities, farmers and other rural citizens, and industrial firms; moreover, the enforcement process can result in clear financial winners and losers (Chen 2008).

The pollution-levy system is a key instrument through which EPB officials enforce emissions standards. Article 18 of the Environmental Protection Law states that 'in cases where the discharge of pollutants exceeds the limit set by the state, a compensation fee shall be charged according to the quantities and concentration of the pollution released' (NPC 1989). The levy system, which is often enforced with the assistance of local courts (X.H. Zhang *et al.* 2010), requires polluting factories to register with their respective EPBs and to disclose their strategies for mitigating the so-called 'three emissions' (*san fei*): air pollution, water pollution and solid waste (H. Wang and Wheeler 2005). However, in most cases, the actual amounts levied against polluting firms are low, which means that the most cost-effective decision for factory managers may be to simply pay the fines and keep polluting. Moreover, since 2003, a new administrative rule has required that pollution fees go directly to the local finance bureau, and firms are no longer guaranteed a partial remission of the fees (Chinese State Council 2003, Lo and Tang 2006).

Attempts have been made to tie the performance evaluations of local government officials more closely to environmental protection through a system known as the 'environmental quality administrative leadership responsibility system'. However, environmental protection counts for only a small fraction of the overall evaluation of political leaders, and evidence suggests that the implementation of this system is lax (Lo and Tang 2006).

China's judiciary constitutes another important channel through which pollution enforcement takes place, although currently less than 1 percent of environmental disputes make their way to the courts (Caijing 2012). Access to the judicial system, like many aspects of political life in contemporary China, is a complex issue. As legal experts point out, litigation is difficult, for a variety of reasons. First, only individuals directly harmed by a given project have standing, the legal right to sue. This means that domestic and international environmental organizations, while they can provide moral, logistical and even financial support, cannot be plaintiffs in a lawsuit. In 2012, the Civil Procedure Law was amended to allow 'government departments and concerned organizations as designed by law' to engage in public interest litigation, and this may result in an increased volume of environmental lawsuits in the near future (Ng 2012, see also A. Wang 2007).

Second, in the case of collective lawsuits involving many parties, each individual party must opt into the lawsuit and provide a copy of his or her national identification card (*shenfen zheng*), which can make collective lawsuits logistically complicated and politically risky. Furthermore, Chinese courts are notoriously difficult to access for those who lack political clout. While lower-level courts are easier for plaintiffs to access, they are particularly vulnerable to the political influence of big players such as state-owned firms or private and shareholder corporations (Stern 2011). As legal scholar Rachel Stern concludes following an analysis of environmental litigation related to pollution damages, 'Despite occasional successes, civil environmental litigation remains a weak tool for environmental protection' (2011, 310).

Because of a general lack of transparency in the court system, it is difficult to track the volume and efficacy of environmental litigation in China. Lü *et al.* (2011) collected longitudinal data on a sample of judicial judgments and orders related to environmental suits in all provincial-level administrative units over a 10-year period. The vast majority were cases within the domain of civil law, with some administrative cases and some criminal cases. Their work highlights the relative differences in political and economic power between polluting firms and the victims of pollution:

> Most plaintiffs in civil environmental cases belong to vulnerable groups that lack the economic capacity to hire attorneys; they tend to rely on themselves, relatives or friends for representation. By contrast, defendants are in generally in a strong position, which undoubtedly increases inequality (Lü *et al.* 2011, 89).

In environmental cases within the civil court system, plaintiffs must demonstrate two things: that they sustained damage to person or property from pollution, and that the damage was caused by a given party. Reliance on scientific evidence from expert witnesses is a key tactic for both plaintiffs and defendants (Lü *et al.* 2011). Once engaged in a civil environmental case, the courts' decisions appear split relatively evenly between plaintiff and defendant, although a 'victory' for the plaintiff may result in a settlement that is only a fraction of the amount requested (Lü *et al.* 2011). In administrative cases, where the defendant is most often a government agency, the courts' judgments are overwhelmingly in favor of the defendant. This is undoubtedly related to the close political ties between the judiciary and the regional government: in line with recent trends toward political and fiscal decentralization (O'Brien 2009), courts rely on local governments for budgetary support. Judicial decisions often represent a balancing act between legal formality (i.e., adherence to law) and individual autonomy (i.e., the power and authority to reach an independent decision) (Stern 2010). This trend is also related to different standards for 'burden of proof' that must be applied in different legal contexts. Based on an analysis of several cases handled by Beijing's CLAPV, Zhang Jingjing (2010) finds that many judges struggle to make distinctions, for example, between 'direct causality' and a 'causal relationship' between pollution exposure and actual harm.

Significantly, recent revisions to China's Water Pollution Law (NPC 2008) place the burden of proof on defendants, who must demonstrate, for example, that pollution emissions from their factory *did not* cause harm (Stern 2010). This represents further steps toward a precautionary approach to pollution management, but it still too early to assess the impact that such statutory changes will have on pollution-related lawsuits.

One growing mechanism for combating pollution is community-based pollution regulation, which involves rural citizens in monitoring pollution levels, alerting EPB officials to violations, and advocating for stricter enforcement. Local government cadres play an important role in environmental management, sometimes acting on behalf of local interests and other times lining their pockets with compensation funds intended for villagers (Van Rooij *et al.* 2012).

On an encouraging note, the central government is actively building institutional capacity to better assess and regulate pollution-related hazards. The central government is currently increasing its efforts to establish a risk management system that would help to predict, address and mitigate environmental hazards through inter-agency collaboration (L. Zhang and Zhong 2010). Within key agencies such as the Ministry of Health (MOH), environmental health issues tend to receive fewer resources and attention than problems such as insurance provision, or maternal and child health; moreover, the emphasis in MOH continues to be on curative, rather than preventive, measures (Fang and Bloom 2010).[5] In 2006, the Ministry of Health and the State Environmental Protection Administration (the predecessor of MEP) established the Combined Offices for Environment and Health, which monitor environmental illness, exchange information, issue public warnings and information, and provide expert advising on pollution standards and anti-pollution legislation (Holdaway 2010, 11). These two ministries jointly drafted the *National Environmental Health Action Plan (2007–2015)* calling for more systematic collaboration between the two agencies (MOH 2006), although a lack of financing, data-sharing and collaboration has made the plan difficult to implement for many local governments (Su 2010).

Conclusions and implications

In this paper, I have attempted to synthesize the growing body of research on industrial pollution in rural China and fashion a framework through which we can think about the topic as a political domain. In regards to the politics of knowledge, an overwhelming majority of recent studies confirm that rural villagers, despite barriers such as poverty and political marginalization, are concerned about the effects of pollution on their health, their families, their communities and their livelihoods. However, the environmental perceptions and values of local people are often embedded in specific cultural, economic and political realities. They tend to be grounded in practical concerns about health and livelihood, rather than in abstract notions of environmental protection. Furthermore, concerns about pollution cannot be separated from political and economic questions about who wins and who loses from industrial development. These findings bring with them an important methodological consideration: while natural and social scientists should rely on macro-level data sets on pollution exposure, where available, they should also roll up their sleeves and engage in community-level field research that allows them to understand villagers' perceptions and experiences in context.

In regards to the politics of action within anti-pollution movements, we have learned a great deal about what motivates rural people to get involved in such movements and what tactics are commonly used. There is widening political space in China today for both formal environmental civil society (such as NGOs or other advocacy organizations) and informal protests and other 'mass incidents'. Anti-pollution lawsuits are also on the rise, although plaintiffs face difficulty in accessing the courts and successfully obtaining redress. The consensus among social scientists is that victories will come slowly and incrementally, and will be achieved by working within, rather than overturning, existing power structures.

In terms of the politics of regulation, central and provincial governments have made tremendous improvements in their institutional capacity in recent years, with the promotion of the MEP to full ministerial status and a steady increase in state funding for pollution control and prevention. However, enforcement of pollution standards remains a key problem, due

---

[5]The New Rural Cooperative Medical Care System (NRCMCS), initiated in 2003, was designed to improve healthcare access among the rural poor. It is a tiered system that allows patients to access local village and township clinics and county hospitals, with heavy subsidies from the central and provincial governments (see Meessen and Bloom 2007).

to lack of resources and a high degree of fiscal dependence on revenues from polluting firms. Recent advancements such as the pollution levy system and an evaluation system that ties political leaders' compensation to environmental enforcement represent steps in the right direction, but the implementation of these policies remains uneven.

Looking to the future, where is this field heading, and what are the most pressing research needs? I have already suggested a methodological approach that combines macro-level analyses of pollution data with community-based fieldwork to understand the perspectives of local people. But what topics should we as researchers prioritize in our work? Although the scope of this paper does not allow for a full exposition of a research agenda, I would like to suggest several areas of critical importance. First, there is a greater need to understand the dynamics of civil society organizations and anti-pollution movements. Such movements increasingly use information technology and social media to disseminate information about environmental contamination, to organize citizen action, and to draw financial and moral support from national and international organizations. Their scope of action, and their influence, will only grow in the years to come. Second, researchers increasingly acknowledge that environmental problems in rural China are part and parcel of the global environmental crisis. As environmental problems – such as transnational pollution, climate change and electronic waste – become more complicated and more difficult to regulate, we will need to broaden our analytical lenses to consider the global ties between producers and consumers and the environmental hazards they co-create. Ideally, such research would involve multi-sited fieldwork in locations as disparate as factories, distribution networks and consumer markets. Finally, researchers will continue to face the challenge of understanding China's environmental governance amidst rapid social, economic and political change. The 18th National People's Congress, held in 2012, ushered in a new generation of political leaders. In a political culture where social stability is the paramount concern, but where the public increasingly calls for fairness and transparency, how will these leaders prioritize environmental concerns relative to other national goals?

As the PRC continues its transformation from an agrarian nation to an industrial powerhouse, rural industry will continue to play a key role in industrial output. Because of its massive scale, geographic dispersion, limited environmental mitigation technology and poor institutional oversight, it is also likely to remain a major contributor to air and water pollution, damaging the health and livelihoods of the nation's 800 million rural residents. However, as this paper has outlined, rural industrial pollution has become a major topic of concern for citizens, a focal point of political and social movements, and a target of increased regulation and enforcement for government agencies. It remains to be seen whether this struggle to balance economic growth with the imperative to protect the environment and human health will be successful.

References

Beach, M., B. Bleish and S. Yang. 2006. *The role of public participation in ecological impact assessment (EcIA) and environmental impact assessment (EIA) in China.* China Environment Series, No. 8. Washington, DC: Woodrow Wilson Center, pp. 117–121.

Becker, J. 1996. *Hungry ghosts: secret famine.* New York: Henry Holt.

Brown, P. 2007. *Toxic exposures: contested illnesses and the environmental health movement.* New York: Columbia University Press.

Brown, P., S. Kroll-Smith and V.J. Gunter. 2000. Knowledge, citizens, and organizations. *In:* S. Kroll-Smith, P. Brown and V.J. Gunter, eds. *Illness and the environment: a reader in contested medicine.* New York: New York University Press, pp. 9–25.

Cai, Y.S. 2010. *Collective resistance in China: why popular protests succeed or fail.* Stanford: Stanford University Press.

Caijing. 2012. Wo guo huanjing qunti shijian nianjun dizeng 29%, sifa jiejue buzu 1% [Environmental mass incidents increase 29% annually in China, judicial settlement accounts for less than 1%] [online]. Beijing: Caijing. 27 October 2012. Available from: http://politics. caijing.com.cn/2012-10-27/112233970.html [Accessed 30 November 2012]

Chan, K.M. 2005. The development of NGOs under a post-totalitarian regime: the case of China. *In:* R.P. Weller, ed. *Civil life, globalization, and political change in Asia: organizing between family and state.* New York: Routledge, pp. 20–41.

Chen, A.J. 2008. Shui wuran shijian zhongde liyi xiangguan zhe fenxi [Stakeholder analysis for water pollution incidents]. *Jiangsu Xuekan [Jiangsu Periodical]*, 4, 169–75.

Chen, A.J. 2009. Zai lun ren shui hexie: Taihu Huaihe liuyu shengtai zhuanxing de qiji yu leixing yanjiu [Further discussion on harmony between humans and water: a study of the turning point and types of ecological transformation in Lake Tai and the Huai River Basin]. *Jiangsu Shehui Kexue [Jiangsu Social Sciences]*, 4, 70–6.

Chen, A.J. and P.L. Cheng. 2011. Aizheng wuren de renzhi yu fengxian yiyngdui: Jiyu ruogan aizheng cun de jingyan yanjiu [Risk perception and response related to cancer and pollution: research on the experience of selected cancer villages]. *Xuehai*, 3, 30–41.

China TVE Yearbook Editorial Committee. 2004. *Zhongguo xiangzhen qiye nianjian [China Township and Village Enterprise Yearbook]*. Beijing: China Agricultural Press.

Chinese State Council. 2003. Administrative regulations on the collection and use of pollutant discharge fees [online]. *China Law and Policy*, 24 January 2003. Available from: www.omm. com/webdata/content/publications/clp030124.pdf [Accessed 1 December 2012.]

Dunlap, R.E. and A. Mertig. 1997. Global environmental concern: an anomaly for postmaterialism. *Social Science Quarterly*, 78(1), 24–9.

Fang, J. and G. Bloom. 2010. China's rural health system and environment-related health risks. *Journal of Contemporary China*, 19(63), 23–35.

Field, J. 2003. *Social capital*. London: Routledge.

Fischer, F. 2003. *Citizens, experts, and the environment: the politics of local knowledge*. Durham, NC: Duke University Press.

French, H.W. 2005. Anger in China rises over threat to the environment. *New York Times*. July 19. Available from: www.nytimes.com/2005/07/19/international/asia/19china.html?pagewanted= all&_r=0. [Accessed 1 December 2012].

Ho, P. 2001. Greening without conflict? Environmentalism, NGOs, and civil society in China. *Development and Change*, 32(5), 893–921.

Ho, P. and R.L. Edmonds. 2008. *China's embedded activism: opportunities and constraints of a social movement*. London: Routledge.

Ho, P. and E. Vermeer. 2006. *China's limits to growth: greening state and society*. Oxford, UK: Blackwell.

Holdaway, J. 2010. Environment and health in China: an introduction to an emerging research field. *Journal of Contemporary China*, 19(63), 1–22.

Hu, T., *et al.* 2012. Que zhi you xiao de wenzezhi: Wo guo shengwu duoyangxing zhizheng nengli chutan [Lack of an effective accountability system: an examination of China's biodiversity governance]. *Huanjing yu Kechixu Fazhan [Environment and Sustainable Development]*, 37(3), 26–30.

Inglehart, R. 1995. Public support for environmental protection: objective problems and subjective values in forty-three societies. *Political Science and Politics*, 28(1), 57–72.

Inglehart, R. 1997. *Modernization and postmodernization: cultural, economic, and political change in forty-three societies*. Princeton, NJ: Princeton University Press.

Jasanoff, S. 2004. The idiom of co-production. *In:* S. Jasanoff, ed. *States of knowledge: the co-production of science and social order*. London: Routledge, pp. 1–12.

Jing, J. 2010. Environmental protests in rural China. *In:* E.J. Perry and M. Selden, eds. *Chinese society: change, conflict and resistance*, 3rd ed. New York: Routledge, pp. 197–214.

Leach, M. and I. Scoones. 2007. *Mobilizing citizens: social movements and the politics of knowledge*. Working Paper. Brighton, UK: Institute of Development Studies.

Li, H.B. and S. Rozelle. 2003. Privatizing rural China: insider privatization, innovative contracts and the performance of township enterprises. *The China Quarterly*, 176, 981–1005.

Lo, C.W.H. and S.Y. Tang. 2006. Institutional reform, economic changes, and local environmental management in China: the case of Guangdong Province. *Environmental Politics*, 15(2), 190–210.

Lora-Wainwright, A. 2009. Of farming chemicals and cancer deaths: the politics of health in contemporary China. *Social Anthropology*, 17(1), 56–73.

Lora-Wainwright, A. 2010. An anthropology of 'cancer villages': Villagers' perspectives and the politics of responsibility. *Journal of Contemporary China*, 19, 79–100.

Lora-Wainwright, A., Y.Y. Zhang, Y.M. Wu and B.Van Rooij. 2012. Learning to live with pollution: the making of environmental subjects in a Chinese industrialized village. *The China Journal*, 68, 106–24.

Lü, Z.M, Z.M. Zhang and X.Q. Xiong. 2011. Zhongguo huanjing sifa xiankuang diaocha: Yi qian fen huanjing caipan wenshu wei yangben [Investigation into the current status of China's judiciary: a sample from thousands of environmental judgments and orders]. *Faxue*, 4, 82–93.

Ma, X.Y. and L. Ortolano. 2000. *Environmental regulation in China: institutions, enforcement, and compliance*. Lenham, MD: Rowan and Littlefield.

Managi, S. and S. Kaneko. 2009. *Chinese economic development and the environment*. Cheltenham, UK: Edward Elgar.

Meessen, B. and G. Bloom. 2007. Economic transition, institutional changes and the health system: some lessons from rural China. *Journal of Economic Policy Reform*, 10(3), 209–32.

Michelson, E. 2007. Climbing the dispute pagoda: grievances and appeals to the official justice system in rural China. *American Sociological Review*, 72(3), 459–85.

MOH (Chinese Ministry of Health). 2006. Guojia huanjing yu jiankang xingdong jihua [National environmental health action plan, 2007–2015] [online]. Beijing: Chinese Ministry of Health. Available from: http://www.moh.gov.cn/open/web_edit_file/20071108173502.doc [Accessed 30 September 2012]

MOH (Chinese Ministry of Health). 2008. *2007 nian wo guo weisheng shiye fazhan tongji baogao* [National health services development statistical report for 2007]. Beijing: Ministry of Health.

Mol, A.P.J. and N.T. Carter. 2006. China's environmental governance in transition. *Environmental Politics*, 15(2), 149–70.

Moore, M. 2011. Chinese lose patience with pollution [online]. London. *The Telegraph*, 6 December. Available from: http://www.telegraph.co.uk/earth/earthnews/8938159/Chinese-lose-patience-with-pollution.html [Accessed 19 December 2011]

Murray, J., T. Primbs, B. Simoneit, B. Tilt and S. Simonich. 2007. Semi-volatile organic compound profiles of urban and rural China, 2002–2003. Poster presented at the annual meeting of the Pacific Northwest Association of Toxicologists. 5 September 2007. University of Washington. Seattle, WA.

Ng, T.W. 2012. Lack of environmental public interest litigation in no one's best interest [online]. Hong Kong. *The South China Morning Post*, 8 November. Available from: http://www.scmp.com/news/china/article/1077382/lack-environmental-public-interest-litigation-no-ones-best-interest [Accessed 10 November 2012]

NPC (Chinese National People's Congress). 1989. *Zhonghua renmin gongheguo huanjing baohu fa [Environmental protection law of the People's Republic of China]*. Beijing: NPC.

NPC (Chinese National People's Congress). 2008. *Zhongguo renmin gonheguo shui wuran fangzhi fa [Water pollution prevention and control law of the People's Republic of China]*. Beijing: NPC.

O'Brien, K.J. 2002. Collective action in the Chinese countryside. *The China Journal*, 48, 139–54.

O'Brien, K.J. 2009. Local people's congresses and governing China. *China Journal*, 61, 131–41.

O'Brien, K.J. and L.J. Li. 2006. *Rightful resistance in rural China*. Cambridge: Cambridge University Press.

Oi, J.C. 2005. Patterns of corporate restructuring in China: political constraints on privatization. *The China Journal*, 53, 115–36.

Pan, Y. 2006. Hexie shehui mubiao xia de huanjing youhaoxing shehui [An environmentally friendly society under the goal of a harmonious society]. *Twenty-first Century Economic Herald*, 34.

SEPA (Chinese State Environmental Protection Administration). 2006. *Zhongguo huanjing zhuangkuang gongbao [Report on the state of the environment in China]*. Beijing: SEPA.

Stalley, P. and D.N. Yang. 2006. An emerging environmental movement in China? *The China Quarterly*, 186, 333–56.

Stern, R.E. 2010. On the frontlines: making decisions in Chinese civil environmental lawsuits. *Law and Policy*, 32(1), 79–103.

Stern, R.E. 2011. From dispute to decision: suing polluters in China. *The China Quarterly*, 206, 294–312.

Su, Y. 2010. Zhongguo huanjing yu jiankang gongzuo de xiankuang, wenti yu duice [Current status, problems and countermeasures in China's environment and health work]. *In:* J. Holdaway, W.Y.

Wang, Y.J. Zhong and S.Q. Zhang, eds. *Huanjing yu jiankang: Kuaxuede shijiao [Environment and health: cross-disciplinary perspectives]*. Beijing: Social Sciences Academic Press, pp. 72–98.

Sullivan, J. and L. Xie. 2009. Environmental activism, social networks and the internet. *The China Quarterly*, 198, 422–32.

Tang, B.S., S.W. Wong and M.C.H. Lau. 2008. Social impact assessment and public participation in China: a case study of land requisition in Guangzhou. *Environmental Impact Assessment Review*, 28, 57–72.

Tang, S.Y. and X.Y. Zhan. 2008. Civic environmental NGOs, civil society and democratization in China. *Journal of Development Studies*, 44(3), 425–48.

Tilt, B. 2006. Perceptions of risk from industrial pollution in China: a comparison of occupational groups. *Human Organization*, 65(2), 115–27.

Tilt, B. 2007. The political ecology of pollution enforcement in China: a case from Sichuan's rural industrial sector. *The China Quarterly*, 192, 915–32.

Tilt, B. 2010. *The struggle for sustainability in rural China: environmental values and civil society*. New York: Columbia University Press.

Tilt, B. and Q. Xiao. 2010. Media coverage of environmental pollution in the People's Republic of China: responsibility, cover-up and state control. *Media, Culture and Society*, 32(2), 225–45.

Van Rooij, B., A. Lora-Wainwright, Y.M. Wu and Y.Y. Zhang. 2012. The compensation trap: the limits of community-based pollution regulation in China. *Pace Environmental Law Review*, 29, 701–45.

Wang, A. 2007. *One billion enforcers*. Washington, DC: Environmental Law Institute.

Wang, H., N. Mamingi, B. Laplante and S. Dasgupta. 2003. Incomplete enforcement of pollution regulation: bargaining power of Chinese factories. *Environmental and Resource Economics*, 24, 245–62.

Wang, H. and D. Wheeler. 2005. Financial incentives and endogenous enforcement in China's pollution levy system. *Journal of Environmental Economics and Management*, 49(1), 174–96.

Weller, R.P. 2005. Introduction: civil institutions and the state. *In:* R.P. Weller, ed. *Civil life, globalization, and political change in Asia: organizing between family and state*. New York: Routledge, pp. 1–19.

Weller, R.P. 2006. *Discovering nature: globalization and environmental culture in China and Taiwan*. Cambridge: Cambridge University Press.

World Bank. 2007. The cost of pollution in China: economic estimates of physical damages. Washington, DC: The World Bank.

Xue, L., E.E. Simonis, and D.J. Dudek. 2007. Environmental governance for China: major recommendations of a task force. *Environmental Politics*, 16(4), 669–76.

Yang, G.B. 2005. Environmental NGOs and institutional dynamics in China. *The China Quarterly*, 181, 46–66.

Yang, G.B. 2010. *The power of the internet in China: citizen activism online*. New York: Columbia University Press.

Zhang, J.J. 2010. Zhongguo huanjing qinquan susong (jiankang sunhai lei) shili fenxi [Case studies in Chinese environmental tort litigation (health damage class)]. *In:* J. Holdaway, W.Y. Wang, Y.J. Zhong and S.Q. Zhang, eds. *Huanjing yu jiankang: Kuaxuede shijiao [Environment and health: cross-disciplinary perspectives]*. Beijing: Social Sciences Academic Press, pp. 170–89.

Zhang, L. 2010. Information disclosure and environmental management: results from recent research. Paper presented at the forum on Health, Environment and Development, Social Science Research Council and Chinese Academy of Science. Beijing, 11 November 2010.

Zhang, L. and L.J. Zhong. 2010. Integrating and prioritizing environmental risks in China's risk management discourse. *Journal of Contemporary China*, 19(63), 119–36.

Zhang, X.H., L. Ortolano and Z.M. Lü. 2010. Litigation law: court enforcement of pollution levies in Hubei Province. *The China Quarterly*, 202, 307–26.

Bryan Tilt is an associate professor of anthropology at Oregon State University. His research focuses on sustainable development, agricultural systems, pollution control and water resources in China and the United States. He is the author of the book *The struggle for sustainability in rural China: environmental values and civil society*, published by Columbia University Press in 2010.

The politics of conservation in contemporary rural China

Emily T. Yeh

Placing conservation within a broad framework of agrarian and environmental politics, this review article argues that natural resource governance is fundamental to rural politics in China. Much of the environmental literature adopts a technocratic approach, ignoring the political nature of the redistribution of access to and control over natural resources, and of knowledge vis-à-vis degradation. Reading the managerial literature with and against the grain of political ecological studies, the essay reviews contemporary environmental issues including Payments for Ecosystem Services and other market-based approaches, the establishment of national parks and resettlement schemes justified through ecological rationales. The first section following the introduction focuses on two of the largest forest rehabilitation schemes in the world. Next, the paper reviews work on China's rapidly growing number of nature reserves, examining their role as enclosures and their entanglement with tourism income generation. This is followed by a discussion of research on the politics of rangeland degradation and property rights. The inclusion of pastoralism within the scope of rural politics is sometimes obscured by the fact that China's extensive rangelands coincide almost completely with its minority populations. The misrecognition of rural politics over resources and the environment as ethnic politics is addressed in the concluding section.

Introduction

With its chemical spills, cancer villages and frequent protests against air and water pollution, China's litany of environmental woes is now well known. Scholars are paying increasing attention to the politics of urban and, to a lesser extent, rural industrial pollution (discussed by Tilt 2013). Yet some long-time observers of China's environment have argued that the 'green' issue of degradation of natural resources such as forests and grasslands are ultimately more threatening to long-term sustainability than 'brown' pollution problems (Edmonds 1999, Smil 2004). Nevertheless, a survey of the literature on Chinese politics reveals more attention to pollution than to natural resources, given the former's close association with violent protests and other visible forms of collective action.[1]

Thanks to Elizabeth Wharton for research assistance, and to three anonymous reviewers for helpful comments on an earlier version of this paper.
[1]One notable exception is Joshua Muldavin's work on the political ecology of agrarian reform, which argues that the 'mining of communal capital' such as reservoirs, irrigation canals and erosion-control structures, and the redirection of investments from communal infrastructure to private agricultural investments have exacerbated rural environmental problems in the reform period. Against the

At the same time, much of the literature that *is* concerned with rural natural resources in China approaches state management policies from a purely technical perspective, rather than exploring the inherently political nature of redistributions of access to and control over resources, or the politics of knowledge at work in the classification of certain areas as being degraded and in need of specific forms of rehabilitation. Though there is a body of work in political ecology on 'green' environmental issues in China (Coggins 2002, Jiang 2005, Sturgeon 2005, Herrold-Menzies 2006a, 2006b), it is dwarfed by the number of studies that hew to a managerial and technocratic approach. Much (though not all) of this work takes as a given that market forms of environmental management – 'selling nature to save it' (McAfee 1999, Robertson 2004) – should be deepened in China.

In this contribution reviewing work on China's forests, nature reserves and grasslands, I adopt a broad framework of agrarian and environmental politics to argue that more attention should be paid to the politics of conservation and natural resource governance, and that these are fundamental to rural politics. Drawing on Kerkvliet's (2009, 227) pithy definition of politics as being about 'the control, allocation, production and use of resources, and the values and ideas underlying those activities', it should be evident that forest and grassland management, the enclosure of land for biodiversity conservation, and processes of deforestation, afforestation, converting cropland to forest and moving pastoralists off rangelands to improve environmental condition, are inherently political. The next three sections of this review follow the common organization of the natural resource management literature into different resource sectors, rather than focusing on political ecological themes across sectors. For each sector – forestry, nature reserves and rangelands – I draw on available political ecology literature while also reading the larger literature in economics, management and remote sensing both with and against the grain.

The second section of the paper examines two of the world's largest forest rehabilitation initiatives, the Natural Forest Protection Program and the Sloping Land Conversion Program, drawing out the politics of their formulation, implementation, and effects. This part of the review also sets these programs in the context of global trends toward Payment for Ecosystems Services as the preferred way to manage natural resources, and the question of the extent to which such schemes are neoliberal. From here I turn to nature reserves and biodiversity conservation, for which there is a larger body of work in political ecology. This section examines the politics of conservation enclosure, both in terms of local access to resources as well as how the implementation of conservation schemes is shaped by broader political economic pressures. The entanglement of parks with tourism development in China and the conflicts that have ensued speak to global trends in the politics of conservation.

The fourth part of the paper turns to rangeland degradation and pastoralism in China, particularly the politics of knowledge about degradation and new forms of property rights and enclosure. Because China's extensive rangelands coincide almost completely with its minority populations, particularly Mongolians, Tibetans and Kazaks, research on rural politics in pastoral areas has made little impact in the mainstream China social science or agrarian studies literatures. Instead, it has been largely been viewed as a minority or peripheral issue, confined to specialized venues on pastoralism. Thus, in the fourth section, I emphasize that pastoral areas are undergoing rapid changes paralleling other rural areas in access to land, environmental conditions, and livelihood strategies. That is,

---

market triumphalist narrative, he suggests the collective period was in some respects better environmentally than what followed (Muldavin 1996, 1998, 2000).

pastoral politics are rural politics, not just 'minority issues.' This is particularly important given that the misrecognition of rural politics as ethnic politics is also used by state authorities as a way to suppress environmental protests by ethnically marked citizens. This observation leads to the final section of the paper, which does not focus on one particular natural resource sector, but rather provides several cases where contentious rural environmental politics have been coded as ethnic problems threatening state stability.

Afforestation and payments for environmental services

Extensive droughts in 1997 caused the lower reaches of the Yellow River to run dry for 267 days. The next year, severe flooding along the Yangtze River caused more than 12 billion USD in property damage and claimed over 3000 lives. Responses to these events – the implementation of the Natural Forest Protection Program (NFPP, sometimes also called the National Forest Conservation Program, or 'the logging ban') and the Sloping Land Conversion Program (SLCP, sometimes called 'Grain for Green') – consolidated China's emergence as an environmental state. The two programs together constitute one of the largest environmental rehabilitation efforts in the world (Jintao Xu *et al.* 2001, 2006, Yeh 2009b).

With an initial investment of 96.4 billion RMB (roughly 12 billion USD), the NFPP was a 10-year program launched in 2000 to completely ban logging over some 30 million hectares of forest in the upper reaches of the Yangtze River and the middle and upper reaches of the Yellow River. Covering 17 provinces, it also called for sharp reductions in commercial harvesting in other forested areas, provided alternative employment for workers in state-owned logging enterprises, and accelerated reforestation and silvicultural treatments. Less research has been done on the effects of NFPP than of SLCP, but there is general agreement that NFPP has increased China's imports of timber, particularly from countries such as Papua New Guinea, Myanmar, Indonesia and Russia, with illegal and unsustainable harvesting practices. That is, the program has in effect exported China's deforestation, though some commentators point out that much of the imported timber is used in China to make products for export (J. Liu *et al.* 2008),

After being piloted in Sichuan, Gansu and Shanxi provinces in 1999, the SLCP was expanded to 25 provinces, with a budget of over 40 billion US dollars. It called for the conversion of over 14 million hectares of cropland on steep slopes of greater than 25 degrees in southwest China and greater than 15 degrees in the northwest, to be converted to forest, or in some areas, grassland. In addition, it called for afforestation of about 17 million hectares of 'wasteland'. Made possible by national grain surpluses, the program gave farmers subsidies of seedlings, grain and cash for five years for those planting 'economic forests' (e.g. fruit and nut trees), eight years for 'ecological forests' (timber species), which were supposed to account for 80 percent of the afforested areas, and two years for grasslands. In 2004, concerns about dwindling public grain reserves led the government to phase out the grain subsidy, which was substituted by a monetary compensation at a constant rate grain price (Yin and Yin 2010). In 2007, the State Council decided to extend subsidies for a second period, but cut them in half and stopped expanding the scope of the program, in part because of concerns about food security and loss of farmland due to urban land conversion (Jintao Xu *et al.* 2006, 2010, Bennett 2008, L. Zhang *et al.* 2008, Yin and Zhao 2012, C. Liu *et al.* 2013).

The majority of published studies about these two programs are based on analyses of several large-scale surveys and government statistics, rather than on-the-ground field studies, leading to what Yin *et al.* (2010) call an emphasis on impact significance (the impacts of implementation on environmental and social goals) rather than implementation effectiveness. The divergence between the generally optimistic view of program accomplishments found in macro-level overviews of the programs (e.g. Jintao Xu *et al.* 2006, J. Liu *et al.* 2008) versus the problems highlighted in local case studies (Démurger *et al.* 2005, Weyerhaeuser *et al.* 2005, Trac *et al.* 2007, Shen *et al.* 2010) suggests questions about the accuracy of data collected in large-scale surveys.

For example, J. Liu *et al.* (2008) take reported forestry statistics at face value to argue that because of NFPP, carbon sequestration has increased, wildlife habitat has improved and soil erosion has been reduced.[2] By contrast, in a case study in Liangshan Prefecture, Sichuan, Trac *et al.* (2007) report that from field visits, it was difficult to tell if anything had been planted at all in several NFPP afforestation sites. J. Liu *et al.* (2008, 9480) state that SLCP has 'directly benefited 120 million farmers in more than 30 million households nation-wide'. However, case studies of SLCP implementation areas suggest that not all farmers receive subsidies, land chosen is often not sufficiently steep, and plant survival and growth are often poor. Despite these realities, however, counties report successful planting and implementation of both programs, with these statistics then used in further aggregation at the provincial and national levels (Démurger *et al.* 2005, Weyerhaeuser *et al.* 2005, Trac *et al.* 2007, Yeh 2009b).

In addition to insufficient funding for local SLCP offices, lack of coordination between county finance bureaus and forestry bureaus makes it difficult for the latter to confirm that compensation reaches the right farmers (Trac *et al.* 2007). Further, reflecting the bureaucratic structure of China's cadre system, cadres have strong incentives to report exaggerated results rather than actual implementation of NFPP and SLCP, because of the way in which they are evaluated for promotion. Thus, Trac *et al.* (2007) conclude based on their field study that the combination of the incentive system and the lack of adequate funding and coordination sets the programs up for failure that is reported as success, while not taking into account the needs of local farmers and herders. Case studies have also paid attention to loss of agrobiodiversity and associated changes in traditional cultural practices and knowledge as a result of SLCP, issues that have been ignored in larger-scale surveys (Shen *et al.* 2010).

Other types of studies have also reported shortfalls in subsidies delivered, with subsidies diverted from farmer compensation by higher levels of government, and villages with connections to local forestry bureaus able to enroll more land than other villages (Bennett 2008, Jintao Xu *et al.* 2010). Because subsidies are applied uniformly over vast regions (with one subsidy level for the Yangtze River watershed and another for the Yellow River watershed), some households are overcompensated relative to their previous crop production while others are undercompensated. Furthermore, insufficient technical support and insufficient budgeting for proper implementation are both widely reported. However, the literature often presents these as technical, rather than political, issues.

Indeed, a number of studies ignore the rural politics revealed in their own findings, most notably in relation to the frequent observation that, counter to the program's explicit principle of being voluntary, most households report having little or no choice about whether and how to participate (Bennett 2008, Jintao Xu *et al.* 2010). For example Jintao Xu *et al.* (2006, 604) note that 'a majority [of surveyed households] stated that they did not have the right to choose which plots and how much of their cropland to be retired, nor

---

[2]They argue that wildlife habitat has improved through the implementation of the program in nature reserves such as Wolong. However, the species planted as part of the program are generally monocrops (as the authors acknowledge), making wildlife habitat improvement questionable.

the right to select the tree/grass species to be planted'. However, they do not explore how and why this still translates into the reported high satisfaction with the program from a majority of participating households (603), an important question in light of other studies that show that farmers who faced government pressure to participate do not welcome the policy (e.g. C. Wang and Maclaren 2012). Moreover, rather than calling into question the violation of rights as specified by the Household Responsibility System, Xu et al. (2006) only comment that this non-voluntary process means 'local people tend not to plant or [to] maintain the tree and grass properly'. That is, the lack of voluntary participation is interpreted as a technical barrier to achieving project outcomes rather than as a political question of land control. Similarly, Jintao Xu *et al.*'s (2010) primary concerns about the lack of voluntary participation are not with the politics of decision-making behind this, but with its effects on the program's cost-effectiveness and the fact that those who do not participate voluntarily may be more likely to return to cultivation at the end of the subsidy period.

In another study, Groom and Palmer (2012) find, counterintuitively, that compulsory participation has a positive impact on incomes relative to quasi-voluntary participation. However, this impact is not due to increases in off-farm income. At the same time, 'purely' voluntary implementation does a better job at increasing off-farm labor market income – that is, of relaxing labor or liquidity constraints to finding off-farm labor. This, they suggest, 'may render the program unsustainable in terms of its effect on both poverty and environment' (51). Here again sustainability is interpreted in an apolitical fashion. Indeed, in the only article among more than 30 reviewed that explicitly mentioned infringement on rights, Mullan *et al.*'s (2010, 324) study of the NFPP states only, '[the fact that] land use rights have been denied without compensation may have implications in terms of equity or in terms of incentives to manage forest … sustainably in the future'.

Another important area of inquiry is SLCP's performance as a PES (Payment for Environmental Services, or Payment for Ecosystem Services) program. The literature on the SLCP almost invariably stresses its significance by noting that it is not only one of China's first PES schemes but also the largest PES in the developing world, and the largest national-level PES scheme in the world (e.g. X. Wang *et al.* 2007, L. Zhang *et al.* 2008, Chen *et al.* 2009, Gauvin *et al.* 2010, Bennett *et al.* 2011, Deng *et al.* 2011, Li *et al.* 2011, Groom and Palmer 2012, Yin and Zhao 2012, C. Liu *et al.* 2013). PES schemes are often considered fundamentally neoliberal, given their embedded assumption that market-based management will allocate conservation resources more efficiently than command and control regulation by states. However, in implementation, very few PES schemes in the developing world conform strictly to a free market model. As McElwee (2012) finds with PES schemes in Vietnam, and McAfee and Shapiro (2010) in Mexico, the hybrid ways in which they are implemented means that most PES schemes are not strictly neoliberal. Indeed, they may instead recentralize state control of resources, or simply be 'old wine in new bottles' (McElwee 2012, 422 on Vietnam).

China has shown considerable interest in adapting international PES experiences to its circumstances (Scherr *et al.* 2006) but, like other developing country programs, SLCP is far from a purely neoliberal PES. At present, the only buyer is the government, and the prices set for compensation bear no resemblance to a free market pricing mechanism. No attempts have been made to calculate the exact value of ecosystem services. As a result, many critiques of the SLCP program in the literature have been of its neglect of market-based instruments (Jintao Xu *et al.* 2006, J. Liu *et al.* 2008), with the assumption that a more neoliberal approach would produce better results. In the PES typology developed by McAfee and Shapiro (2010), SLCP falls between a pro-market, pro-poor approach and a Compensation for Ecosystem Services approach. Unlike the former, it does not focus primarily on a market mechanism, but, unlike the latter, the scheme does not recognize the fundamentally political nature of conservation, nor does it serve to revalue the countryside.

One of the key tensions in the conceptualization and implementation of PES schemes is whether they should be driven primarily by their utility for efficiency in achieving conservation objectives, or whether they can also achieve benefits for alleviating poverty, as the pro-poor and Compensation for Ecosystem Services approaches would have it (McAfee and Shapiro 2010, McElwee 2012). Despite arguments that PES is most effective and best used for conservation goals rather than for alleviating poverty (Wunder 2008), the SLCP is specifically designed to address both goals (Li *et al.* 2011). Thus, studies have focused on evaluating both its environmental and social impacts.

Perhaps the largest potential problem in the program's attempt to achieve its environmental goals is that the relationship between the cultivation of sloping land upstream and the frequency of flooding downstream is not as clear-cut as presumed, as earlier debates over upstream-downstream relationships in the 'Theory of Himalayan Environmental Degradation' also suggested (Thompson *et al.* 1986, Ives and Messerli 1989, Guthman 1997, Blaikie and Muldavin 2004). Moreover, it is not clear that afforestation, even if successful, is the most effective means to reduce erosion (FAO-CIFOR 2005, Bennett 2008). One study of SLCP in an arid region of Shaanxi province suggested that afforestation there actually led to destruction of natural vegetation and exacerbated water shortages, decreasing vegetation cover, soil moisture and number of species relative to simply prohibiting cultivation and grazing (S. Cao *et al.* 2009a).

By contrast, a number of studies do suggest moderately beneficial local environmental effects. For example, in Wuqi County, Shaanxi, households credited SLCP with leading to less severe and frequent soil erosion, less intense sandstorms, improved air quality, and reduced water runoff and flooding after storms. In Tianquan County, Sichuan, SLCP participants reported decreased landslide frequency and intensity (Bullock and King 2011). Other farm-level studies have found improvement in soil quality, microbes and root ecological niches after conversion from cropland to tree cultivation (Z. Liu *et al.* 2005, Peng *et al.* 2005, J.-Y. Xu *et al.* 2007).

Many other studies, however, reveal 'mistargeting' – the implementation of the program on fertile flatlands rather than steep lands, as well as a failure to adequately consider land productivity and environmental heterogeneity in site selection (X. Wang *et al.* 2007, Jintao Xu *et al.* 2010). Site selection is often determined by ease of inspection, as well as minimization of transaction costs. Adequate implementation funding, which could enable more extensive surveys and thus better targeting, is often not available. That is, many non-environmental factors have been used to determine actual implementation; the implementation of these programs is inescapably political. At the same time, it is important to remember that slope steepness alone is not an adequate measure of ecological benefit; it has been used extensively as a crude proxy because other fine-scale data on ecological and hydrological characteristics are not available (Yin and Zhao 2012). A study in Wolong Nature Reserve showed that even though steepness criteria were met, the species planted for reforestation failed to deliver suitable habitat for the giant panda. Consequently, SLCP did not have its desired biodiversity benefit there (J.-Y. Xu *et al.* 2007).

A number of studies have found tradeoffs not only between the conservation and poverty-alleviation aims of SLCP, but also between farmer autonomy and environmental benefits. One study found that households given the right to choose what to plant had improved survival rates, but also planted more economic rather than ecological forests, against the broader aims of the program. At the same time, giving households the right to choose which plots to retire actually reduced survival rates. Furthermore, households with access to off-farm labor markets – those that would benefit more from the program – also produced worse environmental results (Bennett *et al.* 2011). This suggests a fundamental flaw in the assumption that retiring land is necessary for sustainability; instead, households that are more vested in agriculture are also better positioned to transition to sustainable agro-forestry systems (Bennett *et al.* 2011).

Turning to social impacts, a significant focus of studies to date has been on the question of whether off-farm labor opportunities increase with SLCP. This is important for several reasons. First, alternative income sources are a much more sustainable and long-term source of income generation and poverty relief than the program's temporary subsidies. In places where free time is directed toward animal husbandry, net income increase is marginal compared to the income opportunities from off-farm and migrant labor (Z. Xu *et al.* 2004, Uchida *et al.* 2007, J. Liu *et al.* 2008). Furthermore, having an alternative source of income is the only way in which farmers may be convinced not to return to farming on converted land once the subsidies are over. Farmers without alternative sources of income generally plan to return to cultivating their converted land. For example, L. Zhang *et al.* (2008) found in Ningxia that only 8 percent of households stated they would not reconvert their land to cropland once compensation stopped. Studies have also found that the amount of land that households plan to reconvert to crop cultivation is reduced with higher age and off-farm household income, but raised by the number of household laborers and total amount of household land enrolled in the program (Chen *et al.* 2009).

Findings about the effects of SLCP participation on off-farm labor participation and income vary substantially, with some reporting a negative effect on income (C. Wang *et al.* 2007), others no effect, and others a positive effect for off-farm labor participation and income (Uchida *et al.* 2009, Yao *et al.* 2010). A number of studies show mixed results, for example of positive effects on household income but no increase in labor transfer, and different effects for households at different income levels (Li *et al.* 2011). Remarkably, of all of the studies reviewed examining the factors determining income and labor effects of SLCP, only one mentioned local political leadership, in particular local government commitment to implementation, facilitation in finding off-farm labor opportunities, and transparency (Yao *et al.* 2010). Again, the local political dimensions of SLCP have been largely ignored.

Because subsidies provide liquidity, some studies have found that it is easier for participants than non-participants to switch to off-farm labor, but that the effects depend on the initial level of 'human and physical capital' (Uchida *et al.* 2009). In particular, it is the younger and more highly educated who are able to find off-farm labor. In addition, the more liquidity-constrained the household, the more positive the impact of the subsidy from SLCP for finding off-farm employment. Other studies suggest that the ability to switch to off-farm labor is limited by 'social capital', or the social relations necessary to gain access to information, technology, markets and capital (L. Zhang *et al.* 2008). In other words, economically and politically marginalized communities are unlikely to benefit from the program in the sense of being able to develop long-term alternatives to crop cultivation. Instead, it is likely they will be forced to return to crop cultivation once the program ends, whereupon they may be stigmatized for being unable to 'understand' the importance of ecological restoration or properly value the environment (cf. S. Cao *et al.* 2009b).

In sum, while many studies have rightly examined SLCP as the largest PES in the developing world, much more work remains to be done on the politics of the project: the extent to which the environmental assumptions are justified, its political economic drivers at the national level, how project implementation patterns are related to local governance logics, the implications of altering land rights defined by the Household Responsibility System contracts, and the extent to which the program has contributed to broader rural-urban migration streams. These are important both because of the sheer scale of SLCP as well as China's interest in pursuing a suite of other PES programs including the FECP (Forest Ecosystem Compensation Fund) and a newer Rangeland Ecological Protection Reward Mechanism (*caoyuan shengtai baohu buzhu jiangli jizhi*). There will also, of course, be much to study once the subsidies for the program are completely discontinued.

Nature reserves and biodiversity conservation

As with its forestry programs, China's nature reserve initiatives are large in scale, and fraught with similar contentious politics in their implementation. Like conservation efforts around the world, enclosures in China have sparked various forms of resistance as they expropriate access to livelihood resources. At the same time, various forms of nature reserves are also compromised by political economic pressures to generate income.

After the first nature reserve was declared in the People's Republic of China (PRC) in 1956, nature reserve numbers grew slowly to 19 by 1965, 481 in 1987, and then much more quickly, to over 2000 in 2004 and 2500 in 2007, covering almost 15 percent of China's total land area. By 2010, the country had 2541 nature reserves, 208 national scenic areas and 660 national forest parks (Zinda 2012). The eight largest reserves, all found in the western provinces of Gansu, Qinghai, Xinjiang and the Tibet Autonomous Region (TAR), encompass an area approximately equal to the remaining 2000 and some reserves (Harris 2008, 112).

Most nature reserves in China fall into the International Union for the Conservation of Nature (IUCN)'s Category VI classification of managed protected areas, which are supposed to allow for community use. In addition, most also follow the United Nations (UN) Biosphere model of dividing parks into strictly protected core areas, buffer zones where controlled commercial and subsistence land use is allowed, and research or experimental zones (Harkness 1998, Coggins 2002, Herrold-Menzies 2006b). Though many scholars stress the extent to which communities are a part of PRC nature reserves, in contrast to the Yellowstone 'wilderness' model adopted in Taiwan (Coggins 2002, Herrold-Menzies 2006b, Weller 2006), Harris (2008) notes that the 1994 Nature Reserve Law in fact adopts very strict measures according to which timber harvests, livestock grazing, medicinal plant collection and crop cultivation are completely prohibited within reserve boundaries, and people are strictly prohibited from entering core zones of nature reserves. These draconian measures are immediately followed by a caveat in the law that these strict prohibitions may be superseded by other local laws or administrative regulations. Nevertheless, Harris suggests that the vision expressed in the law, particularly for core zones, is not as dissimilar from federally designated wilderness areas in the US as sometimes believed.

These apparent differences in interpretation of China's nature reserve model lie in the focus on implementation versus intent. As Harris (2008, 114) puts it, the 1994 regulations were aspirational in character, 'reflecting an ideal of biodiversity protection that no local administrator was expected to fulfill'. Indeed as of 2004, between 1.25 and 2.85 million people were believed to be residing within core zones of nature reserves around China. Many protected areas are 'paper parks', with at least one-third lacking staff, a management agency and funding (Harkness, 1998, Jim and Xu 2004). The nature reserve law of 1994 did nothing to remove control of the land under protection from the government that had managed it at the time of reserve declaration; moreover, except for national-level reserves, it failed to provide a guaranteed source of funding for reserve administration and staffing. This leads to perverse incentives in which reserve managers' primary goal has become revenue generation rather than biodiversity conservation. Thus, the manager of one biosphere reserve in Jiangsu was praised as a model for his revenue-generating activities, including conversion of habitat for shrimp farming inside the core zone (Harkness 1998). Poaching of lucrative protected species such as musk deer is often conducted by the well-armed guards hired by reserves to protect them (Harris 1996).

The rush to designate nature reserves of all kinds from the 1990s onward can be attributed in part to China's desire to win recognition as a respectable leader on the global stage, and in part to deregulatory strategies which have allowed local governments to play an active role in their designation – often in the hopes of achieving the administrative status, political rewards and tourist income that can accompany reserves (Jim and Xu 2004). However, the results have not only been parks that exist only on paper, but also many reserves that have little significant biodiversity value, or are too small to be ecologically viable. While failing to have significant ecological benefits, nature reserve declaration may be more effective as a spatial strategy by local governments to seek competitive advantage. Moreover, the existence of protected areas rarely trumps the lucrative opportunities presented by satisfying China's large resource demands. Mining operations have been developed in even the highest-level protected areas, including in Shangri-la and Deqin in Yunnan as well as the Sanjiangyuan Reserve in Qinghai, often by companies that are either state-run or have close ties to highly-placed state officials, and against the wishes of local peasants (Jianchu Xu and Melick 2007).

As in other parts of the world, nature reserves in China have often taken away customary access to resources, despite their tendency to follow a biosphere reserve rather than the Yellowstone model. In China, many reserves were declared on land that had already been allocated to individual households under the Household Responsibility System, depriving residents of access to land they had already been given rights to. After a nature reserve is established, local people are sometimes charged a fee for the right to continue traditional practices (such as cardamom cultivation), negatively impacting their livelihoods (Jianchu Xu and Melick 2007). In other cases, households are resettled, a process often accompanied by failure to fully deliver compensation packages, difficulties establishing new livelihoods, cultural disruption and coercion.

Despite clear evidence that nature reserve ineffectiveness in conserving biodiversity is often political-economic in origin, the tendency to blame local people is strong, both among policy-makers and, sometimes, natural scientists who ignore the political dimensions of the environment. In one widely cited study published in *Science*, J. Liu *et al.* (2001) use time series remote sensing data from Sichuan province's Wolong Nature Reserve, a flagship reserve for pandas that has received extensive financial support, to show that rates of habitat loss and fragmentation accelerated after park establishment, becoming more severe than conditions outside of the reserve. Despite the authors' assessment that the thousands of tourists attracted by the reserve every year have 'significantly stimulated the extraction of natural resources such as fuelwood to produce marketable goods', they nevertheless conclude that 'local people in the reserve were the direct driving force behind the destruction of the forest and the panda habitat' (2001, 100). This familiar narrative blames local people by considering only proximate factors, contradicting the study's own finding of the role played by broader political-economic forces such as those driving the development of tourism.

In some cases, of course, local people are directly responsible for environmental destruction within reserves, whether as direct acts of everyday resistance against the loss of their access to resources, in reaction to a general perception of being left behind by the benefits of development (Zackey 2007), or because formerly effective forms of community management have been replaced by weak state enforcement, creating open-access situations. In his study of the reintroduction and conservation of the South China tiger in three nature reserves in Fujian province, Coggins (2002) discusses the considerable antipathy that local residents harbor toward the reserve because of their lack of participation in the planning process. Coggins shows that, ironically, traditional local practices such as the maintenance of fengshui forests and the burning of the landscape may have benefited tiger habitat compared to more recent policies that have increased bamboo forests.

Herrold-Menzies' (2006a, 2006b, 2009) study of the Caohai Nature Reserve in Guizhou takes a closer look at peasant resistance to the enclosure of livelihood resources. A significant area of Caohai Lake was drained during the collective period to reclaim farmland, which after decollectivization was allocated to individual households. These households were not informed of government plans to restore the lake in the early 1980s, a process which caused some households to lose more than 50 percent of their land, depriving them of access to adequate land to feed their families. Despite this, they were paid no compensation and were still responsible for paying agricultural taxes on the submerged land. To compensate, peasants turned to fishing, trapping waterfowl and reclaiming land, but these were criminalized by the new reserve, leading to violent confrontation between peasants and reserve staff. As a result, reserve staff welcomed a program by the US-based International Crane Foundation to provide microcredit and small grants programs, which, Herrold-Menzies found, greatly improved and transformed the relationship between communities and the nature reserve. In fact, farmers have come to see the reserve, rather than the local government, as their main partner for development. She tracks the transformation of peasant interactions with the reserve from incidents of beating up staff members and attempting to drown them, to a much more peaceful relationship in which local residents have instead submitted peaceful petitions for compensation (Herrold-Menzies 2006a, 2006b, 2009). Thus, the provision of training sessions, employment opportunities and loans in this integrated conservation-as-development program has transformed violent protests against a conservation enclosure to action more along the lines of 'rightful resistance' (O'Brien and Li 2006). The reserve itself has also been transformed, becoming somewhat more responsive to local livelihood needs. At the same time, though violence has been eliminated, peasants still continue various forms of covert non-compliance such as fishing during the spawning season, as a form of income generation and everyday (rather than violent, or 'rightful') resistance (Scott 1985).

These dynamics of reserves as conservation enclosures that reduce rural farmers' and pastoralists' access to resources are common. Another case is the Lhalu wetland in Lhasa, TAR, where state programs created local dependence on natural resources, which were subsequently enclosed (Yeh 2009a). Historically, water levels in this wetland were too high for livestock to graze, and peasants living around the wetland were instead engaged in crop agriculture nearby. In the 1960s, failed attempts to convert the 'wasteland', as the wetland was seen by government officials at that time, to productive agriculture significantly drained the wetland, a process that was furthered in the early 1990s by the building of a canal next to the wetland for 'city beautification'. At the same time, villagers' farmland was expropriated for urbanization, and development projects encouraged villagers to adopt new breeds of cows that they could only graze in the now-drained wetland. Less than a decade later, however, it was enclosed and declared a nature reserve that was to be promoted for tourism, and grazing was prohibited. Not surprisingly, local residents were resentful, particularly because it was various previous state-led transformations of nature that led to their dependence on the wetland for grazing in the first place. They engaged in everyday forms of resistance, including continuing to graze their livestock, as well as stealing and reselling parts of the metal fence that had been constructed around the reserve.

Finally, a new model of conservation – national parks – has been studied since the very first one in China, Pudacuo National Park, was introduced in northwest Yunnan in 2007 through efforts of The Nature Conservancy (Zhou and Grumbine 2011, Zinda 2012, 2014, Moseley and Mullen 2014). Intended as a market-based method to combine conservation with community participation by using tourism revenues to benefit both rural residents and the environment, the national park quickly became a site of contestation, as government units with divergent mandates competed for prestige and control. Zinda (2012, 2014) traces how a succession of shifting alliances emerged around efforts to establish Pudacuo and other parks. Local governments competing to expand tourism economies adopted the title 'national park' for upgraded attractions, but prioritized high-volume tourism and lagged on the active conservation management and resident involvement recommended in initial proposals. Line agencies, on the other hand, tried to acquire organizational turf. Pudacuo National Park has ended up becoming an important source of revenue for local governments, but local residents receive less than 3 percent of park revenues and are generally dissatisfied with the results. Thus, though the national park model was introduced as a new conservation model with aspirations of addressing many flaws in China's existing nature reserves, like PES programs for forestry, it has suffered many of the same problems, stemming from fundamental differences in power and inequality. Efforts to reforest and to conserve biodiversity are political processes that have significantly altered rural residents' access to livelihood resources, while political-economic pressures, bureaucratic structures and the system of incentives for cadres and local governments often prevent significant achievement of their environmental goals.

Pastoral politics

While nature reserves have been declared on 15 percent of China's land area, grasslands account for about 42 percent of China's total territory. Thus, the territorial importance of China's pastoral regions is difficult to overstate (Brown *et al.* 2008). The 266 pastoral and semi-pastoral counties of the PRC, mostly located in the five key pastoral provinces of Inner Mongolia, the TAR, Qinghai, Xinjiang and Gansu, are home to 161.5 million people, who herd the world's largest population of sheep and goats, as well as the fourth largest concentration of cattle (Williams 1996, J. Liu 2010). Attempts to reverse rangeland degradation and modernize pastoral production have been the key drivers of rural politics in these pastoral areas over the last two decades. With the implementation of the Household Responsibility System in agricultural areas in the early 1980s, livestock were decollectivized in pastoral areas. The return to the household as the basic unit of production while pasture remained in common use made the ensuing system not unlike traditional common property arrangements before collectivization in the 1950s (Yeh and Gaerrang 2011). However, after the passing of the National Grassland Law in 1985, rangeland use rights also began to be allocated to individual households (Williams 1996, Ho 2000b). Privatization of rangeland use rights began in Inner Mongolia, followed by other pastoral areas such as Ningxia, Xinjiang, and the Tibetan Plateau. On the Tibetan Plateau, eastern Tibetan areas saw implementation in the 1990s, while it began in the TAR only in the 2000s.

The privatization of grassland use rights has been driven in large part by concerns about severe grassland degradation, purportedly caused by overgrazing. Harris (2010) carefully examines the claim, first articulated in 1997 and taken up by the State Council in 2002, that 90 percent of China's grasslands are severely degraded. Searching for the basis of these claims, he finds that the statistics derive from 'undocumented surveys conducted by local-level staff of grassland and livestock bureaus, using criteria originally envisioned for the allocation of pasture following de-collectivization', without use of a baseline and based only on a single quantitative measure (Harris 2010, 3). There is little question of overgrazing and sometimes very severe degradation in localized areas. However, rigorous attempts to quantify the national extent of rangeland degradation have had ambiguous results that further call into question the 90 percent figure. Moreover, there is significant counter evidence from place-based studies in which local herders' knowledge and ecological assessments contradict the narrative of pervasive degradation (Holzner and Kreichbaum 2000, Goldstein and Beall 2002, Yundannima 2012).

Examining official statistics further, Yundannima (2012) found numerous contradictions both within and between reports, which further call into question the credibility of degradation claims. Indeed, an official at the Grassland Monitoring and Supervision Center under the Ministry of Agriculture acknowledged in 2009 that the 90 percent figure for degradation, and other frequently cited figures, are only guesses that are 'neither based on uniform, commonly recognized criteria, nor results of detailed ground surveys and continuous ground observations, lacking scientific validity' (J. Liu 2010 in Yundannima 2012). Despite this, the assertion of pervasive degradation persists in policy statements as well as scientific papers, particularly, though not exclusively, those originating in China (Zhu and Li 2000, Zeng *et al.* 2003, Economy 2004, Ren *et al.* 2007, T. Zhang 2007).

This has led to several studies of the 'politics of knowledge' (see Tilt 2013) about grassland degradation. These studies find that severe and pervasive degradation on China's rangelands has become a kind of 'received wisdom', a narrative that blames local people for environmental degradation in the absence of adequate evidence, which is often used to justify certain interventions, and which is repeated so many times that it becomes common sense within certain scientific and policy communities (Yundannima 2012). Discussing the 'theory of Himalayan environmental degradation', Blaikie and Muldavin (2004) trace the reproduction of perceived wisdom in China to disjunctures between epistemic communities of social and natural scientists, as well as between those working within the Chinese national context and other national contexts, who 'write and read for different journals, speak different languages', have different conceptualizations of sound research and effectively see different landscapes (2004, 541). Also examining the disjunctures between different epistemic communities, Williams (2000, 2002) argues that international, national and local scales of natural scientific practice work together to privilege non-local representations of nature, and that grassland science in Inner Mongolia ultimately functions to reproduce unequal social relations. Remarking on a different epistemic divide, Xu Jun (2010) makes an oblique reference to the highly politicized nature of resettlement policies implemented to remedy degradation (discussed below). She notes, 'western scholars are arguing about the various reasons or goals of China's central government's [policies, while] most Chinese scholars are paying more attention to the harsh living conditions of eco-immigrants', a statement that points to the fraught politics of framing questions about rangeland management in China.

Along with overgrazing-induced degradation, household use rights privatization and other subsequent policies have also been based on an assumption of the tragedy of the commons, the belief that only privatized land-use rights can provide an adequate incentive for households to manage their livestock without causing rangeland degradation, by making herders responsible for matching herd sizes to rangeland resources and for investing in improvements for sustainable management (Ho 2000a, Banks *et al*. 2003, Z. Yan and Wu 2005, Harris 2010). However, since the passing of the National Grassland Law of 1985, implementation of use rights privatization has varied significantly from place to place. For example, in many parts of Xinjiang, use rights were theoretically allocated to the level of the household, but management is often still at the level of groups of households or the village (Banks 2001, 2003, Zukosky 2008). In Machu County of Gansu province, too, use rights were allocated to individual households, but pastures are still often used in common by groups of households. In Nagchu, TAR, the rangeland household responsibility system was not implemented until the 2000s. In many villages, management and use is at the scale of the village rather than the individual household (Banks *et al*. 2003, Richard *et al*. 2006, J. Cao *et al*. 2011a, 2011b, Yundannima 2012).

The rangeland household responsibility system and its accompanying fencing, used for both reserve grass and for demarcating boundaries, have been the subject of different analytical lenses. Some scholars have stressed the governmental logics of fencing, privatization and other more recent interventions (Williams 2000, 2002, Yeh 2005). Others are more focused on the differences between policy and actual on-the-ground arrangements. For example, observing that pastoralists in Xinjiang's Altai Prefecture dismiss the household contracts as being only 'on paper...a kind of writing', Zukosky (2008, 44) suggests that rather than indexing actual patterns of use, the grassland allocation certificates are a kind of poetic form that constitutes the state, producing its 'unqualified identity as a disinterested, developmental state with sovereign power, not a constantly negotiated and contingent representation by a diversity of interested actors' (57). Others simply note that variation results from a lack of strict implementation of the provisions of the National Grassland Law (Richard *et al*. 2006).

Yundannima (2012, 67–70) argues against this interpretation of 'lack of strict implementation'. Instead, he suggests that heterogeneity on the ground is due to ambiguity in the policy framework. In fact, the National Grassland Law of 1985, the Amended Grassland Law of 2002 and the Land Administration Law of 2007 all suggest that grazing land may be contracted either to collectives or individuals, whereas the Rural Land Contract Law of 2002, Property Law of 2007 and a document issued by the State Council in 2002 state that it should be contracted only to individual households. It is this legal ambiguity, Yundannima (2012) suggests, as well as the difference between the allocation of use rights versus actual use patterns (about which the laws are silent), rather than poor implementation that accounts for variation.

Despite this legal ambiguity, however, there has been a strong tendency for local and regional governments to interpret the policies as a mandate to limit land-use rights to the scale of individual households. Even in areas where group tenure was initially established, pressure from local governments, as well as other factors, have produced a trend toward household privatization, sometimes despite herders' stated preferences for group management. The movement of pastoralists to town, either as part of government resettlement programs or to facilitate access to education for children, encourages households to take individual leases so that they may then rent out their land to others (J. Cao 2010, Yeh and Gaerrang 2011).

The resulting division of land and accompanying boundary fencing has led to many documented problems, including household inequality, as rich households that can afford to buy barbed wire fences started enclosing more land than allocated, thus increasing grazing pressure on unfenced land. Favoritism in land allocation, the product of local politics, as well as the sheer difficulty if not impossibility of equitably allocating a patchy and heterogeneous resource, has led to inequalities in terms of households' access to both good pastures and sources of water. This in turn has led to increased household labor and economic burdens, as well as rangeland conflicts (Williams 1996, N. Wu and Richard 1999, Williams 2002, Yeh 2003, Z. Yan and Wu 2005, Z. Yan *et al*. 2005, Zukosky 2008, J. Cao *et al*. 2011b). Moreover, some studies have suggested that the rangeland contract system and its enclosures have increased, rather than decreased, grassland degradation by inhibiting the mobility and flexibility needed for pastoral production, and by concentrating livestock grazing and trampling near settlements (Miller 2000, Z. Yan and Wu 2005, Z. Yan *et al*. 2005, Sheehy *et al*. 2006, Taylor 2006, Z. Wu and Du 2008). In addition, household tenure makes it more difficult for herders to access resources during periodic events such as drought or severe snowstorms, through long-distance migration (known as Otor in Inner Mongolia), increasing herders' vulnerability to livestock loss during such events (Xie and Li 2008, Klein *et al*. 2011).

Despite these problems, the scope of fencing has been expanded by a new policy, *tuimu huancao* (translated as 'converting pastures to grasslands' or 'retire livestock, restore pastures'), which calls for grazing bans in periods ranging from months to 10 years. In the TAR, the effects of the policy have largely been to strengthen the implementation of the rangeland household responsibility system and increase fencing between villages and seasonal pastures. In his study of *tuimu huancao* implementation in Nagchu, Yundannima (2012) finds that in practice, local officials tend to implement the policy in order to satisfy pressures from above rather than because they believe the policy will improve rangeland conditions. As a highly visible and easily quantified target, fencing is an easy way for local government officials to demonstrate their achievement of policy goals (see also Bauer 2005). In addition, fencing is favored by local officials in their efforts to 'chase projects' (*pao xiangmu*) as a way to capture state subsidies (Bauer 2005, Yundannima 2012). Because cadre performance is evaluated based on the presence or absence of fencing on inspection visits, the policy has turned largely into one of fencing installation rather than attention to grassland conditions, and in some places the outcome of this logic is that pastoralists are assigned to patrol fencing material, demonstrating how the political economy of local budgets in contemporary China intersects with environmental protection efforts (Yeh 2009b).

The *tuimu huancao* policy has taken on quite a different meaning and significance in the Sanjiangyuan, or Source of the Three Rivers (Yangtze, Yellow and Mekong) region of Qinghai province. There, it has been implemented together with ecological migration, a program that subsidizes herders to sell their livestock and move to resettlement areas of varying distance away from their former homes, for a period of 10 years (Foggin 2008). According to government reports, some 50,000–60,000 herders have already moved, with plans for a total of 100,000. Studies of this policy have highlighted different types of politics. Some have focused on the rationales and logics of ecological security, internal territorialization and governmentality that underlie the program (Yeh 2005, 2009b, Dell'Angelo 2007, Cencetti 2010), while others have reported findings that highlight local agency, for example, noting that most of those who choose to migrate are those with very few livestock in the first place, or those who are very wealthy and want to move into town to pursue business opportunities (Du 2009, 2012).

Regardless, all studies find that those who resettle face significant challenges. Lack of means to generate a livelihood in resettlement areas has been perhaps the biggest immediate obstacle, coupled with deeper questions of identity, and cultural and linguistic continuity. Because the former pastoralists usually do not have the skills to find employment in town, they become dependent on subsidies, leading to serious questions about what will happen once the subsidies expire. Unlike the SLCP, where farmers can easily reconvert their forested land to cropland, it will be much more difficult for displaced herders separated from their land to return to an entire way of life after a decade away. Another problem is that resettled herders' cash expenditures increase significantly. The subsidies are often insufficient to cover their expenses, leading to a decline in living standards and, sometimes, health conditions. Social problems have also emerged, as settlements are nicknamed 'robber villages' for their reputations among urban residents of becoming places of theft (Dell'Angelo 2007, Foggin 2008, 2011, Sonamkyid 2008, Du 2009, 2012, Wende 2009, Jun Xu 2010).

As a result, pastoralists' agentive maneuvers to maintain their livelihoods have largely worked against the intent of the policy to reduce grazing pressure and move herders off the land. In many cases, pastoralists who participate in the program leave their livestock in the care of relatives rather than selling them, and return to the rangelands to help their relatives as hired labor. Resettled pastoralists also return to the pastures to pick yak dung for fuel or harvest caterpillar fungus, or simply abandon the new settlements and move back to the grasslands (Du 2009, Jun Xu 2010). Moreover, land that is supposed to be 'retired' from grazing often continues to be used by villagers who have not moved. In sum, multiple forms of politics permeate the ecological migration program, from the way it works as a technology to govern unruly minority population to the everyday politics of local government implementation (or lack thereof), and herders' agency in maneuvering around the obstacles it poses to the reproduction of livelihoods and culture.

Contentious environmental politics or ethnic politics?

Because of the coincidence of China's rangelands with its minority populations, the work on pastoralism reviewed above has not found a salient place in the broader literature on rural politics. Similarly, environmental protests that take place in China's minority areas have tended to be treated by both state authorities and scholars as minority issues, rather than part of the larger repertoires of contentious politics available to rural Chinese citizens, such as protests in response to the land seizures of China's 'new enclosure movement' (see Le Mons Walker 2008, Hsing 2010, Sargeson 2013) or everyday resistance against nature reserves (Herrold-Menzies 2006a, 2006b, 2009). For state authorities, labeling protests as 'splittist' (separatist) threats to the nation-state is a highly effective way to quell protests, a political tool that deploys inter-ethnic 'harmony' as a coercive management device (see Bulag 2002). Scholars rightly see that ethnic identity and sometimes struggles over territory play a role in protests in minority areas. However, class and ethnic identity are not either/or issues that can be cleanly disentangled. Rural environmental politics in minority areas, like those in Han majority areas, are fundamentally about access to and control over resources. Thus, a robust study of rural politics of China should include them as well.

These dynamics are notable particularly in cases of mining across the Tibetan Plateau, considered by many local residents to be the gravest environmental problem they face, but the one which they can do least about given the powerful array of interests that benefit from mineral extraction. Virtually all local residents around the Amnye Machen mountain range in Golog, Qinghai, one of the most sacred mountains across the Plateau, are opposed to the copper mining that has been taking place since the mid-2000s. Tibetans believe that mining sacred mountains leads to diseases and disasters as well as a general decline in environmental quality (or what they call the 'nutrition' of the soil). Pastoralists in the area state that water pollution from the mining has caused livestock deaths as well as human illnesses. Local government officials are also said to be unhappy about the mine, operated by a mining conglomerate with a poor environmental record, including a large accident in another operation in Fujian province. However, local leaders are unable to oppose it given its backing by a high-level central leader or, perhaps, according to the suspicions of local pastoralists, to a relative of Hu Jintao's. In 2011, 600 students from the local minority Normal School marched 16 kilometers in protest of the mining. In response, the Central Government ordered an investigation according to the logic that this could not be just an environmental protection issue, but rather must have had some other ethnic motive behind it. An investigation subsequently uncovered some papers printed with the statement 'You must speak Tibetan', in reference to a movement that has spread across the Plateau since 2009 among Tibetans to speak pure Tibetan (rather than the code-switching with Chinese, as has become increasingly common). Government officials declared that telling other Tibetans that they must speak Tibetan was destroying unity. Further, they declared it the political act that was the true cause of the demonstration, denying the validity of local dissatisfaction with mining and its effects on water quality, human health and livelihoods.

Another case began in February 2011 near the village of Abin on the western side of Khawa Karbo, another very important sacred mountain that straddles the provincial border between Yunnan and the TAR. A Chinese mining company negotiated with local officials, but not residents, to open a gold mine, against the wishes of farmers living nearby (K. Yan 2012). When villagers attempted to negotiate with the company, they were met with harassment, attacks and death threats. In response, villagers pushed some 300,000 USD worth of mining equipment into the Nu River. Subsequently, women and children fled to other villages to escape violent retaliation. In a common strategy in rural China of appealing to higher levels of authority, the villagers wrote an open letter to higher levels of government listing their grievances and asking for justice. In addition to detailing the history of the conflict, they point out that the region lies in the United Nations Educational Scientific and Cultural Organization (UNESCO) 'Three Parallel Rivers' World Heritage Site and is a biodiversity hotspot; thus, the villagers state, they have a duty to protect the area. The letter ends in a familiar mode of appealing to the rectitude of higher levels of government, stating 'We love the country, trust the Party and government'. After several more confrontations, the mining company boss reportedly fled and an official ordered the mine shut. This rare apparent village success in protest was reported by a young Beijing-based environmental activist, who happened to come across the case in Tibet while conducting other research, to an international network concerned with sacred lands. However, her optimism proved premature. The miners soon returned in force and, as of the summer 2012, no resolution had been reached.

In most respects, this is a very familiar story of contentious environmental politics: rural protests against environmental damages, an appeal to justice from higher levels of government, a violent response from local state authorities and the miners, and a violent counter-response by villagers. As Le Mons Walker notes (2008, 476), rural outbursts are triggered throughout China by 'incidents in which officials, the newly wealthy or even minor state employees acted with contempt or brutality toward … peasants'. The main difference here was the response that met the young Chinese environmental activist who subsequently tried to publicize the case through Beijing's environmental journalists and non-governmental organizations (NGOs). Not only was she surprised to find her phone tapped, but also that none of the NGOs were willing to touch the case because of its location in the TAR where, they felt in contrast to other contentious cases they had taken on, there was no chance at all of making a difference. Instead, they themselves risked being accused of being accomplices to 'splittism'.

Finally, the case of the Voluntary Association for the Protection of the Environment of Domed Anchung Sengge Namzong is illustrative of how environmental – and other forms of rural politics – are effectively quelled with the label of ethnic politics. This grassroots association was formed in 2003 in a remote, rural area of Chamdo in the eastern TAR by a local organic intellectual, Rinchen Samdrup. Through his interest in the environment, sparked by witnessing deforestation and littering on the local sacred mountain, as well as through a visit by a well-known environmentalist from Qinghai, the more than 1300 adult residents of a cluster of 11 hamlets decided to organize themselves into a voluntary association and drew up a list of rules against hunting and fishing. Over a period of six years, they undertook a large number of environmental protection activities, including planting over half a million trees, monitoring wildlife, patrolling the valley against poachers and fishers, and cleaning up garbage. In addition, they also organized community environmental education events at the annual festival, established a Tibetan language environmental website about their group, and produced an annual Tibetan language journal about the environment, drawing from translations into Tibetan of major environmental news in China and around the world, Chinese national environmental laws, statements by Chinese leaders on the environment, passages from Buddhist and Bön texts, as well as essays by villagers themselves. For its remarkable work, the group caught the attention of environmentalists in Beijing, and won several major national environmental prizes.

However, in 2009, both Rinchen Samdrup and his younger brother, who coordinated most day-to-day activities of the rural group while Rinchen was often in Lhasa, were detained. His brother was sentenced without trial to 21 months in a labor camp on charges of having endangered state security by having 'illegally compiled three discs of audio-visual materials on the ecology, environment, natural resources and religion of Chamdo Prefecture', illegally possessing materials from 'the Dalai clique abroad' and for 'supplying photographs and material for an illegal publication', a reference to their Tibetan-language environmental protection journal. Furthermore, he was accused of breaking the law by assisting Rinchen in applying for registration for their environmental NGO. Rinchen was tried and sentenced the following year to five years of imprisonment on charges of 'incitement to split the nation' for an oblique reference to the Dalai Lama that appeared on their website, in an article that he denied in court having posted.

What actually transpired appears to have been a relatively mundane rural conflict spinning out of control as the villagers sought to assert their rights through appeals to higher levels of government, leading eventually to their environmental activism being labeled 'splittism'. The villagers had a long-standing and mutual antipathy for the Tibetan county head of the Public Security Bureau, who was from a neighboring area with a long-standing land dispute with the hamlets of the association. When the villagers began to prevent outsiders from poaching, this provoked the ire of the official, who reportedly beat villagers for trying to stop outsiders from hunting, and who further sought to provoke them by coming with other cadres to hunt. Villagers tried, unsuccessfully, to take this official to court and, when this failed, they eventually petitioned higher levels of government, including an attempt to petition in Beijing. This angered not only the county Public Security Bureau official and his backers in the prefecture, but also regional-level leaders. Ultimately, the labeling of the group as splittist and as aligned with the 14th Dalai Lama (despite substantial evidence to the contrary, and despite the group's many attempts to state their loyalty to the Party and state) proved a remarkably easy way to both disband and punish them.

Conclusion

In this review of recent scholarship on China's rural 'green' environmental issues, I have argued for expanding our view of rural politics to consider the governance of forests, nature reserves and grasslands, and environmental activism in minority areas. Much of the currently available literature on forest conservation accepts as givens the goals and forms of China's current 'greening' efforts, focusing on technical difficulties rather than the politics of redistributing peasants' access to land resources or the political-economic factors shaping their environmental and social effects. Given the global rise of PES schemes and China's embrace of programs including not only NFPP and SLCP but also new forms of REDD+ (Reducing Emissions from Deforestation and Forest Degradation) forestry programs, much more research is needed on the politics of these programs, from their governmental logics to the question of rights to land.

Like current work on forestry, much literature on biodiversity conservation in China is also highly managerial and technocratic in orientation. However, there is also a more significant body of work from the perspective of political ecology, which has demonstrated that the problems that are to be solved by nature reserves are often created by previous government interventions, such as through the draining of wetlands for agriculture during the collective period. However, it is local people who get blamed for contemporary environmental problems, and it is their access to resources that is generally taken away by conservation enclosures. As with the conservation of forests and grasslands, the governance of nature reserves is heavily shaped by scrambles for funding at the local level. On the one hand, there is rarely sufficient funding to carry out projects as envisioned. On the other hand, local government officials seek to grab projects – whether funding for SLCP or fencing from *tuimu huancao*, or the declaration of nature reserves as a form of place branding and distinction – to capture benefits from higher levels of government for themselves.

In all three sectors of conservation, of forests, nature reserves and grasslands, there are significant questions to be asked about the politics of knowledge with regard to degradation. Areas zoned or labeled as degraded by state authorities and restoration programs may not be degraded in the eyes and knowledge of local rural residents. In some cases, these labels enable interventions that further exacerbate environmental problems or that seek to alter property rights toward a more market-driven system. In grassland areas, tragedy of the commons assumptions combine with exaggerated claims of degradation to promote household boundary fencing, which many studies have shown increases degradation by concentrating grazing and trampling, while reducing the mobility and flexibility necessary for successful and sustainable pastoralism. In all three resource sectors, too, China is moving toward quasi-neoliberal models of governing nature, including Payment for Environmental Services, national parks and privatization of use rights. However, these efforts thus far have often turned out to be 'more of the same', because of entrenched top-down models of implementation, political economic incentives (tourism revenue as a drive for national parks, for example) and legal ambiguity. This is, of course, not an argument that truly neoliberal models would have better results. Instead, it suggests that the push in that direction both from scholars and policy-makers in China will no doubt produce new projects whose assumptions and effects require careful research.

Whereas afforestation in particular has been dominated by economics and remote sensing approaches, work on pastoralism has been more balanced with a significant body of political ecology and common property-theory informed work. However, this has not been in dialogue with mainstream agrarian studies in part because of pastoral regions' location on the (very large) peripheries of the PRC, mostly inhabited by ethnic minorities.

I have argued that class and ethnic issues cannot be separated in the study of rural politics in China. This is particularly important because the state apparatus itself often uses ethnic labels to dismiss and repress environmental protests and problems through discourses of state sovereignty. The exclusion of such issues in the literature mirrors and reinforces these rural peoples' exclusion from political repertoires of resistance available to other citizens, a lesson for agrarian studies scholarship that is surely not limited to China.

References

Banks, T. 2001. Property rights and the environment in pastoral China: evidence from the field. *Development and Change*, 32, 714–40.

Banks, T. 2003. Property rights reform in rangeland China: dilemmas on the road to the household ranch. *World Development*, 31(12), 2129–42.

Banks, T., C. Richard, Li Ping and Zhaoli Yan. 2003. Community-based grassland management in Western China: rationale, pilot project experience and policy implications. *Mountain Research and Development*, 23(2), 132–40.

Bauer, K. 2005. Development and the enclosure movement in pastoral Tibet since the 1980s. *Nomadic Peoples*, 9(1), 53–83.

Bennett, M. 2008. China's sloping land conversion program: institutional innovation or business as usual? *Ecological Economics*, 65(4), 699–711.

Bennett, M., A. Mehta and Jintao Xu. 2011. Incomplete property rights, exposure to markets and the provision of environmental services in China. *China Economic Review*, 22, 485–98.

Blaikie, P. and J. Muldavin. 2004. Upstream, downstream, China, India: the politics of environment in the Himalayan region. *Annals of the Association of American Geographers*, 94(3), 520–48.

Brown, C., S.A. Waldron and J. Longworth. 2008. *Sustainable development in Western China: managing people, livestock and grasslands in pastoral areas*. Cheltenham: Edward Elgar.

Bulag, U. 2002. *The Mongols at China's edge: history and the politics of national unity*. Boulder: Rowman and Littlefield.

Bullock, A. and B. King. 2011. Evaluating China's Slope Land Conversion Program as sustainable management in Tianquan and Wuqi Counties. *Journal of Environmental Management*, 92, 1916–22.

Cao, J. 2010. *Research on grassland management on the Qinghai-Tibetan Plateau* (in Chinese). Lanzhou: Lanzhou University Press.

Cao, J., N.M. Holden, X-T. Lü and Guozhen Du. 2011a. The effect of grazing on plant species richness on the Qinghai-Tibetan Plateau. *Grass and Forage Science*, 66(3), 333–6.

Cao, J., You-cai Xiong, Jing Sun, Wan-Fang Xiong and Guozhen Du. 2011b. Differential benefits of multi- and single-household grassland management patterns in the Qinghai-Tibetan Plateau of China. *Human Ecology*, 39(2), 217–27.

Cao, S., Li Chen and Xinxiao Yu. 2009a. Impact of China's Grain for Green Project on the landscape of vulnerable arid and semi-arid agricultural regions: a case study in northern Shaanxi Province. *Journal of Applied Ecology*, 46, 536–43.

Cao, S., C. Xu, L. Chen and X. Wang. 2009b. Attitudes of farmers in China's northern Shaanxi province towards the land-use changes required under the Grain for Green Project, and its implciations for the project's success. *Land Use Policy*, 26, 1182–94.

Cencetti, E. 2010. Tibetan Plateau grassland protection: Tibetan herders' ecological conceptions versus state policies. *Himalaya*, 30(1–2), 39–50.

Chen, X., F. Lupi, G. He, Z. Ouyang and J. Liu. 2009. Factors affecting land reconversion plans following a payment for ecosystem service program. *Biological Conservation*, 142, 1740–7.

Coggins, C. 2002. *The tiger and the pangolin: nature, culture and conservation in China*. Honolulu: University of Hawaii Press.

Dell'Angelo, J. 2007. *The Sanjiangyuan environmental policy and the Tibetan nomads' last stand: a critical political ecology analysis*. MS thesis. London School of Economics and Political Science.

Démurger, S., M. Fournier and G. Shen. 2005. Forest protection policies: national guidelines and local implementation in northern Sichuan. *China Perspectives*, 59, 2–14.

Deng, H., P. Zheng, T. Liu and X. Liu. 2011. Forest ecosystem services and eco-compensation mechanisms in China. *Environmental Management*, 48, 1079–85.

Du, F. 2009. From eco-refugees to eco-migrants: a research in Maduo County in the source area of the Yellow River (in Chinese). *Chinese Academy of Social Sciences*, (unpublished draft).

Du, F. 2012. Ecological resettlement of Tibetan herders in the Sanjiangyuan: a case study in Madoi County of Qinghai. *Nomadic Peoples*, 16(1), 116–33.

Economy, E. 2004. *The river runs black: the environmental challenges to China's future*. Ithaca: Cornell University Press.

Edmonds, R.L. 1999. The environment in the People's Republic of China 50 years on. *The China Quarterly*, 159, 640–9.

FAO-CIFOR (The Food and Agriculture Organization of the United Nations - Center for International Forestry Research). 2005. *Forests and floods: drowning in fiction or thriving on facts?* Bogor, Indonesia: RAP Publication.

Foggin, J.M. 2008. Depopulating the Tibetan grasslands: national policies and perspectives for the future of Tibetan herders in Qinghai province, China. *Mountain Research and Development*, 28(1), 26–31.

Foggin, M. 2011. Rethinking 'ecological migration' and the value of cultural continuity: a response to Wang, Song and Hu. *Ambio*, 40, 100–1.

Gauvin, C., E. Uchida, S. Rozelle, J. Xu and J. Zhan. 2010. Cost-effectiveness of payments for ecosystem services with dual goals of environment and poverty alleviation. *Environmental Management*, 45, 488–501.

Goldstein, M.C. and C. Beall. 2002. Changing patterns of Tibetan nomadic pastoralism. *In*: W.R. Leonard and M.H. Crawford, eds. *Human biology of pastoral populations*. Cambridge: Cambridge University Press, pp. 131–50.

Groom, B. and C. Palmer. 2012. REDD+ and rural livelihoods. *Biological Conservation*, 154, 42–52.

Guthman, J. 1997. Representing crisis: the theory of Himalayan environmental degradation and the project of development in post-Rana Nepal. *Development and Change*, 28, 45–69.

Harkness, J. 1998. Recent trends in forestry and conservation of biodiversity in China. *The China Quarterly*, 156, 911–34.

Harris, R. 1996. Approaches to conserving vulnerable wildlife in China: does the color of cat matter – if it catches mice? *Environmental Values*, 5, 303–34.

Harris, R. 2008. *Wildlife conservation in China: preserving the habitat of China's wild west*. Armonk, NY: M.E. Sharpe.

Harris, R. 2010. Rangeland degradation on the Qinghai-Tibetan Plateau. *Journal of Arid Environments*, 74, 1–12.

Herrold-Menzies, M. 2006a. From adversary to partner: the evolving role of Caohai in the lives of reserve residents. *Canadian Journal of Development Studies*, 23(1), 39–50.

Herrold-Menzies, M. 2006b. Integrating conservation and development: what we can learn from Caohai, China. *The Journal of Environment and Development*, 15, 382–406.

Herrold-Menzies, M. 2009. Peasant resistance against nature reserves. *In*: You-Tien Hsing and Ching Kwan Lee, eds. *Reclaiming Chinese society: politics of redistribution, recognition and representation*. New York: Routledge, pp. 83–98.

Ho, P. 2000a. China's rangelands under stress: a comparative study of pasture commons in the Ningxia Hui Autonomous Region. *Development and Change*, 31, 385–412.

Ho, P. 2000b. The clash over state and collective property: the making of the rangeland law. *The China Quarterly*, 161, 240–63.

Holzner, W. and M. Kreichbaum. 2000. Pastures in south and central Tibet (China): methods for a rapid assessment of pasture conditions. *Die Bodenkultur*, 51, 259–66.

Hsing, Y. 2010. *The great urban transformation: politics of land and property in China*. New York: Oxford University Press.

Ives, J. and B. Messerli. 1989. *The Himalayan dilemma: reconciling development and conservation*. London: Taylor & Francis Group.

Jiang, H. 2005. Grassland management and views of nature in China since 1949: regional policies and local changes in Uxin Ju, Inner Mongolia. *Geoforum*, 36, 651–3.

Jim, C.Y. and S.W. Xu. 2004. Recent protected-area designation in China: an evaluation of admistrative and statutory procedures. *Geographical Journal*, 17(1), 39–50.

Kerkvliet, B. 2009. Everyday politics in peasant societies (and ours). *Journal of Peasant Studies*, 36 (1), 227–43.

Klein, J.A., E.T. Yeh, J. Bump, Y. Nyima and K. Hopping. 2011. Coordinating environmental protection and climate change adaptation policy in resource-dependent communities: a case study from the Tibetan Plateau. *In*: J.D. Ford and L. Berrang-Ford, eds. *Climate change adaptation in developed nations: from theory to practice*. New York: Springer, pp. 423–38.

Le Mons Walker, K. 2008. From covert to overt: everyday peasant politics in China and the implications for transnational agrarian movement. *Journal of Agrarian Change*, 8(2–3), 462–88.

Li, J., M. Feldman, S. Li and G. Daily. 2011. Rural household income and inequality under the sloping land conversion program in western China. *Proceedings of the National Academy of Sciences*, 108(19), 7721–6.

Liu, C., Sen Wang, Hao Liu and Wenqing Zhu. 2013. The impact of China's priority forest programs on rural households' income mobility. *Land Use Policy*, 31, 237–48.

Liu, J. 2010. Paying attention to pastoral issues, accelerating pastoral development (*zhongshi sanmu wenti, jiakuai muqu fazhan*) [online]. Grassland Monitoring and Supervision Center, Ministry of Agriculture. Available from: http://www.grassland.gov.cn/Grassland-new/Item/2361.aspx. [Accessed February 24 2013].

Liu, J., M. Linderman, Z. Ouyang, Li An, J. Yang and Zhang H. 2001. Ecological degradation in protected areas: the case of Wolong nature reserve for giant pandas. *Science*, 292, 98–101.

Liu, J., S. Li, Z. Ouyang, C. Tam and X. Chen. 2008. Ecological and socioeconomic effects of China's policies for ecosystem services. *Proceedings of the National Academy of Sciences*, 105(28), 9477–82.

Liu, Z., H. Tian and J. Zhang. 2005. Analysis on communities of soil microbes under different models of forest rehabilitation. *Journal of Nanjing Forestry University (Nature Sci Ed)* (in Chinese), 29 (4), 45–8.

McAfee, K. 1999. Selling nature to save it? Biodiversity and the rise of green developmentalism. *Environment and Planning D: Society and Space*, 17(2), 133–54.

McAfee, K. and E. Shapiro. 2010. Payments for ecosystem services in Mexico: neoliberalism, social movements and the state. *Annals of the Association of American Geographers*, 100(3), 579–99.

McElwee, P.D. 2012. Payments for environmental services as neoliberal market-based forest conservation in Vietnam: panacea or problem? *Geoforum*, 43, 412–26.

Miller, D. 2000. Tough times for Tibetan nomads in western China: snowstorms, settling down, fences and the demise of traditional nomadic pastoralism. *Nomadic Peoples*, 4(1), 83–109.

Moseley, R.K. and R.B. Mullen. 2014. Contesting authenticity, conserving biological and cultural diversity: The Nature Conservancy in Shangrila. *In*: E.T. Yeh and C. Coggins, eds. *Mapping Shangrila: Contested Landscapes in the Sino-Tibetan Borderlands*. Seattle: University of Washington Press.

Muldavin, J. 1996. The political ecology of agrarian reform in China: the case of Heilongjiang province. *In*: R. Peet and M. Watts, eds. *Liberation ecologies: environment, development, social movements*. New York: Routledge, pp. 227–59.

Muldavin, J. 1998. Agrarian change in contemporary rural China. *In*: I. Szelenyi, ed. *Privatizing the land: rural political economy in post-Communist societies*. London: Routledge, pp. 92–124.

Muldavin, J. 2000. The paradoxes of environmental policy and resource management in reform-era China. *Economic Geography*, 76(3), 244–71.

Mullan, K., A. Kontoleon, T. Swanson and Shiqiu Zhang. 2010. Evaluation of the impact of the natural forest protection program on rural household livelihoods. *Environmental Management*, 45, 514–25.

O'Brien, K., and Lianjiang Li. 2006. *Rightful resistance in rural China*. New York: Cambridge University Press.

Peng, W., K. Zhang, Y. Chen and Q. Yang. 2005. Research on soil quality change after returning farmland to forest on the loess sloping cropland. *Journal of Natural Resources* (in Chinese), 20(2), 272–8.

Ren, H., W.J. Shen, H.F. Lu, X.Y. Wen and S.G. Jian. 2007. Degraded ecosystems in China: status, causes and restoration efforts. *Landscape Ecology Engineering*, 3, 1–13.

Richard, C., Zh. Yan and G. Du. 2006. The paradox of the individual responsibility system in the grasslands of the Tibetan Plateau, China. In: *USDA Forest Proceedings*. RMRS-P-39. Rocky Mountain Research Station, Fort Collins: US Department of Agriculture, 83–91.

Robertson, M. 2004. The neoliberalization of ecosystem services: wetland mitigation banking and problems in environmental governance. *Geoforum*, 35, 361–73.

Sargeson, S. 2013. Violence as development: land expropriation and China's urbanization. *Journal of Peasant Studies*, 40(6) doi: 10.1080/03066150.2013.865603.

Scherr, S., M. Bennett, M. Loughney and K. Canby. 2006. *Developing future ecosystem service payments in China: lessons learned from international experience*. Washington, DC: Forest Trends.

Scott, J. 1985. *Weapons of the weak: everyday forms of peasant resistance*. New Haven: Yale University Press.

Sheehy, D.P., D. Miller and D.A. Johnson. 2006. Transformation of traditional pastoral livestock systems on the Tibetan steppe. *Secheresse*, 17(1–2), 142–51.

Shen, S., A. Wilkes, J. Qian, L. Yin, J. Ren and F. Zhang. 2010. Agrobiodiversity and biocultural heritage in the Dulong Valley, China. *Mountain Research and Development*, 30(3), 205–11.

Smil, V. 2004. *China's past, China's future: energy, food, environment*. New York: RoutledgeCurzon.

Sonamkyid. 2008. The implementation of a resettlement development project: socio-economic changes accompanying the transition from nomadic to town life in Sogrima Town, western China. MA Thesis, Miriam College. Quenzon City, Philippines.

Sturgeon, J. 2005. *Border landscapes: the politics of Akha land use in China and Thailand*. Seattle: University of Washington Press.

Taylor, J.L. 2006. Negotiating the grassland: the policy of pasture enclosures and contested resource use in Inner Mongolia. *Human Organization*, 65(4), 374–86.

Thompson, M., M. Warburton and T. Hatley. 1986. *Uncertainty on a Himalayan scale: an institutional theory of environmental perception and a strategic framework for the sustainable development of the Himalayas*. London: Milton Ash.

Tilt, B. 2013. The politics of industrial pollution in rural China. *Journal of Peasant Studies*, 40(6) doi: 10.1080/03066150.2013.860134.

Trac, C.J., S. Harrell, T.M. Hinckley and A.C. Henck. 2007. Reforestation programs in southwest China: reported success, observed failure and the reasons why. *Journal of Mountain Science*, 4 (4), 275–92.

Uchida, E., J. Xu, Z. Xu and S. Rozelle. 2007. Are the poor benefiting from China's land conservation program?. *Environment and Development Economics*, 12(4), 593–620.

Uchida, E., S. Rozelle and Jintao Xu. 2009. Conservation payments, liquidity constraints and off-farm labor: ompact of the grain-for-green program on rural households in China. *American Journal of Agricultural Economics*, 91(1), 70–86.

Wang, C. and V. Maclaren. 2012. Evaluation of economic and social impacts of the sloping land conversion program: a case study in Dunhua County, China. *Forest Policy and Economics*, 14, 50–7.

Wang, C., H. Ouyang, V. Maclaren, Y. Yin, B. Shao, A. Boland and Y. Tian. 2007. Evaluation of the economic and environmental impact of converting cropland to forest: a case study in Dunhua county, China. *Journal of Environmental Management*, 85, 746–56.

Wang, X., J. Bennett, C. Xie, Z. Zhang and D. Liang. 2007. Estimating non-market environmental benefits of conversion of cropland to forest and grassland program: a choice modeling approach. *Ecological Economics*, 63, 114–25.

Weller, R. 2006. *Discovering nature: globalization and environmental culture in China and Taiwan*. Cambridge: Cambridge University Press.

Wende, D. 2009. The ecological migration project: the case of Ca Chog, Mtsho Sngon province, China (unpublished paper).

Weyerhaeuser, H., A. Wilkes and F. Kahrl. 2005. Local impacts and responses to regional forest conservation and rehabilitation programs in China's northwest Yunnan province. *Agricultural Systems*, 85, 234–53.

Williams, D.M. 1996. Grassland enclosures: catalyst of land degradation in Inner Mongolia. *Human Organization*, 55(3), 307–13.

Williams, D.M. 2000. Representations of nature on the Mongolian Steppe: an investigation of scientific knowledge construction. *American Anthropologist*, 102(3), 503–19.

Williams, D.M. 2002. *Beyond great walls: environment, identity and development on the Chinese grasslands*. Stanford: Stanford University Press.

Wu, N. and C. Richard. 1999. The privatization process of rangeland and its impacts on pastoral dynamics in the Hindu-Kush Himalaya: the case of Western Sichuan. Conference Paper for People and Rangelands: Building the Future. *In*: D. Eldridge and D. Freudenberger, eds. *Proceedings of the VI International Rangeland Congress*. Townsville, Australia., pp. 14–21.

Wu, Z. and W. Du. 2008. Pastoral nomad rights in Inner Mongolia. *Nomadic Peoples*, 12(2), 13–33.

Wunder, S. 2008. Payments for environmental services and the poor: concepts and preliminary evidence. *Environment and Development Economics*, 13, 279–97.

Xie, Y. and W. Li. 2008. Why do herders insist on Otor? Maintaining mobility in Inner Mongolia. *Nomadic Peoples*, 12(2), 35–52.

Xu, J. and D. Melick. 2007. Rethinking the effectiveness of public protected areas in southwestern China. *Conservation Biology*, 21(2), 318–28.

Xu, Jian-ying, L. Chen, Y. Lu and B. Fu. 2007. Sustainability evaluation of the grain for green project: from local people's responses to ecological effectiveness in Wolong Nature Reserve. *Environmental Management*, 40, 113–22.

Xu, Jintao, E. Katisgris and T. White. 2001. *Implementing the natural forest protection program and the sloping land conversion program: lessons and policy recommendations*. Beijing: China Forestry Publishing House.

Xu, Jintao, Ran Tao, Zhigang Xu and M. Bennett. 2010. China's sloping land conversion program: does expansion equal success? *Land Economics*, 86(2), 219–44.

Xu, Jintao, Runsheng Yin, Zhou Li and Can Liu. 2006. China's ecological rehabilitation: unprecedented efforts, dramatic impacts, and requisite policies. *Ecological Economics*, 57, 595–607.

Xu, Jun. 2010. Challenges: resettlement of nomads in Qinghai Province [online]. Conference Paper for International Association of Tibetan Studies, Vancouver. Available from: http://www.case.edu/affil/tibet/tibetanNomads/books.htm#X [Accessed 24 February 2013].

Xu, Zhigang, M. Bennett, Ran Tao and Jintao Xu. 2004. China's sloping land conversion programme four years on: current situation, pending issues. *International Forestry Review*, 6, 317–26.

Yan, K. 2012. *Tibetan village stops mining near the Nu River*. Berkeley: International Rivers Network.

Yan, Zhaoli and Ning Wu. 2005. Rangeland privatization and its impacts on the Zoige wetland on the eastern Tibetan Plateau. *Journal of Mountain Science*, 2(2), 105–15.

Yan, Zhaoli, Ning Wu, Yeshi Dorji and R Jia. 2005. A review of rangeland privatization and its implications on the Tibetan Plateau, China. *Nomadic Peoples*, 9(1), 31–52.

Yao, Shunbo, Yajun Guo and Xuexi Huo. 2010. An empirical analysis of the effects of China's land conversion program on farmers' income growth and labor transfer. *Environmental Management*, 45, 502–12.

Yeh, E.T. 2003. Tibetan range wars: spatial politics and authoirty on the grasslands of Amdo. *Development and Change*, 34(3), 499–523.

Yeh, E.T. 2005. Green governmentality and pastoralism in western China: 'converting pastures to grasslands'. *Nomadic Peoples*, 9(1–2), 9–29.

Yeh, E.T. 2009a. From wasteland to wetland? Nature and nation in China's Tibet. *Environmental History*, 14(1), 103–37.

Yeh, E.T. 2009b. Greening western China: a critical view. *Geoforum*, 40, 884–94.

Yeh, E.T. and Gaerrang. 2011. Tibetan pastoralism in neoliberalizing China: continuity and change in Gouli. *Area*, 43(2), 165–72.

Yin, Runsheng and Guiping Yin. 2010. China's primary programs of terrestrial ecosystem restoration: initiation, implementation and challenges. *Environmental Management*, 45, 429–41.

Yin, Runsheng, Guiping Yin and Lanying Li. 2010. Assessing China's ecological restoration programs: what's been done and what remains to be done? *Environmental Management*, 45, 442–53.

Yin, Runsheng and Minjuan Zhao. 2012. Ecological restoration programs and payments for ecosystem services as integrated biophysical and socioeconomic processes – China's experience as an example. *Ecological Economics*, 73, 56–65.

Yundannima. 2012. From 'retire livestock, restore rangeland' to the compensation for ecological services: state interventions into rangeland ecosystems and pastoralism in Tibet. PhD dissertation, University of Colorado Boulder, Geography.

Zackey, J. 2007. Peasant perspectives on deforestation in southwest China: social discontent and environmental mismanagement. *Mountain Research and Development*, 27(2), 153–61.

Zeng, Y., Z. Feng and G. Cao. 2003. Land cover change and its environmental impact in the upper reaches of the Yellow River, northeast Qinghai-Tibetan Plateau. *Mountain Research and Development*, 23(4), 353–61.

Zhang, L., Q. Tu and Arthur Mol. 2008. Payment for environmental services: the sloping land conversion program in Ningxia Autonomous Region of China. *China and World Economy*, 16(2), 66–81.

Zhang, Tingjun. 2007. Perspectives on environmental study of response to climatic and land cover/land use change over the Qinghai-Tibetan Plateau: an introduction. *Arctic, Antarctic and Alpine Research*, 39(4), 631–4.

Zhou, D.Q. and R.E. Grumbine. 2011. National parks in China: experiments with protecting nature and human livelihoods in Yunnan province, People's Republic of China. *Biological Conservation*, 144, 1314–21.

Zhu, L. and B. Li. 2000. Natural hazards and environmental issues. *In*: D. Zheng, Q. Zhang and S. Wu, eds. *Mountain genecology and sustainable development of the Tibetan Plateau*. Dordrecht: Kluwer Academic Publishing, pp. 203–22.

Zinda, J.A. 2012. Hazards of collaboration: local state co-optation of a new protected-area model in southwest China. *Society and Natural Resources*, 25(4), 384–99.

Zinda, J.A. 2014. Making national parks in Yunnan: shifts and struggles within the ecological state. *In*: Emily T. Yeh and Chris Coggins, eds. *Mapping Shangrila: Contested Landscapes in the Sino-Tibetan Borderlands*. Seattle: University of Washington Press.

Zukosky, M.L. 2008. Reconsidering governmental effects of grassland science and policy in China. *Journal of Political Ecology*, 15, 44–60.

Emily T. Yeh is an Associate Professor of Geography at the University of Colorado Boulder. She conducts research on nature-society relations, primarily in Tibetan parts of the People's Republic of China, including projects on conflicts over access to natural resources, the relationship between ideologies of nature and nation, the political ecology of pastoral environment and development policies, vulnerability of Tibetan herders to climate change, and emerging environmental subjectivities. Her book, *Taming Tibet: landscape transformation and the gift of Chinese development* (Cornell University Press, 2013), explores the intersection of the political economy and cultural politics of development as a project of state territorialization.

The politics of water in rural China: a review of English-language scholarship

Darrin Magee

Politics is about access and power, and access to freshwater resources in rural China is complicated and understudied. China's massive size and diverse climate make it hard to generalize about freshwater resources in rural areas of the country. On balance, China is not water-scarce, yet geographic and temporal variations in water availability are dramatic, with China's driest areas receiving far less precipitation than the wettest areas. Rural areas are the locus of competition among freshwater users including agriculture, power companies, industry, households and ecosystems. Additionally, while peasants may hold usage rights (not title) to farmland, it is not a given that they will hold rights to water that will guarantee the productivity of that farmland in areas where precipitation is low. Finally, water quality is, unfortunately, an increasingly important factor affecting the availability of 'fresh' water, as is evident in the notion of quality-induced scarcity (*shuizhixing queshui*). This contribution reviews a small but important body of scholarship on rural water politics in China, identifying existing themes and suggesting new directions where scholarship is currently lacking.

1. Introduction: seeking a politics of water in rural China

Politics is about access to resources and exercise of power. It is about institutions, some of which facilitate access to resources, others of which limit that access. It is also about discourse, language that naturalizes and legitimizes certain activities (e.g., 'basin-wide management') while challenging and delegitimizing others (e.g., 'over-abstraction[1] of water'). The politics of water resources in rural China encompasses structures and practices related to a number of uses for freshwater resources, including irrigation, household use (drinking, cooking and hygiene), power generation (hydroelectric and thermal), industry, and in-stream flows necessary to support biodiversity and ecological systems. Yet while rich scholarly literatures exist on agriculture, industry, daily life and, to a lesser extent, rural energy production, much of the scholarship in these areas tends to engage tangentially at best with the fundamental questions on water resources, including questions regarding

I am grateful to my undergraduate research assistant, Ms. Kexun Sun, and to The Henry Luce Foundation for its support of the Asian Environmental Studies Initiative at my institution, Hobart and William Smith Colleges. All errors are, of course, mine alone.

[1]Pinyin: *guodu chouqu*. To abstract water is to extract it from an underground or surface source and convey it to some point of use.

control of and access to those resources. There are, of course, several important exceptions that I highlight below.

It is unclear why western scholarship on rural water politics in China is so limited, but several explanations seem plausible. First, water use in rural China is, for the most part, not metered, just as agricultural use in much of the world is not metered. Excepting perhaps water quantity data that is aggregated at the provincial or watershed scale, reliable data on actual availability of freshwater resources in different areas, let alone costs (economic, ecological and other) of securing and utilizing those resources, are hard to find. In many cases, it is likely that the data simply do not exist. Second, given the dramatic variation in climate regimes across the continent-sized country, generalizing from the micro-scale studies that do exist is difficult if not impossible. Finally, scholars (and humans in general), except those who have spent significant amounts of time in water-poor areas, may to a certain extent simply be blind to the geographic and historical contingency of freshwater resources. Such hydro-myopia is surely evident in other national contexts around the world. In the United States, for example, water users in the desert west gain (and lose) access to water resources based on history, meaning whoever first puts a water resource to use is granted 'first in time, first in right' access to that water. In the eastern United States, however, geography trumps history: those who live on a water resource are granted rights to that water, so long as those rights do not materially impinge upon the rights of downstream users. Climate change introduces another layer of uncertainty as precipitation patterns become more erratic.

Another challenge in studying water from a social scientific perspective lies in the nebulous nature of water itself. Yet this is precisely where important questions about the politics of water arise. Should water be considered a public good or private property? Should it be a tradable commodity? Should ownership or use rights to water be severable from ownership or use rights to adjacent land? What foundations exist in Chinese legal traditions for water rights systems? What relevant lessons, if any, can Chinese policy makers draw from foreign water rights systems? What role have neoliberal institutions such as water rights markets played thus far in China? Have those institutions improved access to water for those who need it most, or limited access to only those who can afford to pay prices that may, at least in the future, reflect the increasing scarcity value of clean water? What institutions exist to moderate conflicts over water? Where do peasants seek recourse when they have been hydraulically wronged through pollution, diversion or other harms to waters they need?

According to the Water Law of the People's Republic of China, all water is the property of the central state or of the collective; thus, the simple answer to the above questions is that water is a public good not tradable for private gain, and not available to be owned by any entity other than the state or its rural extensions (Standing Committee of the National People's Congress 2002). Indeed, the expanse and strength of China's hydraulic bureaucracy led Wittfogel (1957) to famously argue that 'Oriental' rulers possessed a unique ability to mobilize vast populations in the construction of major irrigation projects, giving rise to a particular 'Asiatic' mode of production, concentrating political power in a hydraulic bureaucracy, and birthing a governing model he labeled 'Oriental despotism'. Yet experiments with water rights markets and other neoliberal institutions have been ongoing since the start of the Reform and Opening Period (*gaige kaifang*) in 1978. These mechanisms are ostensibly designed to increase end-use efficiency by channeling water via so-called inter-sectoral transfers toward higher economic value uses such as urban areas and industry, theoretically not at the expense of agriculture (Cai 2008). There are, however, no Chinese water courts for adjudicating water rights cases, and the tendency of the

Chinese judiciary is to side with power rather than with peasants seeking remedies for those who abuse power by using, for instance, publicly-owned rivers and lakes as pollution sinks for private gain (Van Rooij 2010). Yet the glass is not necessarily half-empty. Some cases of water rights trading have proven effective in increasing water use efficiency and access. New ownership and management models for tubewells and other irrigation assets have arisen and show similar promise for increasing end-use efficiency of water in rural areas. Legal aid groups are providing guidance to peasants whose streams, lands and crops are contaminated with the by-products of China's industrial miracle, many of which (such as cadmium, mercury and lead) have extremely deleterious and well-understood effects on human and ecosystem health. And as one scholar of China's legal system recently put it, local courts must balance their role of interpreting the law in the service of the Communist Party on the one hand, and diffusing local discontent on the other (Stern 2010).

The prolific and widely read scholar Vaclav Smil has made a number of contributions on human-environment issues in China over the course of several decades. Much of his work steers clear of the water politics of greatest interest here, though, despite the fact that the alarmist tone he often sounds begs for deeper engagement on policy questions. 'Urban water shortages', he notes, 'tend to attract a disproportionate share of attention, but, as in any other populous Asian country, China's water use is heavily dominated by irrigation requirements, creating a number of extensive environmental impacts' (Smil 1993, 44). Here, the politics of *why* urban water crises attract greater attention is unexplored, as are the impacts of that disproportionate attention. Irrigated food production is a culprit, but the amount of that food production in the direct service of urban centers is not specified, although he does specify later that much of the demand for 'rural' water is due to rising 'urban needs' (Smil 1993, 155). His view on the link between energy production and water (Smil 1993, 1998) is essentially limited to an acknowledgement that energy production (including thermal) requires water, and that China's numerous rivers give it 'global primacy in [a] desirable clean energy source' (Smil 1998), namely hydropower. Smil's work does nonetheless provide detailed, data-rich background to the significant challenges China has faced during the Reform Era in terms of energy and food production, both of which depend heavily on reliable access to fresh water (Smil 1984, 1992, 1993, 2003). Shapiro's (2001) account of the destructiveness of Mao's campaign-style politics on the natural environment provides a similarly valuable primer on the three decades prior to the reforms. To be sure, the discourse of sustainability and environmental stewardship was far less pervasive in the mid-twentieth century than today, and even less likely to be central to national political decision-making at the time.

Below, I begin by briefly outlining China's water situation, including quantity, quality and institutions. China's water challenges are by any measure daunting: falling water tables in the north threaten social disruption and population displacement; contamination of groundwater and surface water sources are severe and widespread; warming resulting from climate change may lead to decreased snowpack and accelerated melting of existing glaciers that feed the country's most important rivers, and the list goes on. My intent is to sketch a background, albeit simplified and incomplete, on which to situate scholarship on rural water politics. In Section 3, I trace several important streams of that scholarship, organized conceptually around competing demands for water resources across three sectors in rural China: agriculture, dams and power generation, and industry. Across these sectors, I highlight the ways in which institutions, power relations and discourse have shaped the politics of water by enabling or constraining access to water resources. The most developed and important body of work revolves around irrigation water management. In a country of over half a billion farmers, where the single largest user of fresh water is agriculture, this is not surprising. A second stream of work examines (primarily large) hydropower and the related spatial politics of water and energy. While rural in location, many of China's largest hydropower projects are decidedly urban in intent; that is, massive hydropower stations on rivers in rural western China provide electric power to urban industrial centers hundreds of kilometers away. Finally, I examine industrial demands on water in rural China and the resultant impact on decision politics regarding access, including access to water bodies as a source for externalizing (and diluting) pollution. Section 4 concludes by connecting the politics of water to the politics of social stability, and suggests directions for further work.

2. Overview of freshwater resources in China

Chinese mythology tells of the Great Yu, the widely revered founding ruler of the Xia Dynasty (~2000–1600 BCE) who, through wisdom and personal sacrifice, engineered away the floods that had devastated the early Yellow River kingdoms. That the story of the Great Yu Controlling the Waters (*Dayu zhishui*) remains significant today is a testament to the fundamental importance of water resources – too little, too much – for China's leaders for millennia. In terms of overall precipitation and availability of freshwater resources, China is not water-scarce. The country does, however, face real challenges in terms of per capita freshwater availability, temporal and geographical water resource shortages, and rapid and widespread declines in water quality. These problems are likely to persist or worsen for decades, and will almost certainly make competition for reliable and useful water across the country more acute. Institutions managing access to water in both urban and rural China, to the extent those institutions exist, will almost certainly find it increasingly difficult to strike a balance between treating fresh water as a basic necessity for human existence and societal development, a saleable commodity capable of generating significant profits, and a pollution sink and energy source – to say nothing of water as a fundamental human right to which all Chinese citizens might be entitled. Many of these challenges are already apparent: as detailed below, the Chinese Ministry of Water Resources and Ministry of Environmental Protection have both published data in recent years evincing widespread water quality and quantity challenges, and the term 'quality-induced scarcity' (*shuizhixing queshui*) has now become common parlance among policy makers, scholars and others concerned about the state of water in contemporary China. The remainder of this section details the current situation and likely trends in China's freshwater resources, sketching a backdrop for contextualizing state-of-the-art scholarship on rural water politics in China.

China receives an average of 645 mm of precipitation per year, comparable to the 715 mm per year of the United States (World Bank 2012). With its population of 1.3 billion, however, China's per capita precipitation endowment is around one-fourth that of the United States, and less than one-fourth the world average. Moreover, China's freshwater availability has wide temporal and geographic variations. Summer monsoons affecting southern and southeastern coastal areas bring massive amounts of precipitation to those parts of the country, contributing to a growing season over 300 days long and to more than 80 percent of the runoff of the entire country's rivers. The northeast and northwest, on the other hand, are out of reach of most of the monsoon rainfall, and suffer from periods of drought ranging from mild to severe. Severe declines in groundwater levels are also widespread in northern China due to over-abstraction of aquifers for agricultural, industrial and municipal use. As McCormack notes, 'China wrestles with a water crisis that is unique and troublesome on two levels: water is scarce in the North while the South suffers from an overabundance that has been ruinous at times' (McCormack 2001).

The central government's response to shortages in the northern parts of the country has involved some efficiency measures, such as encouraging less water-intensive crops, strengthening institutions, and implementing market-based systems that reward end-use efficiency. Smil (1993) gives several examples of industries that saw significant decreases in wasted water during the first decade of reforms, including the petrochemical, steel and paper. At the same time, central water planners have moved forward with grander engineering fixes, most notably the South-North Water Diversion (SNWD; *nanshui beidiao*). The SNWD aims to divert up to 44.8 billion m^3 of water per year from the Yangtze River basin in the south to the drier Yellow River basin in the north through three channels (eastern, middle and western) (Magee 2011), a figure that has grown roughly 50 percent since the project was first detailed at the start of the Reform Period (Smil 1993, 153). As of December 2012, the roughly 1200-km eastern route, designed to supplement water supplies in the Beijing-Tianjin area, is largely complete and utilizes the existing series of waterways collectively comprising the Grand Canal, most of which was constructed during the sixth and seventh centuries. Due to the elevation gradient along the canal route, however, more than a dozen pumping stations are required to lift water from the lower-elevation Yangtze to the higher-elevation Yellow basin, concern that stalled the project in the late 1970s (Smil 1993). These electric pumps will require the conversion of millions of tons of coal per year into electricity in order to move water, with thermal power generation – roughly 80 percent of China's total power generation – being itself an inherently water-intensive activity. Moreover, concerns that water quality will decline over the course of the transfer route have necessitated the construction of hundreds of treatment stations along the route, all of which require electricity to function, bringing further delays to the project (Magee 2011).

In one of his few forays into the politics of water, Smil (1993, 153) notes that the 'revival' of the SNWD plans in 1978 brought about 'China's first, and surprisingly strong, expert and public challenge to a central bureaucratic decision argued on the basis of environmental sustainability'. Opponents argued that efficiencies should be pursued first, and that increased irrigation rates in the north would likely lead to further declines in soil productivity due to salinization and alkalinization. Yet today the project lurches forward, several years behind schedule despite being husbanded directly by the State Council. The project website now even includes a countdown timer for completion of the eastern and central routes.[2] For Zhang Ye, vice-director of the State Council's SNWD Office, there is no option: 'Only the SNWD can solve the north's water resources problem', proclaimed a recent headline on the project webpage (Office of the South-North Water Diversion 2012). Such a sentiment reflects water technocrats' faith in a system of top-down decision-making on major water infrastructure projects, a system in which central planners claim the mantle of science in the name of public good, and where 'scientific' (*kexueguan de*) is often equated with 'sustainable' (*kechixu de*) with regard to water conservancy projects. Writing for a top China politics journal at the end of the 1990s, one of China's best-known water engineers and planners asserted confidently that 'From a scientific point of view, the south-north water transfer scheme is likely to be environmentally sound', hedging slightly by noting that a 'careful and well-considered' Environmental Impact Assessment (EIA) had yet to be done (Liu 1998, 910).

---

[2]See Office of the South-North Water Diversion (n.d.). As of 23 March 2012, completion of the eastern and central routes was expected in 283 days and 19 months, respectively.

The middle route of the project, which begins at the existing Danjiangkou Reservoir in northern Hubei, required increasing the height of the Danjiangkou dam by some 15 meters in order to create a favorable elevation differential between the intake point and the discharge point approximately 1400 km to the north. The larger reservoir is expected to require the resettlement of tens of thousands of villagers; some of these will be secondary migrants, i.e., individuals and families who have already been resettled at least once before, in many cases due to the original construction of the Danjiangkou dam. Given the SNWD's importance at the national level, and the growing severity of water quality and quantity concerns it is expected to help ameliorate, there is little reason to expect that migrants displaced by the project will fare better than their compatriots who have been displaced by large dams. A *South China Morning Post* article noted in 2003 the close connection between one of the central route's architects, Tsinghua University hydraulic engineering Professor Gu Zhaoqi, and one of his former students: Hu Jintao, China's president from 2003 to 2013 (Cheung 2003).

In addition to water quantity concerns such as those that motivate proponents of the SNWD, severe water quality degradation is also widespread, resulting from both point-source (e.g., factory discharge and sewage outfalls) and non-point-source (e.g., pesticide- and fertilizer-laden runoff from agricultural areas) pollution. Water quality degradation represents the most daunting of China's water resource challenges in the short- to medium term; even if supplies of water are sufficient, they are of limited utility for agricultural, industrial or human consumption purposes if heavily polluted. In southern and eastern provinces such as Guangdong, Fujian and Jiangsu, home to some of China's most important manufacturing hubs, point-source pollution emanating from unregulated and untreated discharges represents the primary culprit. The textile, paper and electroplating industries have long histories as some of the worst polluters (Economy 2004). These have been joined more recently by high-tech industries, including those involved in the manufacture of electronic goods and de-manufacture and 'recycling' of obsolescent electronics in the form of e-waste. Country-wide, official statistics reported 21.74 billion tons of industrial wastewater discharged nationwide in 2008, of which 20.22 billion tons (93 percent) met government discharge standards (National Bureau of Statistics 2009). The latter figure is probably inflated; additionally, it is highly likely that the water that did meet standards nevertheless contained high levels of numerous contaminants (including known carcinogens and endocrine disruptor compounds) that are either not yet regulated, untreatable with the technologies currently in place, or both (see, for instance, Gong *et al.* 2009).

According to the *2010 State of the environment report* prepared by the Chinese Ministry of Environmental Protection, all seven of the country's major basins[3] were 'lightly polluted' (*qingdu wuran*), with certain stretches of all seven basins suffering from more severe pollution (Ministry of Environmental Protection of the PRC 2010). The Chinese water quality grading system ranges from I to V, with Grade V being worst and Grades IV and V unfit for direct human contact (see Table 1). A classification of > V is also occasionally used to indicate the most severely polluted water sources. Among the 408 monitoring stations on 203 rivers in the seven basins, the percentages of water falling in grades I–III, IV–V, and worse than V were 57.3, 24.3 and 18.4 percent respectively in 2009. Water quality was ranked good overall in the Zhu (Pearl) and Yangtze, moderately polluted in the Yellow and Liao, and severely polluted in the Hai. Key pollutants reported include permanganates, ammonium nitrates (from fertilizers) and petrochemicals, as well as

---

[3]Yangtze (Chang), Yellow (Huang), Zhu, Songhua, Huai, Hai and Liao Rivers. The report is based on data from 408 monitoring stations on 203 rivers in seven major basins.

Table 1. China's water quality grading system.

Category	Description and designated use(s)
1	Source regions; national natural protected areas
2	Drinking Water Grade 1; ecological needs
3	Drinking Water Grade 2; ecological needs
4	Industrial water, not fit for direct human contact
5	Agricultural water; 'landscape' use
> 5	Worst quality, not suited even for agriculture

Source: Ministry of Environmental Protection of the PRC (2002).

elevated biological oxygen demand (BOD). Even though stretches of the rivers that are highly polluted are often adjacent to other stretches of relatively good water quality (suggesting that the engineer's adage 'the solution to pollution is dilution' holds), population densities along the most polluted stretches of the major basins make the potential for significant socioeconomic impacts non-negligible. Moreover, given that most water treatment technologies are highly energy-intensive, it must be assumed that quality shortfalls are to some extent equivalent to quantity shortfalls.

The link between water quality and quantity of freshwater resources requires little explanation, but the fact that central government water pollution policy now explicitly recognizes quality as a direct threat to quantity via the term 'quality-induced scarcity' (*shuizhixing queshui*) suggests serious concern of central water and environmental quality leaders. As noted below, Ma Jun's groundbreaking work publishing water polluter data on the web has done much to expose polluters to public scrutiny and, arguably, to spur officials to act. Similarly, Greenpeace's Detox Campaign targets water pollution stemming from the global textiles industry and has had some success in securing commitments from manufacturers to reduce or eliminate toxicity in their production lines.[4] In some cases, the causes of quality declines are at best indirectly anthropogenic, such as saltwater intrusion into aquifers or rivers. In groundwater, over-abstraction – that is, abstraction that occurs more rapidly than recharge rates, resulting in a net loss of water from the groundwater source – creates a cone of depression in the aquifer whose negative pressure relative to adjacent water bodies, thereby drawing in water from those adjacent areas. In aquifers near the sea, the likelihood that the water drawn in will be saline is high. The processes in rivers are simpler, but there are numerous pathways to creating water that is too salty to use for agriculture, industry or domestic consumption. River salinization can occur when rivers that normally flow to the sea are overdrawn and either fail entirely to reach the sea, or reach it with diminished volumes that allow seawater to intrude into the river. In areas where brackish or freshwater aquaculture is conducted, such intrusion can have swift and devastating ecological, and therefore economic, impacts. A second process by which the salinity of rivers increases is simply because they flow in channels of stone and soil, much of which contain salts of various sorts that dissolve readily in water, thereby increasing salinity.

3. Politics of uncertainty: competing demands for water in rural China

Scholarly work on the political importance of water is quite rich. Though Wittfogel's (1957) theory of water's role in 'Oriental despotism' has been critiqued, China's vast geography,

---

[4]See Greenpeace (n.d.).

large population and wide disparities in water resources availability have made for a long history of major water projects. Most notable among these is the 2000-year-old Dujiangyan flood control and irrigation works in Sichuan. Geographers have made important contributions outside the China context. Swyngedouw (1997, 2003) has examined the role of water infrastructure in Ecuador and Spain as part of modernist visions enacted (or at least attempted) at various scales of governance, underscoring that flows of water are intrinsically bound to flows of capital and power relations. Bakker directs her attention to discursive constructions of resources subject to commodification, as well as to discourses of scarcity of those resources, a theme Clarke-Sather (2012) further develops with his aleatory political ecology of water. In her early examination of hydropower in the Mekong basin (where China is the upstream riparian among six other countries), Bakker (1999) highlights the need for multiscalar analyses of major water projects, with particular attention to processes of water resources development as compared to scalar analyses aimed at regions defined *a priori*.

Geographer Erik Swyngedouw (2003, 95) argues that

> Hardly any river basin, hydrological cycle, or water flow has not been subjected to some form of human intervention or use; not a single form of social change can be understood without simultaneously addressing and understanding the transformations of and in the hydrological process.

As noted in the previous section, uncertainty with regard to those processes is a characterization that applies to much of China's rural geography. Uncertainty in terms of quality, quantity and access to fresh water for agricultural, (rural) industry and hygiene and household use creates the basis for a rural politics of fresh water. Different users who would make competing demands on the same water resource may seek water that is of different quality grades, and when those users compete for water coming from the same resource, it is not always clear that the highest-value use should have priority. In the paragraphs that follow, I examine important scholarship related to rural water politics in China, organizing that scholarship under rubrics of three key competing uses. Given the importance of irrigated agriculture to China's overall food security, I begin with irrigation and drainage, followed by dams and hydropower, and finally industry and pollution. Arguably the most important category in any consideration of water politics is that regarding water allocated to maintaining ecosystems, the life-support systems of the social systems considered above. Yet due to space considerations, I do not examine the politics of water provision for ecological systems here.

3.1. *Irrigation and drainage*

According to official statistics, agriculture accounted for 61.3 percent of freshwater use in China in 2010 (Ministry of Water Resources of the PRC 2012), and the figure has likely grown since then. One long-standing focus of scholarship on politics and management of irrigation water has centered on the policy implications of water-related threats to agricultural yields. Huang and Rozelle (1995) cite the 'breakdown of the environment' – especially irrigation, soil erosion and soil salinization – as contributing significantly to declines in grain production in the latter half of the 1980s. The authors note that declines in yields occurred despite increases in inputs such as chemical fertilizer and irrigation water. Yet 'there is a point', the authors argue, 'at which production and yields may respond more to the relief of environmental stress than to additional factor intensification' (Huang and Rozelle 1995, 853). More to the question of rural resource politics, Huang and Rozelle note arguments (their own and those of others) that uncertainty about land tenure and inadequate implementation of the Household Responsibility System may have been other factors contributing to declines in effective water use, the first of which would likely have had a direct bearing on farmers' willingness to invest in such things as improved irrigation systems or other water management technologies or practices. In discussing the import of their findings, Huang and Rozelle argue that 'the negative impact on yields due to environmental stress has become an issue deserving close attention by policy makers' (861).

On a somewhat contradictory note, other research shows that the looming water crisis in northern China has *not* resulted in declining agricultural yields in that part of the country (J. Wang *et al.* 2005). Rather, that crisis has generated new neoliberal institutions of de facto ownership of and investment in rural groundwater. This is due primarily, J. Wang *et al.* argue, to more efficient groundwater management resulting from the privatization of tubewells, even in a political economic context in which water increasingly tends to flow towards industry and away from farmers. Beginning as early as the late 1960s, the diversion of water resources toward rural industry created a greater reliance on tubewells on the part of farmers. In turn, the accelerated extraction of groundwater facilitated by millions of tubewells has led to the now commonly-cited problem of falling water tables, sometimes more than a meter per year, and declining groundwater quality in northern China, as well as secondary effects such as ground subsidence. According to J. Wang *et al.*, shifting patterns of tubewell ownership, specifically the privatization of wells that were formerly collectively owned, have made positive contributions to water resources stability and to the development of the agricultural sector in general in northern China. The authors conclude by recommending continued government support of privately-owned tubewells.

James Nickum has written authoritatively for decades on water issues in China, with much of his work focusing explicitly on rural water management. In one particularly relevant article (Nickum 2010), he explores the successes and failures of Self-Funded Irrigation and Drainage Districts (SIDDs), noting that in the south, the ability to drain water from waterlogged farmland may be just as important as the ability of famers in the north to irrigate parched lands. His use of SIDDs as a case study is situated against a broader theoretical question about how (or whether) China's semi-authoritarian state, with its wariness (paranoia) of all institutional arrangements outside the Party-State apparatus, might affect the formulation and implementation of effective water management policy in rural China. Nickum warns against assumptions of a stagnant authoritarianism that might overlook both reforms in China's water sector, and the contextual factors that impede those reforms.

Here again arises the question of water's status as tradable commodity versus state-owned property. In the same article, Nickum also points out that, notwithstanding the fact that all water (like other natural resources) is technically the property of the state, 'there has been a continuous dialectic between state, collective and private (including foreign) ownership and plan-directed and market-directed approaches to the economy, including water policy' (Nickum 2010, 540). Privately-owned tubewells (J. Wang *et al.* 2005) are an important part of that dialectic; the extent to which 'water enclosures' or 'water grabs' can change China's rural groundwater hydrology remain to be seen.

Chinese scholars have also written on this tension, and as Nickum (2010) notes, there is increasing consensus among scholars and policy makers that China needs to institute some form of a water rights system, albeit one that is focused on use rights versus ownership rights (Y. Wang 2002, 2012, Sun 2009).[5] The need for a water rights system came full-

---

[5]These articles are from a special issue of *The International Journal for Water Resources Development* devoted to water rights questions in China.

force onto the national legislative agenda in 2006, with the 11th Five-Year Plan (2006–2010) calling for the first concrete steps in implementing components of that system (Sun 2009). The *International Journal of Water Resources Development* devoted a special issue in 2009 to the question of water rights in China, which included numerous contributions from scholars involved in national water policy formulation in institutions such as the Ministry of Water Resources, the China Institute of Hydropower and Water Resources Research, and influential research universities. Of particular relevance to any discussion of rural water politics in China is the fact that local governments wear two hats, both as agents authorized to exercise state control over water use, and as water users themselves. This situation, essentially *de facto* ownership of water by the local state, can lead not surprisingly to conflicts over use rights and the creation of regional monopolies or fiefdoms over freshwater resources (Qian and Ni 2007). Not unlike other resources and capital in rural areas, actual control of use and ownership rights frequently rests with village heads (*cunzhang* or *cunzhishu*) whose priorities as individuals may or may not always align with those of other users in the village. SIDDs represent one attempt to institutionalize water rights management.

Tsinghua University public policy scholar Wang Yahua has made important contributions on the theoretical and practical implications of tradable water rights resources in rural China, noting in one recent article that while water rights trading between the agricultural and industrial sectors would likely yield significant gains in water savings, those gains are much smaller when rights are traded within the agricultural sector alone (Y. Wang 2012). Moreover, he notes that the agricultural sector in general lacks the capacity to absorb increases in transaction costs and other uncertainties of water markets that would present less of a threat to the industrial sector. Wang's word of caution about low-income groups' vulnerability vis-à-vis water rights markets echoes similar calls elsewhere; a recent Asian Development Bank (ADB) study on the potential for implementing Payments for Ecosystem Services (PES) systems to incentivize pollution reduction in rural areas by through payments from urban users reports very low willingness to pay (WTP) for clean water tariff increases among low-income (< 1000 Yuan/month) urban residents in eastern China (Jiang *et al.* 2010).

The ability of water markets and tradable water rights to bring about real water savings, however, is far from certain. As Li *et al.* (2011) argue, the apparent failure of market-based systems in some areas of China must be understood as a failure of certain Chinese institutional structures to mesh with assumed parameters under which rational choice options like water abstraction charges (WAC) function. They offer the example of Water Abstraction Policies instituted at the end of the Maoist period that worked in areas such as Shanghai, enabled by relatively strong implementing institutions and compact geography, but failed in places like Shanxi, due to weaker institutions and more sizeable land area. Whereas prior to 1988 water had been essentially treated as an open-access resource, the 1988 Water Law sought to institutionalize on a national level its management as a state-owned resource, with government agencies responsible for managing in accordance with local needs, including determining and collecting use fees and permit regimes. Li *et al.* cite the 'bureaucratic friction and lack of clarity in water allocation plans' (Li *et al.* 2011, 955) as obstacles in the successful implementation of the Water Law's management regimes, a point that other authors have made regarding another important use of water that bridges rural and urban China: hydropower (Yeh and Lewis 2004, Magee 2006).

Li *et al.* argue that while the 2002 revised Water Law strengthened the ability and increased the autonomy of the water conservancy bureaucracy at all levels to conduct the business of water management (e.g., levying fees, issuing use permits and enforcing penalties), four key structural obstacles to local government implementation of water allocation policies remain. First, state ownership of water resources is enacted at different scales, ranging from the extremely local scale to the basin scale. For instance, whereas the Water Resources Department in western Yunnan's Nujiang Prefecture exerts some degree of control over surface and groundwater allocation within its boundaries, those same water resources also fall under the jurisdiction of the Yangtze River Basin Commission, a national-level agency employing some 30,000 people and charged with the comprehensive management of a watershed whose main river is some 6500 km long (Magee 2006, Magee and McDonald 2009). This jurisdictional overlap leads to uncertainty about whether local decisions might be trumped by national-level ones involving, for instance, inter-basin transfers (Magee 2011, Li et al. 2011).

A second institutional obstacle revolves around lack of clarity in water abstraction targets, which results in part from imperfect coordination among national-scale, basin-scale, and local agencies, and in part from the simple unavailability of reliable hydrological data that would enable implementing agencies to make better informed decisions (Li et al. 2011).[6] Even where detailed hydrological data may exist, it is frequently considered classified or 'internal' (neibu) and therefore not readily available to water resource planners, especially those employing foreign experts. This is especially true for hydrological data on transboundary rivers and their watersheds. Exacerbating the difficulty of effectively sharing data and establishing coordinated targets among different users within a basin is the fact that the basin commissions themselves, according to Li et al., lack the legal authority to enact processes to facilitate such sharing and coordination. Finally, Li et al. point out a fourth institutional obstacle in China's dual-responsibility system, in which water resource agencies at all levels are subordinate both to their parent agency at the next-highest governmental level, as well as to the local government.

In Chinese, the term tiao represents the vertical authority (i.e., that deriving from the center), while kuai signifies the horizontal (i.e., that deriving from the locality). China scholars have written much about the conundrum the tiao kuai system presents for policy implementation. Writing specifically about environmental policy, Lieberthal (1997, 3) noted 'obvious potential conflict' when central and local interests do not align. Naughton (1999, 182) refers to the tiao kuai arrangement as an 'organizational straightjacket'. Mertha (2005) argues that attempts to reshape tiao kuai relations in order to combat local protectionism have resulted in dysfunction along vertical (functional) bureaucratic lines, such as within the Ministry of Water Resources and its functional dependents, the Water Resource Bureaus at the provincial levels and below. Mertha's more recent examination of advocacy and activism surrounding water projects suggests that it is precisely the gaps and overlaps in administrative authority over water projects that enable conflicting priorities to surface and facilitate the formation of sometimes surprising coalitions to promote those priorities (Mertha 2008).

With irrigation consuming the lion's share of rural China's freshwater resources, the logic of pursuing reforms in irrigation water management is abundantly clear. Competing discourses of collective management versus private ownership offer different models whose

---

[6]This is not a uniquely Chinese phenomenon, of course. The Colorado River in the western United States traverses seven US states before passing through a small section of northwestern Mexico. Yet the US failed to address the allocation of water to Mexico until nearly 20 years after establishment of the Colorado River Compact, which divided an overly generous estimate of the river's flow among the seven states alone.

suitability to local conditions and preferences may vary across such a vast country. Irrigation assets, like many of the built assets in rural China whose ownership became muddied during decollectivization, often fell by default under the control of local officials under pressure to boost local revenues (agricultural and other) through whatever means possible at the start of the reforms. Power relations institutionalized by the *tiao kuai* system continue to leave water bureaus subject to political pressures and priorities both horizontal and vertical, which may conflict with each other.

3.2. *Hydropower and other dams*

A few years after severe flooding in the Yangtze basin in 1998 threatened millions in cities like Wuhan, and in the midst of construction of the massive and controversial Three Gorges Dam designed, among other things, to thwart such floods, McCormack (2001) wrote on two competing paradigms for understanding water resources in China. Mega-scale engineering projects like Three Gorges Dam, the Lancang (upper Mekong) hydroelectric cascade and the South-North Water Diversion (*nanshui beidiao*) are, according to McCormack, emblematic of China's embrace of modernist visions of harnessing rivers to do society's work. Clearly influenced by the World Commission on Dams (WCD 2000) report, which largely criticized on a number of measures the performance of large dams around the world, McCormack argues that China was 'the archetypal proponent of the globally declining paradigm and the determined "modernism" of Beijing's powerful water bureaucrats stands sharply opposed to the emerging 'post-modern global consensus on water' (McCormack 2001, 8).

Now, with the benefit of more than a decade of hindsight, McCormack's assertion that a 'post-modern' global consensus on water that had sustainability (not defined by him) at its core seems to have been overly optimistic. In China, current domestic dam construction efforts are proceeding at a breakneck pace in a discursive context that prioritizes rural electrification, poverty alleviation and such regional development slogans as 'Send Western Electricity East' (*xidian dongsong*) and the Western Development Campaign (Magee 2006). Perhaps more striking, Chinese hydropower companies have in recent years been extremely successful at exporting dam-building expertise around the world, once again motivated by central government exhortations to 'go outward' (*zou chuqu*), especially to Africa and Latin America (see, for instance, International Rivers 2012). Yet at the time McCormack's view seemed plausible; large dams had been roundly condemned on a variety of counts by the WCD report, an unprecedented number of legislators in China's National People's Congress refused to give their blessing to the Three Gorges Project, and international development institutions such as the World Bank and the Asian Development Bank had visibly pulled back their funding for large dam projects (though they continued to support related projects such as feasibility studies and power grid development).

Hydroelectric projects the world over are embedded in complicated networks of rivers, transmission lines and political economic relations, and Chinese hydropower is no exception. China has the world's greatest hydroelectric potential – essentially a measure of how big the rivers are and how far they fall – and boasted just over 200 GW of installed hydroelectric generation capacity at the end of 2010, a figure set to rise to as much as 350 GW by 2020 (Sinohydro 2012).[7] Small hydropower (*xiao shuidian*), which usually refers to

---

[7]The article actually states 420 GW as the 2020 target, but that figure includes 350 GW of traditional hydropower and 70 GW of pumped storage hydropower.

projects under 50 MW in China, is not counted in these figures, and likely contributes another 40 GW (Zhao and Zhu 2004). Many of the largest hydropower projects currently underway are in southwestern provinces such as Yunnan, Guizhou and Sichuan, where large river volumes combine with steep terrain to create attractive development conditions. Of those projects, mega-dams such as the 5850-MW Nuozhadu, 4200-MW Xiaowan on the Lancang (upper Mekong) and numerous others rank in the 99th percentile of large dams in China on a number of measures.

Southwestern China is also home to significant populations of ethnically and economically marginalized individuals who have long lived under regimes of resource appropriation. In the words of one scholar of large hydropower in rural Yunnan, 'dams have become a form of governance, one that builds on previous relations of resource control and the politics of ethnicity and the margin' (McDonald 2007, 88). While large dams there often come with the blessing of state slogans such as poverty alleviation and rural electrification, the negative impacts on local communities near the dams can outweigh the broader, more diffuse benefits, such as marginal increases in power grid reliability, that tend to accrue at a regional or national scale (Brown *et al.* 2008, McNally *et al.* 2009). One reason for this uneven geography of costs and benefits is that large hydroelectric dams, aside from submerging valuable farmland with their reservoirs, are simply not designed to 'plug in' to local grids operating at low voltages and incapable of handling the output of such huge projects. Instead, almost all of them will connect directly to urban industrial load centers in the east via ultra-high-voltage lines operated by the South China Grid, thereby bypassing many of the communities most in need of poverty alleviation and critical infrastructure (Magee 2006). This is not to suggest that all dams are bad, or that China's dam-building or compensation practices are particularly flawed, but rather to underscore that the negative and positive impacts of dams for hydropower and other purposes are unevenly distributed geographically (Scudder 2005, Tullos *et al.* 2010).

Producing hydroelectricity is, of course, not the only purpose of dams. Others include improving transportation by 'flattening' steep stretches of rivers and improving their navigability, controlling floods by capturing storm runoff and releasing it in a more regulated fashion, creating impoundments for irrigation and, as an occasional side benefit, providing opportunities for recreation. Large-scale projects across rural China, or those on important and/or transboundary rivers, may be subject to oversight from high-level organizations such as the State Council, the National Development and Reform Commission, trans-provincial river basin commissions, and national-level power conglomerates that trace their lineages from former ministries (Magee 2006, Lieberthal and Oksenberg 1986, Lampton 1987). Smaller projects, on the other hand, usually fall under the jurisdiction of local administrative units at the provincial level or below.

Whatever the geographic scale of oversight, the distribution of the impacts of hydropower will be uneven and inequitable; some peasants will lose their land and face resettlement, with its inherent risks, to new communities where arable land and other vital resources may already be scarce and welcome mats nonexistent (Padovani 2004, Andrews-Speed and Ma 2008, Brown and Xu 2010). The Three Gorges Project provides a textbook example of how top-down decision-making on major water infrastructure frequently disenfranchises already vulnerable populations – in this case, over a million residents (Edmonds 1992, Andrews-Speed and Ma 2008). For rural residents especially, disrupted social networks resulting from resettlement may represent as great or worse a threat to well-being than loss of land (and therefore livelihood) (Heggelund 2006), especially in areas with high concentrations of minorities with limited formal education

(Brown and Xu 2010). The extent to which conflicts arise among the resulting winners and losers varies (McDonald 2007, Mertha 2008), but protests over dam-related injustices, real or perceived, have gained national and international media attention in recent years. Some scholars (Brown *et al.* 2009, Tullos *et al.* 2009, 2010) have sought to develop theoretical frameworks and decision-support tools to improve assessments and decisions surrounding large dams (cites), but the degree to which such tools could substantially impact the politics of hydropower development in China, where power and (techno-economic) knowledge differentials between developers and members dam-affected communities can be great, is not clear. Much work remains to be done to improve the inequities inherent in dam-building projects the world over, and to understand the true socioeconomic, biophysical and geopolitical costs of those dams, especially in political contexts where decisions about major infrastructure such as dams are often taken behind closed doors and with inadequate consultation of those immediately local to the dam who stand to feel the negative impacts most acutely.

3.3. *Industry and pollution*

As noted in Section 2, water pollution from industry is widespread and serious in both rural and urban China. This is not a new phenomenon; as Smil (1993, 47) noted two decades ago, official figures of 35 Gt of wastewater discharges roughly matched the annual discharge of the Yellow River in a moderately dry year. A few years later, another geographer argued that China faced 'imminent ecological disaster' due to industry's unchecked pollution of urban and rural waterways (Muldavin 1996, 245). Even more troubling is the fact that the geographic extent of surface- and groundwater sources contaminated by heavy metals, synthetic chemicals and other toxins resulting from industrial processes will only expand as factories move farther and farther inland to avoid the rising costs of labor and other inputs in the coastal cities that have been at the heart of China's industrialization to date. Many of the chemicals that contaminate water resources in China are persistent organic pollutants (POPs), toxic compounds that do not degrade easily in nature and which may bioaccumulate and biomagnify in ecosystems, entering food webs and increasing in concentration as they work their way up to higher trophic levels.

One of the most influential authors and advocates on water pollution-related concerns in China is Ma Jun, formerly a journalist at the *South China Morning Post*. His first book, *China's water crisis* (Ma 2004)[8] set in stark relief the geographic extent and seriousness of quantity and quality challenges to much of the country's fresh water. He founded and directs the Institute for Public and Environmental Affairs (IPE; *gongzhong huanjing yanjiu zhongxin*), which maintains an important Internet database of water pollution incidents, including a pollution map and profiles of responsible parties.[9] Ma has gained international recognition in recent years, including winning the Goldman Environmental Prize in 2012, for his advocacy on water pollution control in China, and the IPE's pollution map and related micro- and macro-level data are valuable resources for scholars, policy makers, and others concerned about water pollution. He and others have also sought to reduce water pollution before it happens by 'greening' supply chains for global manufacturers (see, for instance, Ma *et al.* 2010).

---

[8]First published in Chinese in 1999 as *Zhongguo Shui Weiji* by China Environmental Sciences Publishing House.
[9]See Institute for Public and Environmental Affairs (n.d.).

Awareness of polluters and pollutants is not necessarily enough to participate in the politics of water pollution, though. One must also be able to navigate the discourse, institutions and power dynamics of China's legal system. This can be especially daunting for peasants not well versed in the nuances of China's legal codes and procedures, and with limited financial means to hire an attorney. Two critical steps in the process of moving from suffering from the harms of pollution to actually seeking redress in the form of a grievance, according to Van Rooij (2010, 58), are 'naming' and 'blaming'. Here, depth of knowledge of the conditions surrounding and harms resulting from pollution are important in naming, which Van Rooij defines as 'recognizing the seriousness of the pollution', as well as blaming, or 'attributing these damages to a responsible entity' (Van Rooij 2010, 58). Pollution victims may also be reluctant to bring cases against factories on which they are economically dependent (Tilt 2010, Van Rooij 2010), a phenomenon not unique to China but instead found in 'factory towns' around the world.[10]

Certain organizations have arisen that seek to address the power differential between polluting industries and peasants by providing legal aid to peasants who have suffered harm, either physically or financially or both, from those industries, and helping bring cases to court. Operating within a political system that is often skeptical at best about the role of non-governmental organizations, these civil society groups often push the envelope in their work to open up the politics of pollution, which is frequently the politics of water pollution. The Center for Legal Assistance to Pollution Victims (CLAPV) is perhaps the best known. Van Rooij (2010) describes a class-action suit of 1721 litigants in which CLAPV assisted. Even though the litigants won, subsequent struggles among certain factions within the group created problems with distribution of compensation funds among claimants, highlighting the nested nature of pollution politics even in cases where winners and losers seem clearly defined. One important victory for pollution victims generally came with a 2001 Supreme People's Court decision that reversed the burden of proof so that those claiming pollution-related harms are no longer required to demonstrate causation (Stern 2010, Van Rooij 2010). Fortunately, despite continued political pressures on China's courts and the judges and lawyers who staff them, there seems to be both increasing specificity and clarity in China's environmental law corpus, as well as at least a 'fluctuating degree of autonomy' among judges (Stern 2010, 6), given that their decisions will have some impact on the all-important goal of maintaining social stability.

In some parts of the Chinese industrial landscape, *de*-manufacturing may represent environmental and human health threats of equal or greater concern to those brought about by manufacturing. Emblematic of this subsector of industry are the e-waste 'recycling' shops that operate at the economic, geographic and regulatory margins. Guiyu in eastern Guangdong Province is perhaps the most notorious example of the e-waste economy, fueled by the vast quantities of cheap raw materials, including obsolete electronic devices such as cell phones and computers, coupled with the growing demand for industrially valuable materials such as copper, gold, certain plastics and rare earth metals used in microelectronics (Tong and Wang 2004). Recovery of these materials in places like Guiyu is a labor of fire, acid and hammers, in which workers use a variety of chemical and physical means to remove higher-value components from lower-value ones. Discarded wastewater and runoff from precipitation mixing with solid components cause extreme localized degradation of soil and water (Leung *et al.* 2006). Biomedical research, including studies showing massively elevated blood lead levels in children who have grown up in

---

[10]Walker (2010) details this dynamic in the case of Minamata Disease (mercury poisoning) in Japan.

Guangdong's e-waste villages, makes it painfully clear that many of those toxins are making their way into human bodies (Zheng *et al.* 2008).

As in the cases of irrigation and dams, scholarship on the politics of water pollution in China is limited but growing. Increasingly, researchers from a variety of fields from environmental and medical sciences to anthropology and geography are examining the environmental and human health impacts of industrial activities. Some of that scholarship goes a step further to engage in the socio-political questions surrounding water pollution, such as inconsistent (or nonexistent) enforcement of existing Chinese laws designed to limit pollution-related harm to ecological and human communities, or the disproportionate impacts often felt by marginalized groups such as migrant workers and ethnic minorities (e. g., Lora-Wainwright 2010, Tilt 2010). More research needs to be done to build upon scholarship on the politics of reporting and acting on pollution infractions (Yang 2010), especially as international and Chinese media report with increasing regularity egregious violations of water resources integrity through accidental or intentional dumping of harmful chemicals. Examples that have received international attention include a benzene spill into the transboundary Songhua River in 2005, a diesel spill into a Yellow River tributary in 2010 and a cadmium spill into the Liu River in Guangxi in 2012. Countless other smaller spills every day never gain international attention.

4. Conclusion: discourse, power, and institutions and access

Part of a geographer's work is to dismantle facile spatial categorizations such as 'rural' and 'urban' that can mask processes transcending those categorizations. Geographers in recent years have made important contributions to scholarship on water, grappling with its unruly nature, its socio-physical duality and the ability of water-related networks to produce new hydro-spatial realities (see, for instance, Bear and Bull 2011). Much of the 'development' of water resources in rural China is tied to discourses of *rural* poverty alleviation, *regional* electrification, *national* economic development and energy security, and other noble goals, all of which have an unspoken but significant *urban* component as farmers are 'expropriated by default' (Berkoff 2003, 12) to provide *urban* drinking water and food supply, *urban* energy provision, *urban* pollution sink. Given the diversity of hydro-realities across China, ranging from drought to flooding to quality-induced scarcity, and the variety of users involved, generalizing about rural water politics is challenging but a useful lens for understanding rural politics more broadly. As I have shown in this contribution, while politics may cameo in some scholarship on water resources in rural China, much work remains to be done to unpack the dynamics of decision-making on water resources in the countryside.

In this paper I have reviewed scholarship on the institutions, discourses and power relations that shape water politics in rural China and thereby constrain or facilitate access to water. Discourses of development in various forms, including the Western Development Campaign, local-scale promotion of water-intensive but economically valuable cash crops, and prioritization of water-intensive and heavily polluting industries like high-tech, have clear impacts on the water resources of an increasingly extensive area of China. Water bureaucracies, guided in theory by institutions such as the National Water Law and basin-wide development plans, often face overlapping responsibilities and conflicting priorities with pressures applied horizontally from local governments and vertically from the water bureaus at higher administrative levels. And while there is evidence, at least in some areas, that water resource decisions are becoming increasingly pluralistic, the capability of marginalized users least able to mobilize political, economic and technical resources necessary to participate meaningfully in those decision processes remains constrained by power relations premised on differentials in techno-scientific knowledge capital, ethnic hierarchies and socio-political clout.

To the extent that wide swaths of Chinese citizens perceive protests over water-related wrongs as principled examples of rightful resistance (O'Brien 1996, O'Brien and Li 2006), then the politics of water could very well be the politics of social (in)stability. Opaque decision processes, hasty, cut-and-pasted, or nonexistent environmental impact assessments, lack of real consideration of environmental performance indicators in evaluating local officials and party cadres, and inadequate compensation regarding water conservancy projects did not begin or end with the Three Gorges Dam and the 1.3 million people its reservoir displaced, textbook case that it may be. Instead, such projects abound, and while the scale of impacts in terms of individuals displaced, farmland submerged or toxins discharged may vary by an order or magnitude or two, the reality of compromised water resources – in quality, quantity or both – is one nearly every Chinese citizen faces to a greater or lesser extent. Thus, whether the issue of the day is cadmium-polluted rivers and soils resulting from illegal dumping, unjust diversions of agricultural water toward industry and municipalities, hastily-built hydropower projects whose output will make only a small contribution to the reliability of China's electrical grid, or giant diversions whose social and ecological costs will only be understood in the past tense, scholars of China would do well to actively situate the politics in scholarship on water, and the water in scholarship on politics.

References

Andrews-Speed, P. and X. Ma. 2008. Energy production and social marginalisation in China. *Journal of Contemporary China*, 17, 247.

Bakker, K. 1999. The politics of hydropower: developing the Mekong. *Political Geography*, 18, 209–232.

Bear, C. and J. Bull. 2011. Water matters: agency, flows, and frictions (Guest Editorial). *Environment & Planning A*, 43, 2261–6.

Berkoff, J. 2003. China: the South–North Water Transfer Project – is it justified? *Water Policy*, 5, 1–28.

Brown, P., D. Magee and Y. Xu. 2008. Socioeconomic vulnerability in China's hydropower development. *China Economic Review*, 19, 614–27.

Brown, P. *et al.* 2009. Modeling the costs and benefits of dam construction from a multidisciplinary perspective. *Journal of Environmental Management*, 90, S303–11.

Brown, P. and K. Xu. 2010. Hydropower development and resettlement policy on China's Nu River. *Journal of Contemporary China*, 19, 777–97.

Cai, X. 2008. Water stress, water transfer and social equity in Northern China: implications for policy reforms. *Journal of Environmental Management*, 87, 14–25.

Cheung, R. 2003. Water pioneer's dream coming true [online]. *South China Morning Post*. Available at www.scmp.com/article/415482/water-pioneers-dream-coming-true [Accessed 1 December 2012].

Clarke-Sather, A. 2012. *From the heavens to the markets: development, nation, and the mediation of water in northwest China. Geography.* Boulder: University of Colorado.

Economy, E. 2004. *The river runs black: the environmental challenge to China's future.* Ithaca: Cornell University Press.

Edmonds, R.L. 1992. The Sanxia (Three Gorges) Project: the environmental argument surrounding China's super dam. *Global Ecology and Biogeography Letters*, 2, 105–25.

Greenpeace. n.d. The Detox Campaign. Available at http://www.greenpeace.org/international/en/campaigns/toxics/water/detox/intro/ [Accessed 11 February 2013].

Gong, J. *et al.* 2009. Occurrence and environmental risk of endocrine-disrupting chemicals in surface waters of the Pearl River, South China. *Environmental Monitoring and Assessment*, 156, 199–210.

Heggelund, G. 2006. Resettlement programmes and environmental capacity in the Three Gorges Dam project. *Development and Change*, 37, 179.

Huang, J. and S. Rozelle. 1995. Environmental stress and grain yields in China. *American Journal of Agricultural Economics*, 77, 853–64.

Institute for Public and Environmental Affairs. n.d. Homepage. Available at http://www.ipe.org.cn [Accessed 31 January 2013].

International Rivers. 2012. *The new great walls: a guide to China's overseas dam industry*. Berkeley: International Rivers.

Jiang, Y., L. Jin and T. Lin. 2010. *Higher water tariffs for less river pollution – evidence from Min River and Fuzhou City, People's Republic of China*. Economics Working Paper Series. Manila: Asian Development Bank.

Lampton, D.M. 1987. Water: challenge to a fragmented political system. *In*: D.M. Lampton, ed. *Policy implementation in post-Mao China*. Berkeley: University of California Press, 157–189.

Leung, A., Z.W. Cai and M.H. Wong. 2006. Environmental contamination from electronic waste recycling at Guiyu, southeast China. *Journal of Material Cycles and Waste Management*, 8, 21–33.

Li, W., M. Beresford and G. Song. 2011. Market failure or governmental failure? A study of China's water abstraction policies. *The China Quarterly*, 208, 951–69.

Lieberthal, K. 1997. China's governing system and its impact on environmental policy implementation. *China Environment Series*, 1, 3–8.

Lieberthal, K. and M. Oksenberg. 1986. *Bureaucratic politics and Chinese energy development*. Washington, DC: US Department of Commerce.

Liu, C. 1998. Environmental issues and the south–north water transfer scheme. *The China Quarterly*, 156, 899–910.

Lora-Wainwright, A. 2010. An anthropology of cancer villages. *Journal of Contemporary China*, 19, 79–99.

Ma, J. 2004. *China's water crisis*. Norwalk, CT: Eastbridge.

Ma, J. *et al.* 2010. *Greening supply chains in China: practical lessons from China-based suppliers in achieving environmental performance*. Washington, DC: World Resources Institute.

Magee, D. 2006. Powershed politics: hydropower and interprovincial relations under Great Western Development. *The China Quarterly*, 185, 23–41.

Magee, D. 2011. Moving the river: China's south–north water transfer project. *In*: S.D. Brunn, ed. *Engineering Earth: the impacts of megaengineering projects*. Dordrecht: Springer, 1499–1514.

Magee, D. and K. McDonald. 2009. Beyond Three Gorges: Nu River hydropower and energy decision politics in China. *Asian Geographer*, 25, 39–60.

McCormack, G. 2001. Water margins – competing paradigms in China. *Critical Asian Studies*, 33, 5–30.

McDonald, K. 2007. *Damming China's grand canyon: pluralization without democratization in the Nu River Valley*. Berkeley: University of California.

McNally, A., D. Magee and A. Wolf. 2009. Hydropower and sustainability: resilience and vulnerability in China's powersheds. *Journal of Environmental Management*, 90, S286–93.

Mertha, A.C. 2005. China's 'soft centralization': shifting *tiao/kuai* authority relations. *The China Quarterly*, 184, 791–810.

Mertha, A.C. 2008. *China's water warriors: citizen action and policy change*. Ithaca: Cornell University Press.

Ministry of Environmental Protection of the PRC. 2002. Dibiao shui huanjing zhiliang biaozhun [Environmental quality standards for surface water] (GB 3838-2002). Available from: http://english.mep.gov.cn/standards_reports/standards/water_environment/quality_standard/200710/W020061027509896672057.pdf [Accessed 3 February 2013].

Ministry of Environmental Protection of the PRC. 2011. 2010 Zhongguo huanjing zhuangkuang gongbao [2010 China State of the Environment Report]. Available from: http://jcs.mep.gov.cn/hjzl/zkgb/2010zkgb/201106/t20110602_211577.htm [Accessed 13 November 2013]

Ministry of Water Resources of the PRC. 2012. 2010 Nian Zhongguo Shuiziyuan Gongbao [2010 China Water Resources Report] [online]. Ministry of Water Resources. Available from: http://www.mwr.gov.cn/zwzc/hygb/szygb/qgszygb/201204/t20120426_319624.html [Accessed 7 December 2012].

Muldavin, J.S.S. 1996. The political ecology of agrarian reform in China. *In*: R. Peet and M.J. Watts, eds. *Liberation ecologies*. London: Routledge, 227–259.

National Bureau of Statistics, ed. 2009. *Zhongguo tongji nianjian [China statistical yearbook]*. Beijing: Zhongguo tongji chubanshe.

Naughton, B. 1999. *Danwei:* the economic foundations of a unique institution. *In*: X. Lu and E.J. Perry, eds. *Danwei: the changing Chinese workplace in historical and comparative perspective.* Armonk, NY: M. E. Sharpe, 169–194.

Nickum, J.E. 2010. Water policy reform in China's fragmented hydraulic state: focus on self-funded/ managed irrigation and drainage districts. *Water Alternatives*, 3, 537–51.

O'Brien, K.J. 1996. Rightful resistance. *World Politics*, 49, 31–55.

O'Brien, K.J. and L. Li. 2006. *Rightful resistance in rural China.* Cambridge and New York: Cambridge University Press.

Office of the South-North Water Diversion. 2012. Zhang Ye: only the SNWD can solve the north's water resource shortage problems [online]. State Council Office of the SNWD. Available from: http://www.nsbd.gov.cn/zx/rdht/201212/t20121210_251568.html [Accessed 5 December 2012].

Padovani, F. 2004. Les effets sociopolitiques des migrations forcées en Chine liées aux grands travaux hydrauliques: l'exemple du barrage des Trois-Gorges. *Les Etudes du CERI*, 103, 1–37.

Qian, H. and Y. Ni. 2007. The current situation of agricultural water use rights and policy innovation (in Chinese). *Zhongguo Nongcun Shuili Shuidian [China Rural Water and Hydropower]*, 5, 138–141.

Scudder, T. 2005. *The future of large dams: dealing with social, environmental and political costs.* Sterling, VA: Earthscan.

Shapiro, J. 2001. *Mao's war against nature: politics and the environment in revolutionary China.* Cambridge: Cambridge University Press.

Sinohydro. 2012. China will strive for 420 GW installed hydropower by 2020 [online]. Sinohydro. Available from: http://www.sinohydro.com/664-998-605360.aspx [Accessed 4 December 2012].

Smil, V. 1984. *The bad Earth: environmental degradation in China.* Armonk, NY: M.E. Sharpe.

Smil, V. 1992. China's environment in the 1980s: some critical changes. *Ambio*, 21, 431–6.

Smil, V. 1993. *China's environmental crisis: an inquiry into the limits of national development.* Armonk, NY: M.E. Sharpe.

Smil, V. 1998. China's energy and resource uses: continuity and change. *The China Quarterly*, 156, 935–51.

Smil, V. 2003. *China's past, China's future: energy, food, environment.* New York: Routledge.

Standing Committee of the National People's Congress. 2002. Water Law of the People's Republic of China [online]. Order of the President of the People's Republic of China (No. 74). Available from: http://www.lawinfochina.com/law/display.asp?ID=2461&DB=1. [Accessed 11 November 2013].

Stern, R.E. 2010. On the frontlines: making decisions in Chinese civil environmental lawsuits. *Law & Policy*, 32, 79–103.

Sun, X. 2009. Introduction: the development of a water rights system in China. *International Journal of Water Resources Development*, 25, 189–92.

Swyngedouw, E. 1997. Power, nature, and the city: the conquest of water and the political ecology of urbanization in Guayaquil, Ecuador, 1880–1990. *Environment and Planning A*, 29, 311–32.

Swyngedouw, E. 2003. Modernity and the production of the Spanish waterscape, 1890–1930. *In*: K.S. Zimmerer and T.J. Bassett, eds. *Political ecology: an integrative approach to geography and environment-development studies.* New York: The Guilford Press, 94–114.

Tilt, B. 2010. *The struggle for sustainability in rural China: environmental values and civil society.* New York: Columbia University Press.

Tong, X. and J. Wang. 2004. Transnational flows of e-waste and spatial patterns of recycling in China. *Eurasian Geography and Economics*, 45, 608–21.

Tullos, D. *et al.* 2010. Perspectives on the salience and magnitude of dam impacts for hydro development scenarios in China. *Water Alternatives*, 3, 71–90.

Tullos, D., B. Tilt and C. Reidy. 2009. Introduction to the special issue: understanding and linking the biophysical, socioeconomic and geopolitical effects of dams. *Journal of Environmental Management*, 90, S203–7.

Van Rooij, B. 2010. The People vs. Pollution: understanding citizen action against pollution in China. *Journal of Contemporary China*, 19, 55–77.

Walker, B.L. 2010. *Toxic archipelago: a history of industrial disease in Japan.* Seattle: University of Washington Press.

Wang, J., J. Huang and S. Rozelle. 2005. Evolution of tubewell ownership and production in the North China Plain. *The Australian Journal of Agricultual and Resource Economics*, 49, 177–95.

Wang, Y. 2002. *Shui quan he shui shichang: shui guanli fazhan xin qushi* [Water rights and water markets: new trends in development of water management]. *Jingji Yanjiu Cankao [Review of Economic Research]*, 20, 2–8.

Wang, Y. 2012. A simulation of water markets with transaction costs. *Agricultural Water Management*, 103, 54–61.

WCD (World Commission on Dams). 2000. *Dams and development: a new framework for decision-making*. London: Earthscan.

Wittfogel, K.A. 1957. *Oriental despotism: a comparative study of total power*. New Haven: Yale University Press.

World Bank. 2012. Average precipitation in depth (mm per year) [online]. The World Bank. Available from: http://data.worldbank.org/indicator/AG.LND.PRCP.MM. [Accessed 7 December 2012].

Yang, G. 2010. Brokering environment and health in China: issue entrepreneurs of the public sphere. *Journal of Contemporary China*, 19, 101–18.

Yeh, E.T. and J.I. Lewis. 2004. State power and the logic of reform in China's electricity sector. *Pacific Affairs*, 77, 437–65.

Zhao, J. and X. Zhu. 2004. *Private participation in small hydropower development in China: comparison with international communities*. Beijing: UNHYDRO.

Zheng, L. *et al.* 2008. Blood lead and cadmium levels and relevant factors among children from an e-waste recycling town in China. *Environmental Research*, 108, 15–20.

Darrin Magee is Associate Professor of Environmental Studies and Director of the Asian Environmental Studies Initiative at Hobart and William Smith Colleges in Geneva, New York. A specialist on environmental issues in China, he has lived and worked in mainland China, Taiwan and Hong Kong. His research and teaching address water, energy and waste issues, including large-scale hydropower and other water infrastructure. He is also a Fellow for Rocky Mountain Institute's 'Reinventing fire: China' Project.

Index